EXPLORATIONS IN COMPUTING

An Introduction to Computer Science and Python Programming

CHAPMAN & HALL/CRC
TEXTBOOKS IN COMPUTING

Series Editors

John Impagliazzo
Professor Emeritus, Hofstra University

Andrew McGettrick
Department of Computer
and Information Sciences
University of Strathclyde

Aims and Scope

This series covers traditional areas of computing, as well as related technical areas, such as software engineering, artificial intelligence, computer engineering, information systems, and information technology. The series will accommodate textbooks for undergraduate and graduate students, generally adhering to worldwide curriculum standards from professional societies. The editors wish to encourage new and imaginative ideas and proposals, and are keen to help and encourage new authors. The editors welcome proposals that: provide groundbreaking and imaginative perspectives on aspects of computing; present topics in a new and exciting context; open up opportunities for emerging areas, such as multi-media, security, and mobile systems; capture new developments and applications in emerging fields of computing; and address topics that provide support for computing, such as mathematics, statistics, life and physical sciences, and business.

Published Titles

Paul Anderson, Web 2.0 and Beyond: Principles and Technologies

Henrik Bærbak Christensen, Flexible, Reliable Software: Using Patterns and Agile Development

John S. Conery, Explorations in Computing: An Introduction to Computer Science

John S. Conery, Explorations in Computing: An Introduction to Computer Science and Python Programming

Ted Herman, A Functional Start to Computing with Python

Pascal Hitzler, Markus Krötzsch, and Sebastian Rudolph, Foundations of Semantic Web Technologies

Mark J. Johnson, A Concise Introduction to Data Structures using Java

Mark J. Johnson, A Concise Introduction to Programming in Python

Lisa C. Kaczmarczyk, Computers and Society: Computing for Good

Mark C. Lewis, Introduction to the Art of Programming Using Scala

Bill Manaris and Andrew R. Brown, Making Music with Computers: Creative Programming in Python

Uvais Qidwai and C.H. Chen, Digital Image Processing: An Algorithmic Approach with MATLAB®

David D. Riley and Kenny A. Hunt, Computational Thinking for the Modern Problem Solver

Henry M. Walker, The Tao of Computing, Second Edition

CHAPMAN & HALL/CRC
TEXTBOOKS IN COMPUTING

EXPLORATIONS IN COMPUTING

An Introduction to Computer Science and Python Programming

John S. Conery

 CRC Press
Taylor & Francis Group
Boca Raton   London   New York

CRC Press is an imprint of the
Taylor & Francis Group, an **informa** business

A CHAPMAN & HALL BOOK

CRC Press
Taylor & Francis Group
6000 Broken Sound Parkway NW, Suite 300
Boca Raton, FL 33487-2742

© 2015 by Taylor & Francis Group, LLC
CRC Press is an imprint of Taylor & Francis Group, an Informa business

No claim to original U.S. Government works

Printed in Canada on acid-free paper
Version Date: 20140630

International Standard Book Number-13: 978-1-4665-7244-7 (Hardback)

Library of Congress Cataloging-in-Publication Data

Conery, John S.
   Explorations in computing : an introduction to computer science and Python programming / John S. Conery.
      pages cm -- (Chapman & Hall/CRC textbooks in computing)
   Summary: "This text helps beginners develop their own Python programs. Experiments with fully completed programs are provided at the beginning of each chapter, allowing instructors to use the text in CS0 courses where students do not learn programming. Programming projects appear later in each chapter. Students are encouraged either to write the code that implements the functions introduced earlier or extend the existing programs. All the projects push students to explore further on their own"-- Provided by publisher.
   Includes bibliographical references and index.
   ISBN 978-1-4665-7244-7 (hardback)
   1. Computer science. I. Title.

QA76.C58594 2014
004--dc23   2014022784

Visit the Taylor & Francis Web site at
http://www.taylorandfrancis.com

and the CRC Press Web site at
http://www.crcpress.com

Contents

Preface

This book is an introduction to computer science. It is intended for beginning CS majors or students from other fields who want a general introduction to computer science and computer programming.

The main focus is on the "big ideas" in computing. Contrary to what most people expect, computer science is much more than just programming. Computer science students learn to write programs, but the goal is to use programming skills to explore fundamental concepts and computational approaches to solving problems.

The distinguishing feature of this book is a set of tutorial exercises included in each chapter. The idea is to use an interactive programming language to provide an environment where students can type expressions, view the results (both in the terminal window and through algorithm animation), and run experiments that help them learn the concepts.

The analogy I like to make when I use this material in my own classes is to compare the tutorials to lab projects in an introductory chemistry class. There the instructor gives detailed instructions on the materials and methods, and students are expected to follow the instructions as precisely as possible. Students learn by observing as the experiment unfolds and writing lab reports that explain what happened.

For the "computational experiments" in *Explorations in Computing* students are shown the statements that create pieces of data and call functions that implement algorithms. An example of this sort of experiment is the section on the insertion sort algorithm. In an interactive Python session, students type statements to create lists of random numbers, display the lists as a series of bars on a canvas, and call the function that implements the sort algorithm. As the algorithm runs the bars are shuffled on the canvas. Later students are given a chance to write their own implementations of the algorithm, and the end of the chapter has several programming exercises based on simple iterative algorithms similar to insertion sort.

This book is a revised and updated version of *Explorations in Computing: An Introduction to Computer Science*, an introductory textbook I wrote in 2011 (and also published by Chapman & Hall/CRC Press). The new book has two major differences from the previous one. The first, most obvious, difference is the switch from Ruby to Python. Python has been widely adopted as the language of choice for first-year (CS1) computer science courses. By revising the lab software to use Python the hope is that students and instructors will find it easier to make a seamless transition from the introductory projects in this book to the deeper studies in later courses.

The second difference is that this new edition is also an introduction to Python programming. The primary emphasis is still on "computational thinking" and important concepts in computing, but along the way readers are presented with sufficient Python programming skills that they can implement their own programs to explore the ideas.

A Note for Students: How to Use This Book

You have no doubt heard the adage, "What you get out of a course depends on what you put into it." That saying is especially true for learning about computation with this book. You could simply read the book and hope to absorb the material, but to truly learn about computing you need to experience first-hand how the computer solves problems.

Each chapter features a tutorial project that helps you explore a particular problem and ways of solving it computationally. You are strongly encouraged to have your computer open as you read a chapter, and at the end of each section type in the Python statements exactly as they are shown in the tutorial exercise.

The "computational experiments" described in this book typically start by having you run complete programs that have already been implemented in PythonLabs, a set of modules written specifically for this book. These initial experiments include animations that illustrate the basic steps of the computation featured in that chapter. The remaining sections go into details that show how the key parts of the Python programs were implemented. As you work on the tutorials for these sections you will be writing your own programs, using the code in the book as the starting point.

The tutorials are designed so that you should be able to complete them in about the same amount of time you would spend on a lab project in a chemistry class. You could run through the tutorials in less time—about as fast as you can type, or if you are reading the book online, as fast as you can copy and paste—but you should take the time to make sure you understand what your computer is doing as you carry out each step in the tutorial.

At the end of each chapter you will find a set of exercises. These are similar to the questions you would find in a more traditional textbook and are designed to test your understanding of the material in that chapter. If you have completed the tutorial and understood what happened at each step along the way you should be able to answer these questions.

---

tl;dr

This book is designed to be read one section at a time. Each section has very few pages.

The projects at the end of each section are "tutorial" style exercises that tell you exactly what to type. Do these exercises. They will reinforce what you read.

T0. Type this statement to see a message from Python:

```
>>> print("hello")
hello
```

After you complete a chapter try your hand at the programming projects at the end of the chapter. The only way to learn how to program is to write programs. These exercises will get you started.

Notes for Instructors

The book is organized as a set of projects that gives students an opportunity to explore important ideas in computer science. In most cases, the main concepts are algorithms, and the projects are examples of how algorithms provide computational solutions to important problems.

An interactive programming language like Python provides a "computational workbench" where students can experiment with algorithms by typing expressions and immediately seeing the results. The language sets up an environment where students can run computations and explore the effects of changing parameters or modifying operations performed at key steps of the computation.

The topics presented in the book are outlined below. The general pattern for each chapter will be to first introduce the concept presented in that chapter; this introductory section will essentially be an essay that tries to make the case that the idea is interesting and worth understanding in more detail. The main part of the chapter will be the development, through a series of projects, of one or more algorithms that illustrate the idea and provide the student with a chance to experiment.

1 Introduction

The book starts with a general introduction to computation, focusing on the idea that computer science is not just about computers and is not just programming.

2 The Python Workbench

The second chapter is a brief introduction to Python and how it can be used as a "computational workbench" to set up experiments with computations. The project takes the students through the construction of a few simple programs, for example a function that converts temperature from Celsius to Fahrenheit. This chapter introduces the ideas of objects, variables, functions, and conditional execution.

3 The Sieve of Eratosthenes

This chapter introduces the first real algorithm studied in the book. It also introduces a few more practical techniques used later in the book: making lists of numbers and iterating over a list. The project starts with simple expressions involving integers, shows how to make a list of numbers, then how to selectively remove composite numbers, and leads finally to an algorithm that creates a complete list of prime numbers.

4 A Journey of a Thousand Miles

This chapter builds on the basic idea of iteration presented in the previous chapter. The project shows how iteration can be used to solve two common problems, searching and sorting, using linear search and insertion sort. An important idea in computing in this chapter is scalability, and students are introduced to $\mathcal{O}$ notation.

Chapter	CS Concepts	Python Programming
1. *Introduction*	Algorithms, computation	
2. *The Python Workbench*	Interactive computing	IDE, objects, functions, libraries, import, variables, assignment, numbers, strings, Boolean values, if statements
3. *The Sieve of Eratosthenes*	Iteration, containers, algorithm animation	Lists, list indices, for loops, incremental development
4. *A Journey of 1000 Miles*	Linear search, insertion sort, scalability	While loops, Boolean operators
5. *Divide and Conquer*	Binary search, merge sort, exponential growth, $\log_2$	◆ Recursive functions
6. *Spam, Spam, Spam, Mail, and Spam*	Machine learning, Bayesian inference	File paths, text files, string processing, dictionaries
7. *Now for Something Completely Different*	Pseudorandom numbers, permutations, testing	mod operator, namespaces, classes, instance variables
8. *Bit by Bit*	Binary representations, trees, queues, parity bits, text compression	ASCII, Unicode, bitwise operators
9. *The War of the Words*	von Neumann architecture	
10. *I'd Like to Have an Argument, Please*	Natural language processing	String library, more string methods, regular expressions
11. *The Music of the Spheres*	Computer simulation, computational science	
12. *The Traveling Salesman*	Graphs, genetic algorithms	◆ Generators

◆ denotes extra-credit project

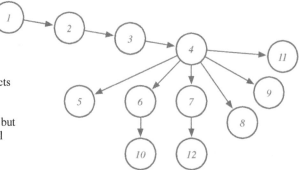

The first four chapters introduce fundamental programming constructs and should be presented in order. Chapters 5 through 7 also present important programming constructs but may be omitted if later projects will not use these constructs.

5 Divide and Conquer

The important idea in this chapter is that a more sophisticated strategy for solving a problem can lead to a more efficient computation. The project shows how binary search takes up to $\log_2 n$ steps instead of n, and merge sort takes at most $n \log_2 n$ steps instead of n^2.

6 Spam, Spam, Spam, Mail, and Spam

The algorithm used for experiments in this chapter is a Bayesian spam filter, a simple example of how "big data" can influence decision making. The chapter also introduces Python constructs for reading text files, Python dictionary (associative array) objects, and has additional exercises in string processing.

7 Now for Something Completely Different

The big idea in this chapter is randomness, and how random numbers can be used in a variety of algorithms, from games to scientific applications. There is an interesting paradox here: can we really generate random outputs from an algorithm? The answer is that random numbers generated by an algorithm are *pseudorandom*, and the project takes students through the steps in the development and testing of a pseudorandom number generator. This is the chapter where students are introduced to module structure in Python and get a chance to write their own simple class definitions.

8 Bit by Bit

The projects in this chapter are related to encoding data: using patterns of binary digits to encode numbers and letters, the number of bits required to encode a set of items, text compression with Huffman trees, and error correction with parity bits.

9 The War of the Words

This chapter introduces the important ideas that functions can also be encoded as a string of bits and that instructions are stored in a computer's memory along with data. The project uses the game of Corewar, which is a contest between two programs running in the same virtual machine; a program wins if it can write a halt instruction over the opponent's code. The projects lead the student through the phases of a processor's fetch-decode-execute cycle and emphasizes how a word that is a piece of data (the constant 0) for one program becomes an instruction (halt) for the other program.

10 I'd Like to Have an Argument, Please

The project in this chapter is based on a Python implementation of Joseph Weizenbaum's ELIZA program, and shows how very simple pattern matching rules can be used to transform input sentences, giving the illusion that the computer is carrying on a conversation. By the end of the chapter students will see how difficult natural language processing is, and how semantics and real-world knowledge are required for effective natural language understanding.

11 The Music of the Spheres

The big idea in this chapter is computer simulation. The project leads to an *ab initio* simulation of the motion of planets in the solar system. The chapter introduces issues related to verification and other topics in computer simulation.

12 The Traveling Salesman

The final chapter introduces the idea of intractable problems, building on ideas of scalability from earlier chapters. The project is based on a genetic algorithm and gives students the opportunity to explore probabilistic solutions. The tutorial has students use predefined code for Map and Tour classes to create random tours, so they can see how tours can be mutated and how collections of tours evolve until an optimal or near-optimal solution is obtained.

Pedagogical Considerations

This book was written primarily for CS0 courses, where the goal is to introduce the key concepts and, as much as possible, give a broad overview of the field. If augmented by additional material on Python programming and further programming exercises it could also be used for a more in-depth introduction as part of a CS1 course.

The projects have been used in courses at the University of Oregon. At UO we cover the first two chapters during the first week, but after that we spend between one and two weeks on the remaining topics chosen for that term. Lectures emphasize material from the first sections of a chapter, describing the problem and how it might be solved computationally, and explaining how that week's lab project gives some experience with the computation. Students have an option of attending a lab session, where an instructor is available to help them work through the material, but many students do the tutorials on their own. Live demonstrations of the tutorial projects, both in lecture and in lab sessions, have proved to be very effective.

At the end of each chapter there is a set of exercises that asks questions about issues raised in the chapter. After the students have completed the tutorial, they are asked to answer a selected set of questions and submit them as a "lab report" that gives them a chance to explain what they learned. Similar questions are given on exams.

Software, Documentation, and Lab Manuals

The software that accompanies this book is a set of modules named PythonLabs. Python-Labs is written exclusively in Python, using only libraries and modules that are part of the standard Python distribution. There is one Python module for each lab project. All of the modules have been collected into a single "egg," which makes it easy to install all the lab software in one step at the beginning of the term. The PythonLabs modules also include data files and sample Python code that students can copy and modify.

A Lab Manual with step-by-step instructions for installing Python and PythonLabs is available from the book website at http://www.cs.uoregon.edu/eic. There is a separate version of the manual for Windows XP, Mac OS X, and Linux. The manual also includes tips for editing programs and running commands in a terminal emulator.

Acknowledgments

This book is the result of many years of teaching introductory courses, and the material has evolved considerably over that time. The students and teaching assistants involved with these courses, and my colleagues at Oregon and elsewhere, have had a major influence on the topics and exercise presented here. I am very grateful for their comments and feedback.

I would like to thank Tom Cortina, Dilsun Kaynar, and Roger Dannenberg from Carnegie-Mellon University and the students in their Principles of Computing course who offered to "test drive" preliminary versions of this Python edition.

Randi Cohen, my editor at CRC Press, was instrumental in getting the first book published. I never would have considered writing a new version of *Explorations in Computing* without her patience and timely encouragement.

As always, I am eternally grateful for the support of my wife Leslie and my daughter Kathleen, who bears at least part of the responsibility for the Monty Python references in the section titles. I love you both.

John Conery
Eugene, Oregon

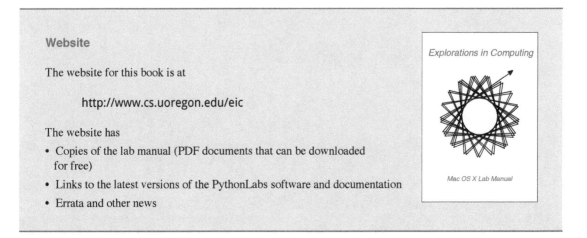

Website

The website for this book is at

      http://www.cs.uoregon.edu/eic

The website has

- Copies of the lab manual (PDF documents that can be downloaded for free)
- Links to the latest versions of the PythonLabs software and documentation
- Errata and other news

Explorations in Computing

Mac OS X Lab Manual

Chapter 1

Introduction

In the summer of 1821, an English mathematician named Charles Babbage was working with his friend, the astronomer John Herschel, to create a book of mathematical tables. Before computers and calculators were available, tables were used by people who needed to solve mathematical equations. To find the value of a function one would search through a table in a reference book like the one shown in Figure 1.1. These books were essential for navigators, architects, merchants, bankers, and anyone else who used math in their profession. There were tables for interest rates, currency conversion, liquid and dry measures, and just about any other quantity that could be expressed with numbers. Not surprisingly, there was a strong demand for accurate tables, and a tremendous amount of time and effort went into creating and checking values printed in reference books.

Babbage and Herschel were working on a nautical almanac, a book of tables containing the positions of the Moon, planets, and stars. These almanacs were used by navigators to determine their location at sea. More than any others, these tables needed to be accurate, as there were concerns that errors in tables could lead to longer routes than necessary, or even shipwrecks if the crew veered off course. Babbage and Herschel would meet periodically to check the tables being made by a group of people they had hired to work through the tedious steps required to fill the rows. At one of their meetings, when reviewing the latest results, Babbage showed his frustration with the large number of errors by exclaiming, "I wish to God these calculations had been executed by steam!"

Of course Babbage wasn't attributing any special powers to water vapor. The year 1821 was the height of the Industrial Revolution, when machines powered by steam engines were beginning to automate tasks previously carried out by humans. Babbage was simply using the terminology of his time to express his wish that the calculations should be done automatically, by a machine, so the results would be more accurate and reliable.

The quote from Babbage brings up an interesting question. Steam power was helping transform physical labor, and machines were beginning to be widely used to augment, or even replace, human effort. But what about mental labor? What made Babbage think steam engines could help him solve mathematical problems?

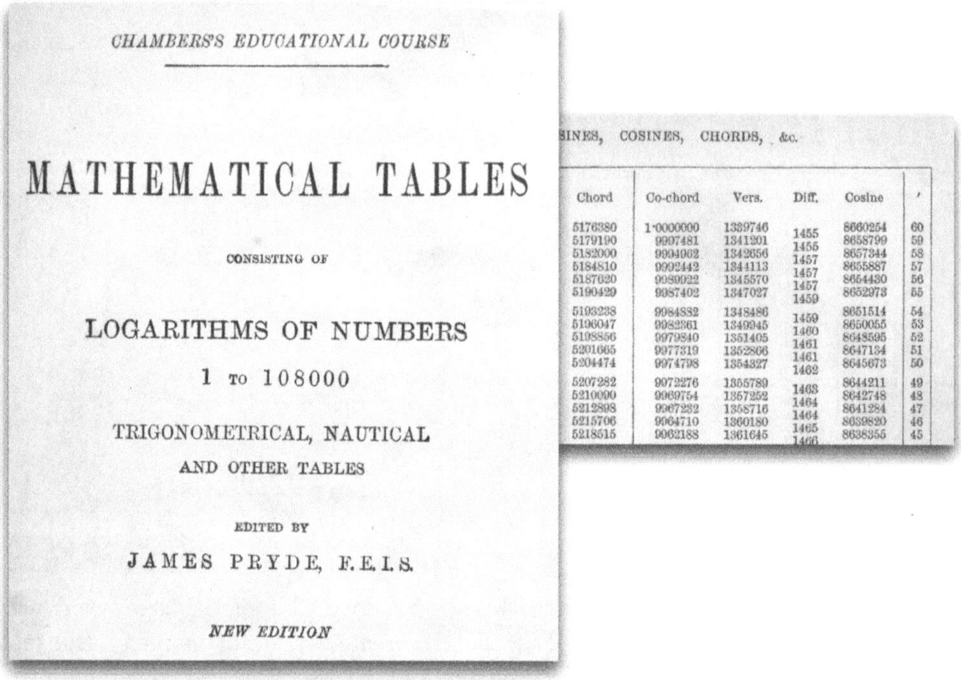

CHAMBERS'S EDUCATIONAL COURSE

MATHEMATICAL TABLES

CONSISTING OF

LOGARITHMS OF NUMBERS

1 TO 108000

TRIGONOMETRICAL, NAUTICAL

AND OTHER TABLES

EDITED BY

JAMES PRYDE, F.E.I.S.

NEW EDITION

SINES, COSINES, CHORDS, &c.

Chord	Co-chord	Vers.	Diff.	Cosine	′
5176380	1·0000000	1339746	1455	8660254	60
5179190	9997481	1341201	1455	8658799	59
5182000	9994962	1342656	1457	8657344	58
5184810	9992442	1344113	1457	8655887	57
5187620	9989922	1345570	1457	8654430	56
5190429	9987402	1347027	1459	8652973	55
5193238	9984882	1348486	1459	8651514	54
5196047	9982361	1349945	1460	8650055	53
5198856	9979840	1351405	1461	8648595	52
5201665	9977319	1352866	1461	8647134	51
5204474	9974798	1354327	1462	8645673	50
5207282	9972276	1355789	1463	8644211	49
5210090	9969754	1357252	1464	8642748	48
5212898	9967232	1358716	1464	8641284	47
5215706	9964710	1360180	1465	8639820	46
5218515	9962188	1361645	1466	8638355	45

Figure 1.1: Chambers's Mathematical Tables, New Edition, *London, 1901. Before there were computers to calculate mathematical functions, a person who wanted to know the value of a trigonometric function would look in a table of sines and cosines. For example, to find the value of* $\cos 30°20'$, *find the page for* 30°, *then scan down to the row for* 20′ *and look in the column labeled "cosine."*

The answer is that the idea of **computation**—solving a complex problem by repeated, systematic execution of a series of simple and straightforward operations—was already well established by the nineteenth century. Mathematicians had developed techniques for calculating entries in tables using using only the most basic operations of arithmetic, such as addition and subtraction, where the value in one row of the table can be determined using values from rows filled in previously. Many tables, including the ones in the book being prepared by Babbage and Herschel, were produced by groups of people who had no advanced mathematical skills, but were hired and trained to fill in a table by doing a specific sequence of additions and subtractions. Prior to the middle of the twentieth century, the word "computer" was a job title, referring to any person who was engaged in systematic calculation of values like those found in mathematical tables. Babbage realized that the simple operations carried out by human computers were mechanical in nature, and he dreamed of one day building a machine that would be able to carry out the steps in a computation automatically.

Today computation is so familiar we take it for granted. A navigator, surveyor, architect, or anyone else who needs to know the value of a mathematical function simply enters a number in a calculator and presses a button labeled with the name of the function. Computation is also at the heart of computer applications that help us with common tasks that seemingly have little or nothing to do with mathematics, such as using a word processor to write an essay, organizing a music library, or playing recorded music.

This book, *Explorations in Computing*, is a book about computation. The focus is on the nature of computation, with the goal of showing how executing a series of simple steps can eventually lead to the solution of a complex problem.

The book is also an introduction to computer programming. A series of programming projects based on Python, a simple but powerful programming language, will give readers a chance to implement the algorithms introduced in each chapter. The projects are "computational experiments" that will help bring the idea of computation to life.

1.1 Computation

Charles Babbage was not the first person to dream of using a machine to help him with his math. The idea of solving a complex problem by systematic execution of simple and straightforward operations is thousands of years old. Ancient Greek, Egyptian, and Chinese philosophers all discovered many important facts about numbers and their relationships. These mathematicians developed methods that are still used today to determine whether a number is prime or how to find the largest common denominator of a pair of numbers. They also devised many different tools to help them perform their calculations; in fact, the word "calculate" comes from the Latin for the pebbles that were used in an abacus or similar device.

Many famous mathematicians in post-Renaissance Europe, including Johannes Kepler (1571–1630), Blaise Pascal (1623–1662), and Gottfried Leibniz (1646–1716), dreamed that one day there would be machines to carry out the steps in a computation automatically. As Leibniz once wrote,

> *Astronomers surely will not have to continue to exercise the patience which is required for computation. It is this that deters them from ... working on hypotheses and from discussions of observations with each other. For it is unworthy of excellent men to lose hours like slaves in the labor of calculation which could safely be relegated to anyone else if machines were used.*

Pascal invented a mechanical calculator that was able to perform additions and subtractions of numbers up to six digits long. Leibniz designed a calculator that could also do multiplications and divisions. These early machines were able to assist human computers, in much the way an abacus or other device might. It wasn't until the nineteenth century that designs for fully automated machines started to appear. Babbage himself played a key role. He designed a machine he called the Difference Engine, coming up with ingenious ideas that later influenced the builders of some of the first successful mechanical computers in the middle 1900s.

The first electronic computers were developed during World War II. Soon after the end of the war, the idea of using machines to automate the calculations in science and business began to spread, and several companies started manufacturing computing machines. The machines were very big and very expensive, however, and were found only in the largest corporations, government bureaus, or university research labs. They were used for such diverse tasks as computing the trajectories of rockets and missiles, predicting the weather, and business data processing applications for payroll and accounting.

Fast forward fifty years, and computation has now become an essential part of modern life. Every day we write mail, share photographs, play music, read the news, and pay bills using our personal computers. Engineers use computers to design cars and airplanes, pharmaceutical companies use computers to develop new drugs, movie producers generate special effects and in some cases entire animated films using computers, and investment firms use computational models to decide whether complex transactions are likely to succeed. Not surprisingly, given the role astronomy has played in the history of computing, modern astronomers also rely heavily on computation. Organizations like the Jet Propulsion Laboratory use computers to carry out the calculations that track the locations of planets, asteroids, and comets with the goal of keeping an eye out for potential threats from impacts, such as the one involving the comet Shoemaker-Levy and Jupiter in 1994.

Computation plays a much more extensive role in modern science than the straightforward "number crunching" involved in the calculation of orbits. The phrase *computational science* refers to the use of computation to help answer fundamental scientific questions. Here the word "computational" is an adjective that describes how the science is done. Computational science is, like the more traditional approaches of theoretical science and experimental science, a way of trying to solve important scientific problems. Computational physicists use computers to study the formation of black holes, investigate theories of how planets form, and simulate the predicted collision, three billion years from now, of our Milky Way galaxy with Andromeda.

As we look at the wide variety of problems that are being solved with the help of computation, several questions naturally arise. Do these problems have anything in common? Is there some aspect of a problem that would lead one to believe it can be solved computationally? For that matter, what does it mean to "compute" something?

The generally accepted definition of a **computation** is that it is a sequence of simple, well-defined steps that lead to the solution of a problem. The problem itself must be defined

---

Squaring without Multiplying

You might think that to compute the value of n^2 you would need to know how to multiply $n \times n$. But there is an easy way to square a number using only additions.

The diagram at right shows how to compute 5^2. At the top of a piece of paper write the first 5 odd numbers. Add the first two numbers and write the sum below them. Then add the next odd number to the sum computed on the previous step. Keep adding numbers from the top row, and after adding the last number you'll have the value of 5^2.

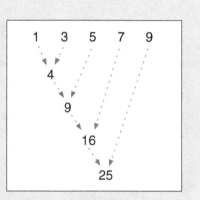

This process is known as the "method of differences" and it is similar to the process Babbage and Herschel taught their workers. In 1823 Babbage started work on a mechanical computer called The Difference Engine that was based on this technique for evaluating polynomials.

exactly and unambiguously, and each step in the computation that solves the problem must be described in very specific terms.

As a simple example of a computation, suppose an ecologist wants to compare the lengths of fish found in two different streams. One way to do this is to calculate the average length of the fish samples from each stream. To solve this problem, the ecologist first needs to specify exactly what is meant by "average" length. The most common definition is the arithmetic mean. Given this definition, the computation starts by summing the lengths of the fish found in the first stream and computing the mean by dividing the sum of lengths by the number of fish (Figure 1.2). Next find the mean for the second stream by adding the lengths of those fish and dividing the sum by the number of fish in that stream. This is an example of a computation because each step (counting, addition, division) is very straightforward, and there is a clear specification of the starting point and the end result.

Note that there is nothing in the description of computation that involves the word "computer." A computation is a process, a sequence of simple operations that leads from an initial state to the desired final result. The process can be carried out entirely by a person, or by a person using the help of mechanical or electronic calculators, or completely automatically by a computer.

The choice of which technology would be most effective depends on the situation. For a very small number of fish, an ecologist could write all the numbers in a single column on a sheet of paper, add them up, and then do the division using paper and pencil. For a larger sample, a person would likely want to use a calculator or abacus; the state of the device will show the running total, so that after the last length is added the final step is to divide the total by the number of fish. For very large samples—for example, using information in a database made by storing videotaped observations of fish passing through a ladder next to a dam—it would be most efficient to use a computer and program it to add the lengths and do the division.

No matter what technology is used, these situations all involve the same basic computation: in each case the average length is determined by calculating the sum of lengths and then dividing by the number of fish. Note also that the process of computing the mean of a set of numbers applies to other situations as well. This same basic computation, of finding the sum of values in the set, and then dividing by the size of the set, can be applied to (almost) any set of numbers.

1.2 The Limits of Computation

In looking to the future, a natural question is whether the astonishing growth in the use of computation will continue. Most people who work in fields related to computing expect technology to continue to improve, with the components used to build computers getting smaller, faster, cheaper, and more energy efficient. The phrase "ubiquitous computing" refers to the prediction that computers will be so small and so inexpensive they will be everywhere, perhaps even in the clothes we wear. With this sort of explosion in the availability of computing power, it's natural to wonder what sorts of tasks these computers might be asked to perform. One way to address the question is to take a closer look at some of the areas where computers are being used today and ask whether there are any limitations on what might be done in the future.

Length (cm)	Average (Mean) Length:
Stream A	
40, 39, 69, 57, 50	(40 + 39 + 69 + 57 + 50) ÷ 5 = 51
Stream B	
44, 62, 59, 58	(44 + 62 + 59 + 58) ÷ 4 = 56

Figure 1.2: *To compare the average length of fish in two different streams, an ecologist would compute the mean length of fish seen in each stream. The mean length is defined as the sum of lengths divided by the number of fish.*

Often the limits to what we can do with a computer are **technological barriers** that could be overcome by using a more advanced piece of hardware. In principle, one could type an essay for a literature class on a cell phone or create a full-length animated feature using a laptop. But it would be more productive to write a paper on a computer with a mouse, a keyboard, and a screen large enough to show a full page of text. Because each frame in an animation is the result of hours of computing on a high-speed supercomputer, a personal computer is not a practical choice for making an animated movie.

There are many important problems that could be solved by computers but are limited by current technology. To take just one example, equations that model changes in weather are very well understood by meteorologists, but the accuracy of weather predictions is currently limited by computing power. With more powerful computer systems (and more detailed measurement of current conditions) it may be possible to make short-term weather forecasts with almost perfect accuracy.

In many cases, however, the limits of what a computer can do are things that cannot be overcome by moving to a more advanced piece of hardware. These barriers might be characterized as **computational limits**: the difficulty lies in specifying how to solve a problem computationally rather than in the hardware used to carry out the computation. Here are some problems that would be very difficult, if not impossible, to solve by a computer:

- When we want to send a text or e-mail to someone, we need to know their contact information. If the person is in our address book, it's trivial to look up their e-mail address. But if we don't know the person we're stuck. In some cases we can do a web search, perhaps to find the name or e-mail address at a company we want to correspond with, but we can't ask a computer to find the e-mail address of someone we just met at a coffee shop.

- When using a computer to manage finances, we can connect to a bank or credit card company to download a list of transactions, and we might be able to have a program do some calculations for different investment options, based on interest rates or projected earnings. But we know the computer can't choose the perfect investment because it can't predict which companies will succeed or future interest rates.

- A computer can help find mileage estimates for different types of cars, or admission and enrollment statistics for different colleges. But a computer can't determine whether the fun of driving a flashy convertible outweighs the practicality of an all-wheel drive station wagon, and people usually find it difficult to quantify all the attributes of different colleges and universities that would allow a computer to make a perfect decision on the best school to attend.

- It is common to talk with a computer via telephone to make a train reservation or arrange air travel. But these "conversations" are very limited, and generally we are restricted to one-word sentences that relate to a specific reservation. We would not get very far if we asked the computer for general travel advice, for example, "What's the weather like in Los Angeles this time of year? Should I go to Hawaii instead?"

In the previous section, computation was defined as a process that follows a series of well-defined steps that lead from an unambiguous starting state to an equally well-defined final result. With this definition in hand, let's take a closer look at the situations described above and some other difficult problems.

Ambiguous Problems

The problem of sending a message to a person we meet at a coffee shop is simply too vague to be solved computationally. It lacks a well-defined starting point, and there is no sequence of steps a computer can carry out to solve it. It's worth noting this is a problem that cannot be solved by a human, either—a friend won't be able to help any more than a computer unless we supply a lot more information.

Some attributes students use when deciding which college to attend are well defined. The cost of tuition, living expenses, and the average GPA and SAT scores of entering freshmen are important factors. If these easily quantified items are the only criteria that are important, a computation could lead to a decision of which college is best.

For most people, however, intangible qualities like geographical location and quality of life are important, and these are hard to put into terms that could be used in a computation (Figure 1.3). People do solve problems that involve intangibles, and it might seem like this is the sort of thing a person could do better than a computer. But the "solutions" obtained by people aren't like the solutions to the problem of computing the mean length of fish: they are recommendations, not unambiguous and reliably correct answers.

Figure 1.3: *In order to use a computer to choose a college to attend, each attribute would have to be well defined and quantifiable.*

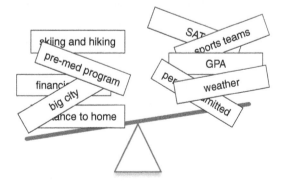

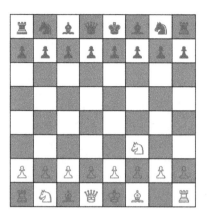

Figure 1.4: *If a computer tries to analyze every possible sequence of moves in response to this opening in a game of chess, it will have to consider over 10^{43} different games.*

Natural Language

It would be nice if we could open a cell phone and say, "Send a message to Aleah, Katie, and Erica to see if they want to come over to study calculus." We could then work on something else while the phone composes a text message, sends it to our friends' phones, and negotiates with the other phones to find a time when everyone wants to meet. This sort of interaction is not possible with current computers, but researchers in a field known as **artificial intelligence** are trying to understand what is involved in these types of communications and developing computational methods to carry them out. A human personal assistant can accomplish this task, so this is an example of the kind of problem that humans solve but computers (currently) do not. It is an open question whether this problem is beyond the limits of computation. It very well might be possible for some future personal digital assistant to accomplish this task.

Intractable Problems

It's tempting to think a computer could easily win a chess tournament by considering all possible moves starting from a given board configuration, then examining each possible response its opponent could make, and eventually choosing the move that leads to a certain victory (Figure 1.4). This is an example of a problem that should be solvable by a computer. The problem has a well-defined input (the starting configuration of a chessboard), output (configurations where each move leaves the opponent's king in "check"), and very specific and well-defined set of operations (the legal moves for each chess piece).

Before we try to write a program that uses this strategy, however, we should do some back of the envelope calculations first. The number of possible chess games has been estimated to be more than 10^{43}. Even if this computation were run on a supercomputer far more powerful than the fastest computers available today, on a hypothetical machine that could somehow compute 1 trillion (10^{12}) alternative board combinations per second, it would require $10^{43}/10^{12} = 10^{31}$ seconds, or roughly 10^{21} years. To put this in perspective, the universe is only 10^{13} years old. So even though this is a well-defined computation, it is one no computer will ever complete, and in that sense it is well beyond the limits of computation. It goes without saying that no human will ever do this computation either. When humans and computers play chess successfully, they are using strategies other than simple "brute-force" exploration of all possible moves.

Computer scientists refer to this sort of problem as **intractable**. Small portions of the problem, such as different opening strategies or endgames, can be analyzed by considering every option, but evaluating every move in a full game is beyond reach. Chapter 12 explores another intractable problem, one that is often faced by organizations that need to do a substantial amount of scheduling or planning.

Unsolvable Problems

Mathematicians in the 1930s made the startling discovery that some problems are simply unsolvable. In logic, these problems are called *undecidable*. The group of unsolvable problems includes determining whether paradoxes, like the familiar statement "this sentence is false," are true or false. The statement can't be true, because that would imply it is false, and likewise it can't be false, because that would make it a true statement about itself. In mathematical terms it is simply undecidable.

The computer science equivalent of the undecidable problems are called **noncomputable functions**. The most well-known, the Halting Problem, asks whether it is possible to examine a running program and decide whether that program will ever terminate. Imagine a situation where an application is running on a laptop, and an icon appears on the screen that says the program is busy, so it will not respond to any keystrokes or mouse clicks. After five minutes we might start to wonder whether the application is progressing very slowly or has crashed. It would be nice to be able to run another program that would examine the first one and say, "Be patient, it will terminate" or "It crashed, you need to kill it and restart it." A fundamental result in theoretical computer science tells us that this problem is undecidable and that it is impossible to write such a "halt-checking" program.

Uncomputability is a different type of limitation than intractability, the limitation encountered by the chess-playing program. The chess player will compute the perfect game of chess if we are patient enough to wait 10^{21} years, so it is only unsolvable in a practical sense. The halt-checker requires us to evaluate an undecidable function, so it is beyond the limits of computation in that we know it is impossible to write a general-purpose program that could carry out a sequence of steps that will let it determine whether or not another program will terminate.

1.3 Algorithms

Let's return to the simple example of a computation presented in Section 1.1, where an ecologist wants to know the average length of fish observed in a stream. We now know what the computation involves: sum up the lengths and divide by the number of fish. The next question is, how do we describe the computation in sufficient detail so the steps can be carried out by a machine?

First consider how the ecologist might enlist the aid of a human research assistant. If the assistant has taken a statistics class, the ecologist can just give the assistant the data and expect him to compute the mean lengths. But if the assistant does not know how to compute a mean, the ecologist needs to describe the operation in detail: write the list of numbers on a piece of paper and then cross them off one by one as they are added to a running sum; after adding the last piece of data, divide by the number of fish.

A detailed description of how to solve a problem by first specifying the starting conditions and then how to follow a set of simple steps that leads to the final solution is known as an **algorithm**. An algorithm is characterized by

- A precise statement of the starting conditions, which are the inputs to the algorithm;

- A specification of the final state of the algorithm, which is used to decide when the algorithm will terminate;

- A detailed description of the individual steps, each of which is a simple and straightforward operation that will help move the algorithm toward its final state.

In short, an algorithm is a specification for how to carry out a computation. Although the word *algorithm* can be used to refer to any method for systematically solving a problem, and algorithms were widely used long before anyone thought of building a machine to perform the steps in a computation, today the term generally refers to a method that will be carried out automatically by a computer.

In the description of an algorithm, the steps must be simple enough to be "understood" by a machine. One way to think of what a machine is capable of doing is to think in terms of symbols, such as numbers or letters. The steps in an algorithm are basically **symbol manipulations** like simple arithmetic operations or comparisons that determine which words come before others in the alphabet. By putting together a large number of simple symbolic operations, a machine can do very complex tasks, such as sorting long lists of names, counting millions of votes cast in an election, or using words to build an index of web pages gathered from the Internet.

As an example of how a task can be described as a sequence of symbol manipulations, the process followed by the research assistant to compute the mean length of a group of fish is shown in Figure 1.5. On the left is the algorithm, showing what needs to be done at each step. The different stages of the computation are shown on the right. To initialize the computation, the number 0 is entered into a calculator, and all the values in the data set are written out on a sheet of paper. Each time the assistant removes a number from the list, it is added to the running total on the calculator and crossed off the list. By the time

---

A Brief History of Algorithms

The world *algorithm* comes from the name of a Persian mathematician, Mohammed ibn Mûsâ al-Khwârizmî (*ca*. 780–850), whose book on the use of Indian numerals introduced Europeans to the numeral 0. The book was translated into Latin as *Algoritmi de numero Indorum* ("al-Khwârizmî Concerning the Hindu Art of Reckoning").

The earliest algorithm, now known as Euclid's Algorithm, dates from at least 300 BC and is still used today to find the lowest common denominator of two numbers. Other ancient algorithms include the Sieve of Eratosthenes, a method for making lists of prime numbers (we'll explore this algorithm in Chapter 3), and methods used by Sun Tzu and other Chinese mathematicians around AD 200.

(a)

Input: L, a list of numbers

Initialize S (the sum) to 0

Let N be the number of items in L

Repeat until L is empty:

    • Add the first item in L to S

    • Remove the first item from L   ★

Output: S divided by N

(b)

S	L
0	40, 39, 69, 57, 50
40	~~40~~, 39, 69, 57, 50
79	~~40~~, ~~39~~, 69, 57, 50
148	~~40~~, ~~39~~, ~~69~~, 57, 50
205	~~40~~, ~~39~~, ~~69~~, ~~57~~, 50
255	~~40~~, ~~39~~, ~~69~~, ~~57~~, ~~50~~

255 / 5 = 51

Figure 1.5: *(a) An algorithm for computing the mean of a set of numbers. (b) An example of a computation specified by the algorithm. The first line shows the initial state, with the sum set to 0 and the full list of numbers. The remaining lines show how the sum is updated by removing the first number from the list and adding it to the running total. After the last number has been removed from the list, the sum is 255, so the final result is computed by dividing 255 by 5.*

the last number has been crossed off, the value displayed on the calculator is the sum of all the lengths. Each step in the algorithm is a symbol manipulation, where the list becomes shorter by one item and the sum has been updated through simple arithmetic operations (which could easily be done by hand as well as a calculator). The final output is a single number, again the result of a simple arithmetic operation, in this case a division.

Often the descriptions of the steps of an algorithm are given in English or another human language, as in the algorithm for computing the mean value of a set of numbers shown in Figure 1.5. This notation, which is sometimes called **pseudocode**, is sufficient for talking about the algorithm, for describing the process to another person, or for trying to understand whether or not the algorithm works. But in order to run the algorithm on a computer, the steps must be written more precisely as statements in a programming language. If the ecologists find they are spending too much of their time computing means by hand, they might decide to invest some time in implementing their algorithm in the form of a program and running it on their computer.

The idea that one can solve a problem through the use of an algorithm is the central concept in computer science. Computer scientists analyze the theoretical properties of algorithms, develop programming languages used to implement algorithms, and design computer systems to automatically execute the steps in algorithms. As technology opens the doors for new applications, computer scientists work with researchers in other fields to find ways to solve important real-world problems, either by inventing new algorithms or improving and adapting existing algorithms.

1.4 A Laboratory for Computational Experiments

The main goal for this book is to explore the fundamental idea of computation and how it can be used to solve some interesting and important real-world problems. A typical chapter will introduce a problem, explain why it is important, and give an overview of one or more algorithms that have been used to solve the problem.

The heart of each chapter is a Python programming project based on the algorithms described in that chapter. These projects are laid out in the form of a **tutorial** that begins in the first section and continues through the remaining sections in that chapter. In the first few chapters, the tutorials introduce basic building blocks of Python programming, showing readers how to create simple pieces of data and write statements that manipulate the data. Later chapters explore more advanced concepts in computing, but still include detailed instructions for how to develop a Python program to implement the algorithms from that chapter. At the end of each chapter there are suggestions for extending the programs plus additional ideas for independent programming projects.

A useful analogy for these tutorials are projects in an introductory chemistry course. Instructors design lab projects by selecting materials and methods that are accessible to beginners, and students follow detailed instructions to carry out experiments that help them learn fundamental concepts about chemical processes. The computational projects in this book are similar: the programs and data have been prepared ahead of time, and the tutorial guides students step by step to gain some insight into how an algorithm works.

The laboratory for these computational experiments can be any small desktop or laptop computer. The software we will use is an interactive environment for developing and testing Python programs. An interactive environment works much like a calculator: users type in expressions, and the system performs a calculation and prints a result.

In the jargon of computer programming, Python supports a style of programming known as **object-oriented** programming. These languages use the word "object" not only to refer to pieces of data, such as numbers and text, but also to collections of data, such as lists of numbers, and to things like functions that can be applied to other objects. In each chapter we will be using Python as our computational laboratory, setting up a small virtual world where we create objects and carry out operations designed to experiment with algorithms related to the main topic of that chapter.

Here are some examples of how an interactive environment based on an object-oriented programming language will be used to set up and carry out experiments with computations:

- The first nontrivial algorithm presented in this book is the Sieve of Eratosthenes, a very old algorithm that has been used since the time of the ancient Greeks to make lists of prime numbers. The name of the algorithm is a hint to the basic idea: create a list of numbers and then sift out those that are not prime. It is easy to set up a straight-forward program that repeatedly works its way through the list. After experimenting with the method, we will find that it is not necessary to do as many sifting operations as one might think, and the insight we gain from the initial experiments will be used to implement a more elegant version that does far less work.

- One of the early milestones in artificial intelligence was a program named ELIZA, which gave the appearance of carrying out a conversation by playing the role of a psychiatrist. A user would type a sentence on a computer terminal, and ELIZA would respond. For

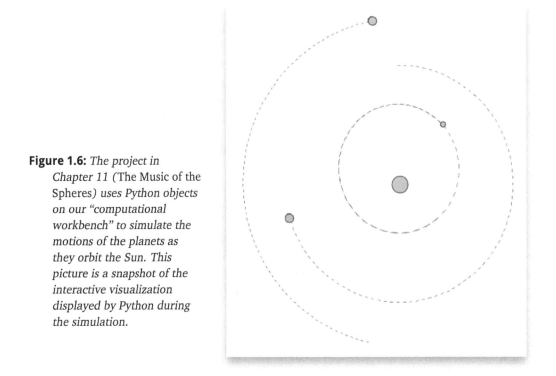

Figure 1.6: *The project in Chapter 11 (The Music of the Spheres) uses Python objects on our "computational workbench" to simulate the motions of the planets as they orbit the Sun. This picture is a snapshot of the interactive visualization displayed by Python during the simulation.*

example, if a person typed "I don't like computers," the program might print "Do computers worry you?" What was fascinating about ELIZA was how well it seemed to participate in a conversation, in spite of the fact that it only did very simple syntactic transformations on input sentences. Our project will create objects that represent the transformation rules and run experiments that apply the rules to test sentences.

- In a modern-day version of the computations supervised by Babbage and Herschel, we will use Python to simulate the motions of the planets as they orbit the Sun (Figure 1.6). The Sun and each of the planets will be represented as Python objects. We will see how to run the simulation and watch the motions of the planets in an interactive visualization. We will also be able to change the simulation parameters, for example, to see what would happen if there were two large objects the size of the Sun.

- A classic problem from the world of mathematics, known as the Traveling Salesman Problem, has the same basic structure as several important real-world problems that require efficient schedules. We will use what is known as a "genetic algorithm" to experiment with one way of solving the Traveling Salesman Problem. Each object in this project will represent a complete tour. We will create a set of tours and put them in a "virtual Petri dish," then sit back and watch as the tours mutate and evolve, eventually giving rise to an efficient solution to the problem.

1.5 Summary

Computation is a very old idea that was well established by time of the Industrial Revolution. Mathematicians had long dreamed of making "thinking machines" that could carry out the steps of a computation, and perhaps inspired by the machines starting to be used in manufacturing, transportation, and other fields, visionaries like Charles Babbage began to think of ways of building machines for computation.

Today computing machines are so widely used they are taken for granted. But as marvelous as they are, modern machines are doing computations the same way human computers did for thousands of years, following detailed instructions specified by an algorithm.

This book is an introduction to the field of computer science. We will look at a wide variety of interesting problems as a way of learning about different kinds of algorithms. Given the name "computer science" one might think the field could be characterized as "the study of computers," but as we have seen the idea of computing has been around a lot longer than there have been machines to do the computations. In the words of one influential computer scientist, Edsger Dijkstra (1930–2002), "Computer science is no more about computers than astronomy is about telescopes." Computer hardware plays a huge role, of course. Much of the motivation for studying computer science comes from the fact that computations are run on machines that perform a wide variety of essential tasks. For many people, a large part of the satisfaction of working in computer science derives from the fact that abstract ideas can be turned into programs that run on real computer systems and address important real-world problems.

Each of the remaining chapters will explore some of these problems. The chapters will introduce algorithms that can help solve the problems, and tutorial projects will give you a chance to experiment with the algorithms in order to gain a deeper understanding.

As you read the chapters it will be tempting to skim the tutorial exercises without actually doing each step on your computer. Most of the steps are very simple, and the outcomes may appear obvious. However, you are strongly encouraged to start an interactive session with Python and to carry out all the steps in each "computational experiment." By seeing how Python executes statements, and by watching the interactive visualizations that accompany most of the lab projects, you will gain valuable first-hand experience. Most importantly, you can use these projects as staring points for your own explorations.

Explore Further

Three books on the history of computing that provided much of the background for this chapter are

- *The Computer from Pascal to von Neumann* was written by Herman H. Goldstine, one of the pioneers in the field who worked with John von Neumann on one of the first electronic computers.

- *When Computers Were Human*, by David Alan Grier, tells the story behind several large-scale computing projects in the days before computing machines.

- *The Difference Engine: Charles Babbage and the Quest to Build the First Computer*, by Doron Swade, describes the nautical almanac project supervised by Babbage and Herschel.

The Computer History Museum (http://www.computerhistory.org) has a variety of materials on early computers, including an online exhibit of Babbage's Difference Engine with a video of a complete reproduction of the machine in action.

Important Concepts Introduced in This Chapter

computation	A sequence of simple, well-defined steps carried out to solve a problem
algorithm	A description of how to solve a problem computationally; an algorithm includes a precise statement of the problem (the input), the desired solution (the output), and the order in which the steps will be executed during the computation
limitations	A problem might not be solvable by computation because it is ambiguous, it requires too many steps to complete, or it is mathematically impossible

---

Exercises

1.1. Consider some of the tasks you do with the help of your computer. What are some of the limitations to what the computer can do? Could you do a better job if you had more powerful technology? Or are the limits related to the nature of problem, something that makes the computation inherently difficult to specify or carry out?

1.2. Below is a list of fields where computers have been used. In some cases computers are well established, and people who work in that area rely heavily on computation, but in other cases the use of computers is still very tentative. Pick an area that interests you and write a short paper on how computers help solve problems in that field. Start by writing down some initial impressions and then do some research on the Internet to see what progress is being made by computer scientists and their colleagues from the problem domain. Questions to ask yourself as you do your research might include, "What are the barriers to the use of computers in this field? Are those limits technological or computational? What are some of the social impacts and ethical issues arising from the use of computers?"

    a) Medicine: Does your doctor use a computer in his or her practice? Can computers diagnose illnesses or prescribe medicines? What does the phrase "personalized medicine" refer to? Will computers play a role in personalized medicine?

    b) Pharmacology: What is "rational drug design"? What role does computation play in the development of new drug treatments?

    c) Engineering: What role do computers play in the design and construction of new cars? airplanes? bridges? How has computing changed the way engineers work?

    d) Architecture: How are computers used to plan new buildings? How do they help architects come up with energy-efficient designs?

    e) Meteorology: Are computer models being used to generate weather forecasts? Do they track hurricanes and other storms? How well do these models predict weather 24, 48, or 96 hours in advance?

    f) Art and Entertainment: How have computers had an effect on music, video, or other artistic endeavors?

    g) Libraries: What impacts are computers having on your school library or local community public library?

    h) Banking and Finance: Do you do your banking online? Do you purchase and pay for any items using the Internet? What is the field of "computational finance" about?

i) Journalism: How does your local paper or school paper use computer technology? How are blogs and social networking sites changing journalism?

j) Government: What role does computer and information technology play in local government in your area? Do you live in a place where electronic voting technology is used?

1.3. Which of the following methods for finding a book at a library could be considered an algorithm? In each case you can assume you have a precise specification of the book you want to find. You know the title, author, and date of publication, and you also know the library owns the book. The desired outcome of your search is that you either find the book or you learn the book has been checked out. Which of the following methods provide an effective set of steps for obtaining the book?

a) Walk up to the first person you see, ask them where the book is.

b) Find a librarian, ask them where the book is.

c) Wait by the book return until the book you want is returned.

d) Use an electronic catalog to find where the book is shelved, then use a map to find the shelf.

e) Start at the shelf nearest the door, then look systematically, shelf by shelf, through all shelves in the library.

f) Pick a shelf at random, see if your book is there; if not, pick another shelf at random and repeat.

g) Recruit ten friends; divide the library into ten regions; assign each friend to a different region; ask them to search every shelf in their region and report back to you.

1.4. Use a spreadsheet to make a table of squares of numbers using only addition, following the method shown in the sidebar on page 4. Start by putting the number 1 in cells A1 and B1. Then use the "fill down" command to tell the spreadsheet that new values in column A should be the result of adding 2 to the value above; after you do this, the value in row i, column A should be the ith odd number. Next fill each cell in column B with the sum of the value in the cell to the left and the cell above. Is the value in row i, column B the value of i^2?

1.5. Suppose you need to make a table of squares of numbers using the same technique as the previous problem, but you don't have a spreadsheet application on your computer. Can you describe the method for computing n^2 in enough detail that a friend can make the table? Use a format similar to that in Figure 1.5 to describe the input and the sequence of operations.

Chapter 2

The Python Workbench

Introducing Python and an environment for interactive experiments

The two main goals for this book are to introduce readers to important ideas in computer science and to teach the Python programming skills needed to explore the ideas in a series of computational experiments. This chapter is a brief introduction to Python, where we will see how programs are organized and learn how to create and test parts of programs.

The algorithms behind the programs in this chapter are trivial—for example, a program that converts temperatures from Fahrenheit to Celsius only requires Python to compute the value of a single equation—but going through the steps of implementing the program in Python is a good way to learn about the language and how we can use it for experiments.

Lab Manual

The instructions for setting up a "computational workbench" on your computer are in a Lab Manual that can be downloaded from the Explorations in Computing website:

http://www.cs.uoregon.edu/eic

There are versions of the manual for Microsoft Windows, Linux, and Mac OS X. Each has detailed instructions for installing and running the software you will use for the tutorial projects.

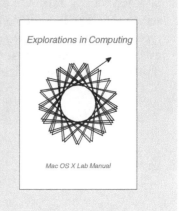

Explorations in Computing

Mac OS X Lab Manual

2.1 Interacting with Python

When we think of a "program," we usually imagine an application that has a main window, a set of operations that can be selected from a menu, and buttons that perform various functions. An application for converting temperatures from Fahrenheit to Celsius might look something like the one shown in Figure 2.1. This program has a window with two boxes for entering numbers. To convert 80°F to Celsius, the user types the number 80 in the Fahrenheit box and clicks the "Convert" button. The program will read the number from the Fahrenheit box, do the conversion, and display the result in the Celsius box.

More complicated programs might have additional controls. For example, the application could have a menu to let the user select degrees Kelvin or some other scale, or a slider to control the number of digits of accuracy for the output. The collection of all of these parts of a program that get input data and present results define the program's **user interface**. Most programs that run on tablets, laptops, or desktop machines have a **graphical** user interface, or **GUI**, that consists of a set of windows, menus, buttons, and other controls.

A graphical user interface hides all the details of what happens as a program is running. Users don't want to see what's going on inside the application when they click a button or select a menu item; they just want the computer to perform some function. For the projects in this book, however, we are interested in the computations carried out behind the scenes. We want to be able to monitor the progress being made by an algorithm as a computation progresses. We need a way to experiment with various parts of a computation, and then be able to put the parts together to see how they work as a whole.

The "computational workbench" we will be using for experiments in this book is based on an older type of user interface. Before computer systems had mice and other pointing devices, and before touchscreens and large high-definition color monitors, computers were connected to a teletype or an electric typewriter. Users typed a command on the keyboard to tell the computer what to do, and any output from that command was printed on paper. Teletypes and typewriters were later replaced by video display terminals (Figure 2.2), but the method of interacting with the computer remained the same: users typed commands, and the computer displayed results.

This style of interacting with a computer is known as a **command line interface**. On a modern computer, we can run software that has a command line interface by using an application called a **terminal emulator**. The application displays a window on the computer's screen that acts like the console on an old-fashioned video terminal. Whatever the user types is displayed in this window, and as soon as the user hits the "return" key the computer executes the command that was just typed. If the command generates any output it is also displayed in the terminal window.

Figure 2.1: *A graphical user interface, or GUI, for a program that converts temperature values. Users enter a number in one text box and click the button. The application will convert the temperature and display the result in the other box.*

Figure 2.2: *Video display terminals, like these from Tektronix, allowed users to type commands on a keyboard and see the output on the display.*

Figure 2.3 shows a terminal emulator program where the user has typed a command telling the system to run a program that converts temperatures from Fahrenheit into Celsius. The command is the first line in the display, and the result generated by the program is on the line just below the command. The window shown in the figure is a snapshot from a Mac OS X system, but Microsoft Windows, Linux, and other operating systems also have terminal emulators and are capable of running programs with command line interfaces.

Several early programming languages, such as LISP and Basic, exploited this style of interface. A person using an **interactive** programming language could type statements based on the syntax of the language. The computer executed the statement and printed any output it generated. This style of computing was very helpful when writing a new program. Programmers would try different versions of statements they wanted to include in a program. When a statement was working, it was copied into the program so it could be executed later when the program was invoked from the command line.

Many modern programming languages, including Python and Ruby, can also be used interactively. The simplest way to use Python as an interactive language is to launch a

Figure 2.3: *A terminal emulator allows us to run software with a command line interface. Here the user typed the command* celsius *80. The output, the result of converting 80° F to Celsius, is shown on the next line.*

```
% celsius 80
26.67

%
```

Choosing an Integrated Development Environment (IDE)

Python programmers have several high-quality open-source IDEs to choose from. The default, included with every Python installation, is named IDLE. The screenshot below was taken from an application named IEP that is also free to install and works on a variety of operating systems.

Because details vary from one operating system to another, the discussion of how to install and use an IDE have been moved to the companion Lab Manual. Versions of the Lab Manual for different operating systems can be downloaded from the book website (`www.cs.uoregon.edu/eic`) and will have updated instructions for each system and a variety of IDEs.

The shell window in IEP, showing how Python evaluates the arithmetic expression that converts 80°F to Celsius.

terminal emulator application and then start Python so it runs inside the terminal emulator. Any Python statements we type into the emulator will be executed immediately, and any outputs those commands produce will be displayed in the terminal window.

Although we could do our experiments by running Python in a terminal emulator, for the projects in this book we will use a slightly different approach. We will use a type of application known as an **Integrated Development Environment**, or **IDE**. The word "integrated" refers to the fact that an IDE combines the abilities of several different applications that all play a part in software development. A typical IDE for Python includes a terminal window so we can enter statements and see what they do. There will also be a text window that allows us to edit programs, and an IDE may include several other panels that serve different purposes during the software development process.

An integrated development environment provides just what we need for our computational workbench. At the beginning of a project we will type commands that test small parts of a computation. Then we will combine the pieces into larger units, and eventually we will run the final program to observe the complete algorithm in action. With this sort of control over the Python programs we will also be able to make small changes to see what effect they have, or to run the algorithms several times, with different inputs each time.

A Note about Displayed Text

Throughout this book there will be examples of Python statements you can type in your IDE or shell window. Immediately following these statements you will see the expected output from Python. These interactions with Python are typeset in a fixed-width (monospace) font.

To help distinguish between text you type and responses from Python, everything you type will be shown in slanted green letters, and everything printed by Python will be shown in black. As an example, the temperature conversion project asks you to type an expression that has Python convert 212°F to Celsius. The text in the tutorial exercise looks like this:

```
>>> celsius(212)
100.0
```

The `>>>` on the first line is the prompt printed by the IDE. The string following that is the statement you type to ask Python to calculate the temperature conversion equation. The text on the second line is what Python should print as the response to your input.

When you start your IDE (see the sidebar on "Choosing an Integrated Development Environment" on page 20) you will see a window where you can enter Python statements. This window is known as the **shell window**, and it typically has a title like "Python Shell." When the system is ready for us to enter a Python command it will display a **prompt** in the shell window. The standard prompt is three greater than signs, so this is what you will see in your shell window when Python is waiting for your input:

```
>>>
```

Working with an interactive programming language programming language is similar to using a calculator: we enter an equation and the computer displays the result. When we run Python interactively, we type a statement, which might be a command to load some software, or a request to evaluate an arithmetic expression, or a number of other operations. Python will execute the statement, and any output it generates will be displayed in the shell window.

Programmers refer to this cycle of typing a statement and waiting for a result as a **read-eval-print loop**. The terminology goes back to the days when output was printed on paper, and we still say the computer "prints" something on the screen, even though it might be more accurate to use the word "display" instead. Python, Ruby, Java, and other modern languages have kept the old terminology. Python statements that generate output are still called "print" commands, even if the output eventually ends up in a window on a screen or in a file saved on disk.

To have Python evaluate an arithmetic expression, simply enter the expression in the shell window and hit the return key. Python will then evaluate the expression and print the result. For example, to ask Python to compute the sum of 5 and 6, just type 5 + 6 and hit return. This is what you will see in the shell window:

```
>>> 5 + 6
11
```

Python and other programming languages use an asterisk as the symbol for multiplication and a slash (/) for division. Here are some examples showing multiplication and division:

```
>>> 5 * 10
50
>>> 6 / 3
2.0
```

In Python arithmetic symbols such as + and * are known as **operators**, and the numbers in these expressions are **operands**. Python has several other operators but we will postpone our discussion of these other symbols and what they mean until we need them for a project.

When an expression has more than one operator, Python applies the operators according to their **precedence**. Because multiplication should be performed before addition, the result of $3 + 4 \times 5$ is 23:

```
>>> 3 + 4 * 5
23
```

If we want Python to evaluate the operations in a different order we can use parentheses:

```
>>> (3 + 4) * 5
35
```

After you complete the following tutorial project we will move on to the next phase in the development of our temperature conversion program.

Tutorial Project

Your Lab Manual has instructions for installing Python and a discussion of different integrated development environments (IDEs). Once you have started your IDE and see the >>> prompt in its shell window you are ready to start the tutorial project.

T1. Type a simple expression. Enter 13 + 2 in the shell window and hit the return key. Python should print the result:
```
>>> 13 + 2
15
```

T2. Try some simple expressions involving other operators:
```
>>> 6 - 3
3
>>> 3 * 7
21
>>> 8 / 4
2.0
>>> 7 + 2 - 3
6
>>> 7 * 6 * 5 * 4 * 3 * 2 * 1
5040
>>> 1 / 3
0.3333333333333333
```

T3. Try some expressions with and without parentheses:

```
>>> 3 + 4 * 5
23
>>> (3 + 4) * 5
35
>>> 8 - 4 / 2
6.0
>>> (8 - 4) / 2
2.0
```

Make sure you understand the output in each of these examples. Is Python printing what you expected?

T4. Does Python care if you include spaces in the middle of your expressions? Type this expression as it is shown, with no spaces:

```
>>> 3+4
```

Type it a second time, with spaces before and after the plus sign:

```
>>> 3 + 4
```

Did you get the same result?

T5. What happens if you leave out an operand? What does Python do if you type 3 + * 5 instead of 3 + 4 * 5?

T6. Type an expression that mistakenly uses a symbol instead of a number:

```
>>> 3 + x
NameError: name 'x' is not defined
```

The error message[1] has some unfamiliar terminology, but at this point you can glean some information: the message has the phrase "not defined," and "x" is enclosed in quotes, so there's a good chance Python was complaining about the x in that expression. We'll see how to assign values so symbols can be used in expressions later in the chapter.

In grade school you might have learned a mnemonic like "My Dear Aunt Sally" to remember the precedence of arithmetic operators (multiplication, division, addition, subtraction).

Python and other programming languages have a slightly different rule: multiplication and division have the same precedence, as do addition and subtraction. If an expression has two operators with the same precedence, the one on the left is applied first.

T7. Type the following expressions to see how Python applies these precedence rules:

```
6 / 3 * 4
8 * 3 / 4
5 - 4 + 2
7 * 4 / 2
9 - 4 - 2
```

Would any of these expressions have a different value using the "dear aunt sally" rules?

T8. What do you think Python will do if you enter 8 / 2 + 5 * 3? Type the expression in your shell window. Did you get the result you expected?

---

[1] Error messages often have many more lines of output. To save space, only the last line of a message will be displayed in the examples in this book.

2.2 Numbers, Functions, and the Math Module

The numbers we used in the previous section were all **integers**, whole numbers with no fractional parts. We can write expressions that use real numbers as well. The general rule is that if we want Python to treat a number as a real number we have to include a decimal point when we write the number.

To be more precise, in programming languages, a number like 5.0 is a **floating point number**, not a real number. Some common real numbers, like $1/3$, $\sqrt{2}$, and π, have an infinite number of digits. Because numbers are stored in a finite amount of space inside a computer, we have to use an approximate value. Floating point numbers are approximations of real numbers, usually accurate to around 12 digits.

You may have noticed in the previous section that the numbers printed by Python as the result of a division operation included decimal points. Python automatically converts the result of every division to a floating point number, even if both operands are integers:

```
>>> 10 / 5
2.0
```

Python has an alternative operator that can be used for algorithms that expect the result of a division to be an integer. This operator is written as two consecutive slashes, with no space between them. In the following examples, notice that Python truncates the results. Any remainder is discarded, and Python does not to the nearest integer:

```
>>> 10 // 3
3
>>> 19 // 10
1
```

Floating point numbers are very common in math and science where numbers represent values like temperatures that are continuously varying. The equation for converting Fahrenheit to Celsius is $C = (F - 32) \times 5/9$. We can use Python to convert 86°F to Celsius by typing the following expression:

```
>>> (86 - 32) * 5 / 9
30.0
```

In this example the output could have been printed as an integer, but in other cases the result has a fractional part:

```
>>> (80 - 32) * 5 / 9
26.666666666666668
```

Mathematically the value of this equation should be $26\,2/3$, but we can see how Python is forced to use an approximate value for $2/3$.

So far all of our expressions have used simple arithmetic operators like + and *. Python also has an extensive library containing mathematical functions, including trigonometric functions, logarithms, and other functions commonly found on a scientific calculator. In addition to math functions, the library contains thousands of other functions that can be used to interact with the operating system (e.g., move a file to a new folder), connect to the Internet, play audio files, and much more. Because there are so many different functions the developers of Python have organized its library into a set of **modules**.

Before we can use a function defined in a module we need to fetch the module from the library. The following statement shows how we tell Python we want to use a function named sqrt (which computes the square root of a number) from the math module:

```
>>> from math import sqrt
```

When we want to use the function, we write its name and then the values it operates on enclosed in parentheses. For example, this is how we ask Python to calculate the square root of 16:

```
>>> sqrt(16)
4.0
```

Computer programmers use a variety of different words to describe what is happening when the system evaluates an expression that contains a function, and the terminology often varies from one programming language to the next. In Python, we usually say the system "calls" a function in order to evaluate the expression. The item in parentheses following the function name is called an **argument**. When the expression is evaluated, we say the argument is "passed" to the function, and the function "returns" a value.

Using this new terminology makes it easier to explain what is going on as Python evaluates more complex expressions. Here is an expression that contains two function calls:

```
>>> sqrt(100) + sqrt(25)
15.0
```

First, Python passes 100 to the sqrt function, which returns the number 10. Then it calls sqrt again, this time to compute the square root of 25. Finally, the two values returned by the different calls are added together and the result is printed.

Floating Point Numbers

Real numbers are stored inside the computer in a format based on scientific notation. A number like 141.42 is stored internally as 1.4142×10^2. This format makes it easier to represent very large numbers (1.4142×10^{23}) or very small numbers (1.1412×10^{-10}) by letting the decimal point "float" to the left or right as the exponent changes.

$$1.4142 \qquad 141.42$$

Multiplying by 100 shifts the decimal point two places to the right

When Python prints a very large number it will use something that looks like scientific notation. For example, the mass of the Earth is 5.9736×10^{24} kg. Python doesn't use superscripts, so it prints the exponent as an e followed by the exponent value: 5.9736e+24.

The same notation is used for small numbers very close to 0. When these numbers are written in scientific notation the exponent is negative. For example, the number 0.00000003 is written in scientific notation as 3×10^{-8}, which Python prints as 3e-08.

Tutorial Project

T9. Use Python to evaluate a simple expression such as 3 * 5 where both operands are integers:

```
>>> 3 * 5
15
```

T10. Repeat the previous expression, but use floating point numbers:

```
>>> 3.0 * 5.0
15.0
```

Can you see the difference between the outputs for these two expressions?

T11. Do you think Python will allow you to mix integers and floating point numbers? What will happen if you try to multiply 3 by 5.0, or add 1.0 to 6? Try these examples and a few other expressions that mix integers and floating point numbers. Does there seem to be a general rule for determining what kind of output is printed?

T12. Use Python to evaluate the expression that converts 100°F to Celsius:

```
>>> (100 - 32) * 5 / 9
37.77777777777778
```

Is this result what you expected, given the rules Python uses for evaluating arithmetic expressions? Is it an accurate calculation of the temperature in Celsius?

T13. Repeat the previous exercise, using Python to convert these temperature values to Celsius: 90°F, 70°F, 212°F, 32°F.

T14. The formula for converting from Celsius to Fahrenheit is $F = C \times 9/5 + 32$. Use this formula to convert the following temperatures to Fahrenheit: 0°C, 10°C, 20°C, 30°C, 100°C.

T15. Type the following command to import the sqrt function from the math module:

```
>>> from math import sqrt
```

T16. Use sqrt to compute the square root of 25:

```
>>> sqrt(25)
5.0
```

T17. The math module also defines some important constants. This command will import the definition of π so you can use it in the current session:

```
>>> from math import pi
```

T18. Verify pi has been imported:

```
>>> pi
3.141592653589793
>>> pi/2
1.5707963267948966
```

The actual value of π has an infinite number of digits. As you can see here, the value used in Python is a floating point number with a finite number of digits.

T19. A module named random has functions for generating random numbers. The randint ("random integer") function produces a random value between a specified minimum and maximum value:

```
>>> from random import randint
>>> randint(1,10)
3
```

You will probably see a different result than the number shown above. Repeat the call to randint(1,10) a few times. Are you seeing a different result each time randint is called? Are all the values it returns between 1 and 10?

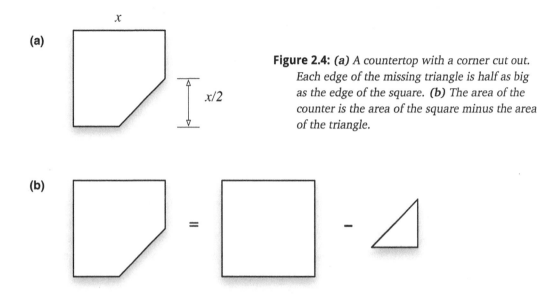

(a)

(b)

Figure 2.4: *(a) A countertop with a corner cut out. Each edge of the missing triangle is half as big as the edge of the square. (b) The area of the counter is the area of the square minus the area of the triangle.*

2.3 Variables

Suppose we want to calculate the area of a countertop shaped like the one shown in Figure 2.4a. The counter is a square with one corner missing, and the sides of the missing triangular piece are half as big as the edge of the square. Figure 2.4b shows one strategy for computing the area: calculate the area of a square and the area of a right triangle, then subtract the area of the missing triangle from the area of the square. The formula for the area of the countertop is thus $x^2 - (x/2)^2/2$, where x is the length of one edge of the square.

To use Python to compute the area, the first step is to calculate x^2, the area of the square. If the length of the edge is 109 cm, we just have to type this expression:

```
>>> 109 * 109
11881
```

A slightly simpler form uses Python's exponentiation operator, which is written with two asterisks in a row, so 109^2 is written this way:

```
>>> 109 ** 2
11881
```

Similarly, we can calculate the area of the missing triangular piece, which is $(x/2)^2/2$:

```
>>> ((109 / 2) ** 2) / 2
1485.125
```

If we want to do the calculation in a single expression, we can put the two pieces together into one Python statement:

```
>>> (109 ** 2) - (((109 / 2) ** 2) / 2)
10395.875
```

This final expression is rather complicated, which makes it hard to type correctly. It would be very easy to leave out a parenthesis, or enter 190 instead of 109, or type a single asterisk instead of a double asterisk. Complicated expressions are not only difficult to type, they are hard to read, and it takes much more effort to try to track down problems.

As a first step in simplifying this expression, we can use a **variable** to stand for the length of the edge. To introduce a variable named x to represent the width of the countertop, we simply type

```
>>> x = 109
```

An expression like this is known as an **assignment statement**. An assignment has a variable name, an equal sign, and the value we want the variable to represent.

It may not look like anything happened because Python does not give us any immediate feedback when we type an assignment statement. However, working behind the scene, Python has created a variable and given it the value we specified.[2] One way to verify the assignment was executed is to type an expression that contains the variable name. The simplest expression is just the name itself, with no operators:

```
>>> x
109
```

Here is the expression that computes the area of the countertop, rewritten to use the new variable:

```
>>> (x**2) - (((x / 2)**2) / 2)
10395.875
```

---

[2]If you are using IEP or a similar IDE you can open a separate window called the "workspace" that will show all the variables that have been defined for the current session.

Layout Rules, Part I: Expressions

Python has very few hard and fast rules for how expressions can be written. We are allowed to use spaces almost anywhere we like.

Figuring out where to put spaces in an expression is mostly a matter of personal preference. Often a statement with a single operator is easier to read if there are spaces around the operator.

When an expression has several operators, it may be easier to understand if higher precedence operators are written without spaces.

```
>>> x**2
9

>>> x ** 2
9
```
spaces here make it easier to see the operator

```
>>> x ** 2 + y ** 2
25

>>> x**2 + y**2
25
```
*emphasize the fact that ** has a high precedence*

What's in a Name?

Mathematical equations typically include single letters like x and y, but variable names in Python can be complete words like **square** and **triangle**. Because most Python projects involve several variables it's important to choose meaningful names in order to keep everything straight.

While Python gives us a lot of freedom in creating names, there are some restrictions:
- All names must start with a letter.
- The remainder of the name can have a mix of upper and lowercase letters, digits, or an underscore: **squareSide**, **square_side**, or **sq123**. are all legal variable names.
- Case is important: **a** and **A** are two different names in Python.

Although names can include uppercase letters, the convention is to use all lowercase when choosing the name for a variable. We'll see other types of names later in the book, where the convention is to use all uppercase or names that are capitalized (uppercase for the first letter, lowercase for the rest).

This version is more readable, but not by much, as it still has a lot of parentheses. We can simplify the expression even further by introducing two more variables, one to stand for the area of the square and the other for the area of the triangle. We could use simple names like y and z for the two new variables, but Python allows us to use complete words as variable names. In the following example, the new variable names are $square$ and $triangle$, which makes it easier to remember what we want each variable to represent:

```
>>> square = x ** 2
>>> triangle = ((x / 2) ** 2) / 2
```

Now we can use the new variables to compute the area of the countertop:

```
>>> square - triangle
10395.875
```

There is another advantage besides readability in using expressions with variable names. Suppose, after computing the area the first time, we double-check the measurements and find the edge is actually 107 cm. To recompute the area, we can type another assignment statement that has x on the left side:

```
>>> x = 107
```

The old value of x is erased and the new value replaces it.

It is important to note that updating x does not automatically update any values that depend on x. At this point we have to tell Python to recompute the values of $square$ and $triangle$ using the new value of x, and then subtract $triangle$ from $square$ again to get the updated area.

The following tutorial project provides some experience with assignment statements and reevaluating expressions entered previously. Development environments support a process called "command line editing" that can save a lot of typing when doing these exercises. Be sure to read the section on command line editing in your Lab Manual.

Tutorial Project

T20. Type these two expressions to define variables named x and y:
```
>>> x = 6
>>> y = 5
```

T21. Try out a few expressions using these variables:
```
>>> x + 3
9
>>> x * y
30
>>> (x + 3) * y
45
```
Did you get what you expected?

T22. Change the value of x:
```
>>> x = 2
```

T23. Repeat the three expressions above that used x. Did the values of these expressions change after you changed the value of x?

T24. Set x to the length of the long edge of the countertop:
```
>>> x = 109
```

T25. Define the area of a square that is x centimeters on each side:
```
>>> square = x ** 2
```

T26. Ask Python to show you the value of this new variable:
```
>>> square
11881
```
Is the result what you expect, the value of x^2?

T27. Define the area of the missing right triangle that is $x/2$ centimeters on each leg, and ask Python to show you the value of this variable:
```
>>> triangle = ((x / 2) ** 2) / 2
>>> triangle
1485.125
```
Is this the correct value?

T28. Enter an expression that computes the area of the countertop:
```
>>> square - triangle
10395.875
```

T29. Change the value of x by entering a new assignment statement:
```
>>> x = 107
```

T30. Recompute the values of `square` and `triangle` using the new value of x and then compute the new area.

With command line editing you just need to hit the up-arrow key on your keyboard a few times until you see the statement you want to execute again. This is a technique we will use often throughout the book so it's worth learning how it works in your development environment.

2.4 Defining New Functions: def Statements

Now that we know how to define variables and evaluate arithmetic expressions we are ready to write our first small Python programs. These initial programs will simply contain a few function definitions. To test the functions we will load them into an interactive session in the shell window of our IDE and type expressions that call the functions. Later in the chapter we will see how we can turn such a collection of functions into a complete program with its own command line interface.

Our first program is going to be an implementation of the function that converts temperatures. After we define this function we want to be able to pass it a temperature value in Fahrenheit as an argument and get back the corresponding temperature in Celsius:

```
>>> temp = celsius(86)
>>> temp
30.0
```

To define this function we need to write only two statements in Python:

```
def celsius(f):
    return (f - 32) * 5 / 9
```

The first line has the word def (short for "define") followed by the name of the function being defined. In parentheses following the name of the function we give a list of **parameters**. These are names of variables that we can use when writing statements inside the main part of the function. Because we're going to pass only one argument to our new celsius function the def statement shows only one parameter, but in general we can put any number of names here (including zero).

The second line is the statement that will be executed when the function is called. In this case we're telling Python to evaluate the expression that converts a temperature value in a variable named f into Celsius and to return the result as the value of the function. In more complicated functions there will be more than one statement here, but in this simple example we only need to evaluate one expression. This group of statements is called the **body** of the function.

Reserved Words

The words def and return used in function definitions have a special meaning to Python. Words like these are known as **reserved words** or **keywords** in a programming language. They are reserved because we are not allowed to use them as the names of variables or new functions.

For example, if you try to define a variable named return Python will print an error message:

```
>>> return = 5
return = 5
      ^
SyntaxError: invalid syntax
```

Layout Rules, Part II: Functions

Most programming languages give programmers a great deal of freedom in how they line up the statements in their programs. Python, however, has some very strict rules.

The main things you need to remember are
- The first line (with the word def and the name of the function) must end with a colon.
- The lines in the body of the function must be indented.
- The same number of spaces must be used at the front of each indented line.

```
def celsius(f):
    return (f - 32) * 5 / 9
```

You can use as many spaces as you like at the front of indented lines, but the Python standard recommends four spaces.

The parameter f is the "link" between the function and the rest of Python. When the function is called, for example celsius(86) on the right side of the assignment statement shown above, the value passed as an argument is stored in the variable named in the parameter list. In this example, Python will set f to 86 before it starts to execute any of the expressions in the body of the function. The return statement tells Python to return the value of the equation to the place where the function was called. Because the function was called from the right side of an assignment statement the returned value is saved in the variable named temp.

A very important fact about parameters like f in this example is that they are **local variables**. That means the name f is valid only within the body of this function definition. The same is true of any variables defined inside the body of a function. If any assignment statements in the body introduce new variables, those names become local variables and they can only be referred to by other statements in the body of the function.

After Python is done carrying out the instructions in the body of a function, and the value it computes is returned to the place where the function was called, all the local variables disappear. The names are no longer valid, and they cannot be used outside the body of the function. The exercises at the end of the tutorial for this section will emphasize this fact by showing what happens when we call a function that creates local variables and then try to refer to those variables after the function returns.

The definition of the celsius function is our first introduction to an important detail about Python programs. Notice how the statement in the body of the function has four spaces at the front of the line. Python is very fussy about how lines in a program are indented. In the case of a function definition, all of the statements in the body of the function must be indented with respect to the def statement (see the sidebar "Layout Rules, Part II: Functions").

Now that we know what a function definition looks like, the next question is how to load it into Python so it can be used in an interactive session. It is possible to type a function definition into an interactive shell session (the process is explained in the Lab Manual if you want to try it) but it is far more convenient to use the program editor window in an

Editing Programs with an IDE

Although it is possible to enter function definitions directly into a shell session, most programmers save definitions in a file. If a definition needs to be updated (perhaps because a bug was found during tests in an interactive session) they can edit the file and load the updated definition.

In this screenshot from a session with IEP the definition of a function named celsius has been saved in a file named celsius.py (the panel at the lower left). A command from the Run menu loaded the function into an interactive shell, where it can be tested (as shown in the top panel). The Workspace panel shows the names that have been defined in the current session.

integrated development environment. Creating a function definition with an IDE is similar to writing a document with a word processor application. Use the File menu to create a new empty document, type in the definition, and save the document. After the file has been saved we can use a command that will load the definition into an interactive session.

The main advantage of using an IDE is that it is far easier to modify a Python program if we need to make some changes. Instead of having to retype the complete definition we just have to edit the file and load the updated definition. Throughout the book we will develop complex programs in stages. The first version of a function might be a simple prototype that we enter and test just to make sure the outline is correct. Then we will gradually add more statements to the body of the function, testing as we go. The IDE allows us to edit the program text, load each new version of the function into a Python session, and type expressions in the shell window to carry out the tests.

When programs are saved in files it is important to include additional information that describes how the program works. For a short one-line program like our celsius function it's pretty clear what the function computes. But as we write more complex programs with several different parts we will want to include some documentation that explains what each piece does.

```
1   def celsius(f):
2       "Convert temperature f from Fahrenheit to Celsius."
3       return (f - 32) * 5 / 9
```

Download: workbench/celsius.py

Figure 2.5: *The final version of the function that converts temperatures from Fahrenheit to Celsius. Line 2 is the "docstring" that will be printed if a user types help(celsius) in an interactive shell.*

The simplest type of documentation is a **comment**. Whenever Python sees a hash mark (#) in a file it will ignore all the remaining characters on that line. That means we can type anything we want, including plain English (or any other language) sentences to explain what we expect the program to do. Some comments are short notes that appear at the end of a line. For example, to the right of an assignment statement that creates a new variable there might be a comment that explains what the variable will be used for. It is also quite common to include several full-length comments, where the # symbol is the first character on the line and the entire line explains something about the program.

Python programs can also include a special type of comment known as a **docstring**. A docstring is written between the def statement that begins a new function definition and the first statement in the body of the function. Also, instead of using a comment character, docstrings are surrounded by quote marks. All the characters between the beginning and ending of the docstring are treated as one long comment.

When a function includes a docstring Python saves the documentation as the function is loaded into an interactive session. If you ever want to read the documentation, call a built-in function named help. For example, if we include a docstring with our celsius function, we can ask Python to print the documentation by typing this expression:

```
>>> help(celsius)
celsius(f)
    Convert temperature f from Fahrenheit to Celsius.
```

All of Python's library functions have docstrings. After loading the sqrt function from the math module we can find out what it does by asking to see its docstring:

```
>>> help(sqrt)
Help on built-in function sqrt in module math:

sqrt(...)
    sqrt(x)
    Return the square root of x.
```

Figure 2.5 shows the final definition of our celsius function, including its docstring. This function can be loaded into an interactive session, where we can call it whenever we want to convert a temperature value from Fahrenheit to Celsius. It can also be used as part of another application. In future chapters we will see how functions we write for one project can be imported and used by other programs. In the next chapter we will see how to create a command line interface to our program so we can run it from the terminal without having to start an interactive Python session.

Tutorial Project

For the project in this section you will write two new function definitions and save them in files. As you work on the projects, have the documentation for your IDE handy so you can figure out how to use the text editor to create new files and how to load files into interactive shell sessions.

T31. Use your IDE to create a new file named celsius.py. Type in the following two lines, exactly as they are shown here (including the four spaces before the word return), and save the file:

```
def celsius(f):
    return (f - 32) * 5 / 9
```

T32. Use your development environment's "run" command to load the program. In IDLE the command is called "Run Module" and it is under the Run menu. In IEP the command is "Run File" in the Run menu.

Even with such a simple program there are several places where things can go wrong. If the IDE says you have a syntax error when you load the program, make sure there is a colon at the end of the def statement and that the statement in the body is indented four spaces.

T33. In the shell window you should now be able to enter expressions that call your new function:

```
>>> celsius(86)
30.0
>>> celsius(212)
100.0
```

If you are not getting the correct answers, check to make sure the equation is correct. Did you include the parentheses around the subtraction operation?

T34. Add the docstring to your program, save the file, and select the Run command again. Your final program should look like the one in Figure 2.5.

T35. Now that your function has a docstring Python should print a helpful message when you ask to see its documentation:

```
>>> help(celsius)
celsius(f)
    Convert temperature f from Fahrenheit to Celsius.
```

T36. Create a new file named countertop.py and type in the code shown in Figure 2.6 (if you prefer, you can download the code from the book website).

T37. The program in Figure 2.6 shows an alternative format for a docstring. If the documentation is too long to fit on one line we can split it over multiple lines. The first and last lines of the documentation should have three quote marks in a row, and in between we can write as many lines as we'd like.

T38. Use your IDE's Run command to load the file into a shell session. Test the new function:

```
>>> countertop(109)
10395.875
>>> countertop(107)
10017.875
```

Do these results agree with your previous experiments, where you entered the expressions directly in the shell?

The variables named square and triangle are defined in the body of countertop, which means they are local variables that can only be used inside the function.

```
1    def countertop(x):
2        """
3        Compute the area of a square countertop with a missing wedge.  The
4        parameter x is the length of one side of the square.
5        """
6        square = x ** 2                     # area of the full square
7        triangle = ((x / 2) ** 2) / 2       # area of the missing wedge
8        return square - triangle
```

Download: workbench/countertop.py

Figure 2.6: *The Python definition of a function that computes the area of a countertop like the one shown in Figure 2.4. This definition shows the multi-line format for docstrings.*

T39. To verify the fact that `square` is a local variable, call `countertop` and then type an expression that uses `square`:
```
>>> countertop(100)
8750.0
>>> 2 * square
NameError: name 'square' is not defined
```

T40. What would Python have printed for the value of 2 * square if square was not a local variable, that is, if the value was still available after returning from the call to countertop?

2.5 Conditional Execution: `if` Statements

The `countertop` function in Figure 2.6 has three simple steps, and Python evaluates the same three equations each time the function is called. Other functions, however, require computations where different statements might be executed for different input values. A good example is the way income taxes are calculated in U.S. states that have a variable tax rate. In these states, a person's tax rate depends on their income level: if their income is below a certain amount they pay one rate, and if it is above that amount they pay a different rate.

Figure 2.7 shows how we can write a Python function to compute a tax rate that depends on an input value. This function is for a hypothetical state where residents who earn less than $10,000 pay no income taxes and income above $10,000 is taxed at 5%.

The statement on line 3 of this function is called an **if statement**. It has the keyword `if` followed by an expression that determines whether or not the lines below it will be executed. In this case the program is checking to see if the value of `income`, a parameter passed to the function, is less than 10,000.[3] Note how there is a colon at the end of the `if` statement, and the statement on the next line is indented. In this example there is only one line in the body of the `if` statement, but Python allows us to include as many lines as we want, as long as they are all indented the same number of spaces.

---

[3]Notice that Python, like other programming languages, does not use commas in large numbers. The number 10,000 is written 10000, 1,000,000 is written 1000000, and so on.

```
1   def tax_rate(income):
2       if income < 10000:
3           return 0.0
4       else:
5           return 5.0
```

Figure 2.7: *A function to compute a tax rate that depends on income level. Residents who earn less than $10,000 will have a rate of 0%, but anyone who earns $10,000 or more is taxed at a 5% rate.*

Download: workbench/taxrate.py

The word else on line 5 of Figure 2.7 is part of the if statement. This line and the statements that follow it are called an **else clause**. As you might expect, these statements mean that if the condition in the if statement is not met the line(s) following else should be executed instead.

Not every if statement needs to have an else clause, but when there is one, the same rules apply: the keyword else is followed by a colon, and all the statements in the body of the else clause are indented the same number of spaces. One final detail: Python insists that the indentation level of the else clause matches the indentation of its if statement, that is, there must be the same number of spaces at the start of both lines.

The expression following if is known as a **Boolean expression**. Like the arithmetic expressions we saw earlier in the chapter, a Boolean expression is a combination of values, variable names, and operators. The difference is that operators in Boolean expressions, which are usually (but not always) comparison operators, result in one of the logic values True or False instead of a numeric value. Here are some more examples using other comparison operators:

```
>>> 10 > 5
True
>>> 10 > 15
False
>>> 10 <= 10
True
```

Returning to the tax rate example, many states have an income tax formula with three or more different tax rates, where higher rates are applied to higher incomes. For example,

```
1   def marginal_tax_rate(income):
2       if income < 10000:
3           return 0.0
4       elif income < 20000:
5           return 5.0
6       else:
7           return 7.0
```

Figure 2.8: *A progressive tax is based on higher rates for higher incomes. In this example, residents will pay either 0%, 5%, or 7%.*

Download: workbench/marginaltaxrate.py

Boolean Operators

The table at right shows how to compare the value of two variables in Python. Notice that to write the mathematical equation $vx \leq y$ in Python we have to use two symbols and write x <= y.

Note also that to see if $x = y$ we have to use a double equal sign and write x == y. That's because a single equal sign is Python's assignment operator, and if we write x = y Python will think we want to assign y to x.

x < y	True if x is less than y
x <= y	True if x is less than or equal to y
x > y	True if x is greater than y
x >= y	True if x is greater than or equal to y
x == y	True if x is equal to y
x != y	True if x is not equal to y

there might be no tax for incomes up to $9,999, a 5% tax on incomes between $10,000 and $19,999, and a 7% tax on incomes of $20,000 or more. Figure 2.8 shows how we can implement a function that computes this type of tax rate. The statement on line 5 starts with a new keyword, elif, which is a combination of else and if. Like else, it only makes sense in the context of an if statement.

At first it might seem like our new marginal_tax_rate function does not meet the specifications. The rule spelled out in the previous paragraph is that income between $10,000 and $20,000 is taxed at 5%, but the Boolean expression on line 5 is only testing one of these conditions, to see if the income is less than $20,000. The program is correct, however, because if the income is less than $10,000 the expression on line 3 would be true and the system would never reach the statements below the first branch in the if statement. When an if statement has multiple elif clauses, Python will try all the clauses in order and will execute the body of the first one that has an expression that evaluates to True.

Tutorial Project

Use your IDE to create the two files shown in Figures 2.7 and 2.8 (or download them from the book website). Start an interactive shell to experiment with Boolean expressions and to test the two functions.

T41. Define some values to use in tests of Boolean expressions:

```
>>> x = 7000
>>> y = 17000
```

The value of x is 7,000 but recall that Python does not use commas in large numbers.

T42. Enter some expressions to see how Python's comparison operators work.

```
>>> x < 10000
True
>>> y < 10000
False
>>> x == y
False
```

Note that to see if x and y have the same value we write x == y with two equal signs.

T43. Test your definition of `tax_rate` by calling it a few times.

```
>>> tax_rate(5000)
0.0
>>> tax_rate(75000)
5.0
```

Is the function behaving the way you expect?

T44. Try some calls to `marginal_tax_rate`:

```
>>> marginal_tax_rate(1500)
0.0
>>> marginal_tax_rate(15000)
5.0
>>> marginal_tax_rate(150000)
7.0
```

Are these values correct?

T45. Try passing an income of $10,000 to `marginal_tax_rate`. Does the value returned match the specification? What about $19,999? $20,000?

T46. What do you think would happen if you passed a negative number to `marginal_tax_rate`? Test your prediction by having Python evaluate `marginal_tax_rate(-10)`. Is the result an error?

♦ When we use Python as a calculator and ask it to evaluate an arithmetic expression it prints the value. We can also save the value in a variable if we put the expression on the right side of an assignment statement:

```
>>> 6 * 7
42
>>> answer = 6 * 7
```

Do you suppose we can do the same thing with a Boolean expression? What do you think Python would do if you entered the following statements? Make a prediction, then check your hypothesis by typing the statements in your IDE's shell window.

```
>>> 6 < 7
True
>>> q = 6 < 7
```

2.6 Strings

Needless to say, computer programs are not limited to "number crunching" applications. People compose e-mail and a variety of other documents, and store names, addresses, product descriptions, and many other types of text in address books and other databases.

When writing programs that deal with text we will be working with pieces of data called **strings**. A string is a sequence of characters, where a character can be a letter of the alphabet, a digit, a punctuation mark, or a symbol such as an arithmetic operator or a currency symbol.

To create a string in Python, simply enclose a sequence of characters in quotes. Here are some examples of assignment statements where variables are given string values:

```
>>> s = 'Hello'
>>> t = "Who's there?"
```

It doesn't matter whether we use either single or double quotes as long as the same type of quote character is used at the beginning and end of the string.

Strings can be used in expressions much the same way we use numeric values. For example, if the operands are strings, the + operator connects the two strings to make a longer string, and the * operator repeats a string a specified number of times:

```
>>> s + "?"
'Hello?'
>>> s * 3
'HelloHelloHello'
```

Another useful operation is specified by the word in. This is a Boolean operator, meaning it evaluates to True or False. The expression x in y will be True if the string x appears anywhere inside the string y. Here are some examples using the two strings defined above:

```
>>> 'll' in s
True
>>> 'here' in t
True
>>> 'who' in t
False
```

The last expression is False because string comparison operations are case sensitive. The letters "Who" with a capital "W" appear in the string t, but "who" with a lowercase "w" is not there.

A function named len (short for "length") will tell us how many characters are in a string. len is a built-in function, which means we do not have to import it from a library. These examples will count the number of characters in the two strings defined previously:

```
>>> len(s)
5
>>> len(t)
12
```

Note that when we asked for the length of t Python counted every character in the string, including spaces and punctuation marks.

In addition to built-in functions like len Python has a collection of basic string functions in a module named str. This module is automatically included with every program, so we do not have to import it to use it. Suppose we want to know how many space characters occur in a sentence. The str module has a function named count, and when we call it, we tell it which string to use and what it is we want to count:

```
>>> sentence = "It was a dark and stormy night."
>>> str.count(sentence, ' ')
6
```

That's the result we expected because there are six space characters separating the seven words in the sentence.

Compound names like str.count are common in Python. The identifier to the left of the period is a module name, and the identifier to the right is the name of a function defined in the module. Names that include both module and function names separated by periods are known as **qualified names**. We'll see several examples of qualified names throughout the rest of the book.

For string operations, Python programmers usually prefer an alternative syntax that lets us avoid writing both the module and variable names. What we do is take the first argument out of the parameter list and write it in place of the module name. Thus

```
str.count(x,y)
```

is transformed into

```
x.count(y)
```

Here is how a Python programmer would typically write the call to count shown above:

```
>>> sentence.count(' ')
6
```

This form of function call can be used for other types of data besides strings, as we will see in the next section. Although this alternative syntax may seem awkward at first it will soon become natural.

For the tutorial project in this section we will write a simple function like the interest rate calculator from the previous section, but our new function will be based on strings instead of numbers. Our goal will be to write a function named plural that will take a common English noun as an argument and return the plural form of the noun. For most words we just need to add a single "s" to the end of the word: the plural of "duck" is "ducks," and "car" becomes "cars." But for other words we need to add "es" to the end. For example, the plural of "fish" is "fishes."

This project is also our first chance to use the PythonLabs library. PythonLabs is a collection of modules that was written specifically for the projects in this book. We will use a PythonLabs function that returns a part of a word, and to test our plural function we will use a list of common nouns that comes with the library.

Tutorial Project

T47. Define a string to use in the first set of experiments in this section:
```
>>> s = 'Here'
```

T48. Type some expressions using string operators and string functions:
```
>>> len(s)
4
>>> s * 2
'HereHere'
>>> (s + '! ') * 2
'Here! Here! '
```
Make sure you understand what Python did to produce that last result.

```
1    def plural(w):
2        "Return the plural form of word w."
3        if w.endswith('h'):
4            return w + 'es'
5        else:
6            return w + 's'
```

Figure 2.9: *This preliminary version of plural simply checks to see if a word ends with an "h". If so, form the plural by attaching "es", otherwise just attach "s".*

Download: workbench/pluralv1.py

T49. How many characters are in the string produced by the third example above? Verify your answer by enclosing the expression in a call to the len function:

```
>>> len((s + ' ') * 2)
12
```

Do you see why there are twelve characters in this string?

T50. Create a longer string and type some expressions using count:

```
>>> s = "What is the airspeed velocity of an unladen swallow?"
>>> s.count(' ')
8
```

T51. Create a couple of strings that will be used to test the plural function:

```
>>> test1 = 'duck'
>>> test2 = 'fish'
```

T52. To look at the letters at the end of a word to see if we need to add "s" or "es" we can use a function defined in Python's str module. The function is named endswith, and it returns True or False. Try calling this function a few times using the test strings:

```
>>> test1.endswith('h')
False
>>> test2.endswith('h')
True
>>> test1.endswith('ck')
True
```

That last test shows we can check as many letters as we want at the end of a string. Try making up a few more examples on your own until you are sure you know how to use endswith.

T53. A simplified version of the plural function is shown in Figure 2.9. Download the code from the book website or create a new file with your IDE and type the code as it appears in the figure.

T54. Load your new plural function into an interactive session. Type the following expressions to test your program:

```
>>> plural(test1)
'ducks'
>>> plural(test2)
'fishes'
```

This first version seems to work—it passes its first test, at least. If you try it on a few more words, you'll quickly find it does the wrong thing on a lot of common nouns ("business" and "growth" to name just two). English has too many special cases and odd words for us to deal with all of them in a simple program, but we can extend the program so it deals correctly with a few more common types of words.

The first extension is to tell the function to add "es" if a word ends in either "h" or "s". Instead of adding an `elif` clause to the `if` statement, however, we will change the Boolean expression. Python has a Boolean operator named `or` that tests for two separate conditions and returns `True` if either condition is `True`.

T55. Define a new test string:
```
>>> test3 = "walrus"
```

T56. Change the Boolean expression on line 3 of your `plural` function so it looks like this:
```
if w.endswith('h') or w.endswith('s'):
```
Save the program and load the new version into your interactive session.

T57. This new version should add "es" to any word that ends with either "h" or "s":
```
>>> plural(test2)
'fishes'
>>> plural(test3)
'walruses'
```

The second extension will deal with words like "charity" that end with a "y". The common rule here is to replace the "y" with "ies". To help implement this rule we will use a function named `prefix` that is defined in PythonLabs.

T58. Type the following command in your shell window to import two functions from PythonLabs into your interactive session:
```
>>> from PythonLabs.Tools import prefix, suffix
```

T59. Enter the definition of a new test string:
```
>>> test4 = "pony"
```

T60. Try calling `prefix` a few times until you're sure you understand what it does:
```
>>> prefix(test1)
'duc'
>>> prefix(test4)
'pon'
```

T61. The `suffix` function returns the character at the end of a word:
```
>>> suffix(test1)
'k'
>>> suffix(test4)
'y'
```
We won't need `suffix` for this tutorial but you may want to use it in one of the programming projects described at the end of the chapter.

T62. To have your function apply this new rule, add an `elif` clause to the `if` statement:
```
elif word.endswith('y'):
    return prefix(word) + 'ies'
```

T63. Because your program will be using the `prefix` function from the PythonLabs Tools library you need to add the statement that imports the function. The final version of your `plural` function should now look like the one in Figure 2.10.

Can you see how the statement in the body of the `elif` clause tells Python to turn a word like "pony" into "ponies"?

```
1    from PythonLabs.Tools import prefix
2
3    def plural(w):
4        "Return the plural form of word w."
5        if w.endswith('h') or w.endswith('s'):
6            return w + 'es'
7        elif w.endswith('y'):
8            return prefix(w) + 'ies'
9        else:
10           return w + 's'
```

Figure 2.10: *The final version of* plural *uses one rule for words that end with "h" or "s", a second rule for words that end with "y", and a third rule for every other word.*

Download: workbench/plural.py

T64. Test the final version of your program on the test string that ends with "y":

```
>>> plural(test4)
'ponies'
```

T65. Repeat the tests you made earlier. Does your program pass all four tests?

If you want to try some more experiments PythonLabs has a list of common nouns you can use for test cases.

♦ Import a function named randnoun (the name is based on Python's randint function, which generates random integers):

```
>>> from PythonLabs.Tools import randnoun
```

♦ Call randnoun a few times to see what it does:

```
>>> randnoun()
'expense'

>>> randnoun()
'condition'
```

Your output will be different, but you should see a random word printed in your shell window each time you call the function.

♦ Run some experiments using random nouns from the word list:

```
>>> plural(randnoun())
'negatives'

>>> plural(randnoun())
'receptions'

>>> plural(randnoun())
'wifes'
```

As you can see, there will still be several cases where the function does the wrong thing. Note that the program is "correct," in the sense that it is doing exactly what we told it to do, but it has rules for only a few of the many special cases in the English language.

2.7 Objects

Python provides extensive support for a style of programming known as **object-oriented programming**, or **OOP**. The ideas behind object-oriented programming were first developed in the 1960s and have been widely adopted in a variety of programming languages since that time.

The word "object" in this context basically means "a piece of data." Every data item in Python is an object. Integers, floating point numbers, and strings are simple kinds of objects, and there are dozens of others, some built into the language and some defined in library modules. In Chapter 7 we will learn how to create our own kinds of objects.

Python has adopted terminology from the world of object-oriented programming, and we will be using these terms throughout the book:

- The word **class** refers to a type of data. Integers, floating point numbers, Booleans, and strings are all classes in Python.

- An individual data object is an **instance** of a particular class. The string 'hello' is an instance of the string class. We also sometimes say an object "belongs to" a class.

- A function that has been defined as part of a class is called a **method**. The functions named count and endswith are examples of string methods.

In the previous section we saw how to use an alternative syntax when we wanted to count spaces in a string. Because the count function is defined as part of the string class, which is named str, to count the number of spaces in a string s we can either write

```
str.count(s,' ')
```

or, using the alternative syntax,

```
s.count(' ')
```

The latter form, where we write a variable name, a period, and then a function name, is the standard way of calling a method in an object-oriented programming language, and it is the syntax that will be used in the rest of the book.

When Python sees an expression like s.count(' ') it does a lot of work behind the scenes. First, it asks, "what is s"? When it learns that s refers to a string, it looks in the string class (str) to see if it contains a method named count. When it finds there is such a function, Python collects all the remaining parameters and passes them in a call to str.count.

Note the similarity between how Python handles a call to a method and how it evaluates expressions. When we ask Python to evaluate an expression like

```
x + y
```

Python looks up the values of x and y. If both are numbers the two are added to produce the result. As we saw in the last section, if the variables refer to strings, they are concatenated. A new string is produced, having the characters in x followed by the characters in y.

(a)   **(b)**

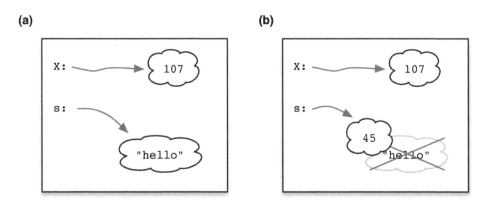

Figure 2.11: *(a) The computer's memory (the "object store") after defining variables named x and s. (b) If we decide to use s to hold an integer the old object referred to by s is discarded.*

Python follows the same process when we ask it to evaluate

```
x.count(y)
```

If x refers to a string, Python calls the count method defined in the string class. But if x is some other type of object Python will call the count method for that class.

How Python chooses to represent objects inside the computer's memory is usually not important for the projects in this book, so objects are shown as abstract quantities inside clouds in drawings like the one in Figure 2.11. Programmers often refer to the memory that holds data items as "object storage," or in a abbreviated form, the **object store**.

A useful way to visualize how Python manages variables is to imagine the object store as a large whiteboard. When a program is first launched, or when you start a new interactive session, the whiteboard will be blank. When Python executes an assignment statement, it first finds space for an object to represent the value on the right side of the statement, and then it writes the variable's name and records the fact that the variable refers to the new object. Figure 2.11a shows what the whiteboard might look like after defining two variables, an integer named x and a string named s.

To see why the whiteboard analogy is helpful, let's do a small experiment that should also help cement the ideas introduced throughout this chapter.

Tutorial Project

T66. Defining the two objects shown in Figure 2.11a:
```
>>> x = 107
>>> s = 'hello'
```

T67. Python has a built-in function named type that will tell us what sort of object a variable refers to:
```
>>> type(x)
<class 'int'>
>>> type(s)
<class 'str'>
```
The output tells us x belongs to the class named int, its name for the class of integers, and s belongs to the class named str, the name for the string class.

T68. Use the two variables in expressions where the operations are valid for each type of variable:

```
>>> s ** 2
11449
>>> len(s)
5
```

T69. The * operator does not have any inherent meaning all by itself. The operation performed by * depends on the types of the values in the expression:

```
>>> x * 2
214
>>> s * 2
'hellohello'
```

We saw earlier that the value of a variable can change—we updated the value of the variable x by associating it with a new number. It turns out we can also change what kind of object is associated with a variable. Python doesn't complain if, after defining s to be a string, we now type an assignment statement that makes s refer to a number.

T70. Enter a new assignment statement, this time setting s to an integer value:

```
>>> s = 9 * 5
```

T71. Verify s now refers to an integer by typing expressions that use s:

```
>>> s * 2
90
>>> len(s)
TypeError: object of type 'int' has no len()
```

This experiment shows Python was perfectly happy to throw away the string object that used to be the value of s, create a new object representing the number 45, and associate s with the new object, as shown in Figure 2.11b. Several conclusions can be drawn from this experiment:

- Variable names are simply labels attached to objects, and Python does not impose any restrictions on what types of names are given to any type of object.

- Whether or not an expression is valid is determined by the types of the values used in the expression: len(s) is valid if s refers to a string but not if it refers to an integer.

- Python figures out whether an expression is valid when it tries to evaluate the expression: the same expression len(s) was evaluated successfully when s was a string but generated an error later after s was updated to refer to a different type of object.

Throughout the rest of the book we will see abbreviated descriptions of how Python evaluates expressions. Instead of writing "if s refers to a string object" or "variable x is a reference to an instance of class y" we will simply say "if s is a string" or "x is a y."

2.8 Summary

Our approach to studying computation in this book is to set up interactive experiments on a "computational workbench" so we can explore algorithms and see how they work with various inputs. In later chapters we will experiment with a variety of interesting algorithms, including searching and sorting, generating random numbers, modeling the solar system, carrying on a conversation in English, and many others. For this chapter the goal was simply to introduce the programming environment we will use and learn how it can manage interactive experiments.

The best way to experiment with Python programs is by typing expressions in an inter-active shell window that is part of an integrated development environment, or IDE. Typing Python commands into the shell window is similar to using a calculator. We enter an ex-pression, and then Python evaluates the expression and prints the result.

The project introduced the two main types of numbers in Python: integers and floating point numbers. Floating point numbers are finite approximations of real numbers. Other types of data described in this chapter were strings, which are sequences of characters used to represent text data, and Boolean (true/false) values.

Numbers, strings, and other types of data are represented inside a computer as objects. Because we typically don't need to know the details of how pieces of data are represented in memory, we can think of them as abstract entities, and we draw them as clouds in figures that show the state of the system. Up to this point we have been describing objects simply as data, but in computer science the term "object" has a richer connotation. For the first few chapters in this book, however, we will simply use the word to mean "a generic piece of data."

Variables are symbolic names associated with objects. When Python evaluates an expres-sion that contains a variable name, it looks in the "object store" for the current value of the variable. The value is substituted in place of the variable so Python can compute the value of the expression. Variables are created by assignment statements that consist of a variable name, an equal sign, and an expression that defines the value for the variable. We can change the value of a variable at any time simply by reusing it in another assignment statement.

Simple operations such as addition and multiplication are written using arithmetic op-erators, such as + for addition and * for multiplication. More complicated functions are specified in the form of calls to functions. Python's Math module has functions for comput-ing square roots, trigonometric functions, and several other operations. For operating on strings, several functions are available as part of a built-in module named `str`, but there are many more collections of functions that can be imported from a variety of different string libraries.

We also saw how to define our own functions. A function that converts temperature values from Fahrenheit to Celsius requires only two lines of code:

```
def celsius(f):
    return (f - 32) * 5 / 9
```

An example of how we can call this function from in interactive shell window is

```
>>> temp = celsius(90)
```

In this call to `celsius`, Python creates a variable named `f`, assigns it the value 90, evaluates the expression in the body of the function, and returns the value of 32, which is then saved as the value of the variable `temp`.

One way to understand what is going on inside the system when we type an expression in a shell window is to think of how we simplify an expression in algebra. Suppose $x = 4$ and $y = x + 5$, and we are asked to find the value of $z = \sqrt{y}$. To solve this problem, we use the value of x to rewrite the second equation as $y = 4 + 5$, which is then simplified to $y = 9$. Now the 9 can be substituted for y in the third equation to give $z = \sqrt{9}$, and one last rewrite gives $z = 3$.

Concepts and Terminology Introduced in This Chapter

terminal emulator	An application that mimics the actions of an old-fashioned computer terminal
IDE	Integrated development environment, an application that combines a terminal emulator, text editor, and other program development tools
integer	A whole number, a number with no fractional parts
floating point	A technique for storing approximate values of real numbers in a computer
string	A short piece of text used as data in a program
Boolean	A type of data used for logical values (true or false)
object	A generic piece of data; an object can be an integer, floating point number, string, or one of the more complex types of data introduced later in the book
variable	A name associated with an object
assignment statement	A Python expression that sets the value of a variable, creating a new variable if the name has not been used before
local variable	A variable that can only be used within the function where it is defined
method	A function associated with a particular data type, usually called by writing a variable name, a period, and the function name.
module	A collection of functions and data; Python's library is organized as a set of modules, and the PythonLabs software developed for this book is also organized into modules

---

Python operates in much the same way. It will do a series of operations that are the equivalent of the equation rewriting steps you would do to perform the same calculation. Suppose we type an expression like

```
>>> a = countertop(x + 2)
```

Python will look up the value of x and use it to calculate x + 2. The result will be passed to the countertop function, and whatever value is returned by the function is stored in a.

In later chapters we will encounter several examples of Python expressions that, at first, may seem very complex. The main thing to remember is that to evaluate an expression, Python starts by gathering all the values it needs. If it sees the name of a variable, it looks up the current value of the variable and substitutes that value in the expression. When it sees the name of a function, it calls the function and then substitutes the value returned by the function into the original expression.

Exercises

The exercises below ask you what Python would print as the value of an expression. To make sure you understand how Python works you should try evaluating the expressions yourself first, and then using the shell window in your IDE to check your answers.

2.1. Suppose you start an interactive session and enter the following three assignment statements:

```
>>> x = 4
>>> y = 7
>>> z = 11
```

What will Python print as the value of the following expressions?

a) x * 2

b) x ** 2

c) x * y

d) x * z

e) y / x

f) y // x

g) sqrt(x)

h) x * y + z

i) x + y * z

j) (x + y) * z

k) 2 * sqrt(x + z + 1)

l) x < y

m) x + y <= z

n) x + y > z

o) x != y

2.2. For this question, assume the celsius and countertop functions shown in Figures 2.5 and 2.6 have been loaded into an interactive session. What will Python print as the result of these expressions?

a) celsius(60)

b) 2 * celsius(212)

c) countertop(10)

2.3. Do you think Python will print anything different if you pass a floating point number to celsius? What would Python print for this expression? Explain.

```
>>> celsius(60.0)
```

2.4. Do you think the celsius function can deal with negative numbers? What will Python do with this expression?

```
>>> celsius(-3)
```

2.5. One of the expressions in the body of the countertop function calculates the area of the missing triangular piece:

```
triangle = ((x / 2) ** 2) / 2
```

Are all the parentheses required in this expression? Explain what would happen if the statement was rewritten in one of the following forms:

```
triangle = (x/2) ** 2 / 2
triangle = x/2 ** 2 / 2
```

2.6. What will happen if you pass a negative number to the `countertop` function? You will get back a number, but what does it mean? Can you explain what Python did to get this result?

```
>>> countertop(-10)
87.5
```

2.7. Suppose the following strings are defined in an interactive session:

```
>>> s = "goal"
>>> t = "In Gaelic football a goal is worth 3 points."
```

What will Python print as the value of the following expressions?

a) `len(s)`

b) `s * 3`

c) `(s + '!') * 5`

d) `s in t`

e) `t.count('g')`

f) `t.count(' ')`

g) `t.endswith('?')`

Programming Projects

Each programming project for this chapter can be completed by writing a single new function definition. Create a new file for each function. A common practice is to give the file the same name as the function but with ".py" at the end. For example, a function named `fahrenheit` would be saved in a file named `fahrenheit.py`. Each project includes test cases showing how the function should be called and the expected results, but you are encouraged to run some additional tests of your own.

2.8. Define a function named `fahrenheit` that will convert temperature values from Celsius to Fahrenheit, using the equation $F = C \times 9/5 + 32$.

```
>>> fahrenheit(25)
77.0
>>> fahrenheit(100)
212.0
```

2.9. You may have seen "speedometer test sections" on interstate highways in the United States. These sections have a series of signs posted exactly one mile apart. To see how fast you are going, in miles per hour, you can calculate $3600/t$, where t is the number of seconds it takes to drive between two signs. Define a Python function named `mph` that will compute the speed of a car given the number of seconds it takes to travel one mile.

```
>>> mph(60)
60.0
>>> mph(48)
75.0
```

2.10. Look on the Internet to find a formula for computing body mass index (BMI) as a function of height and weight. Define a Python function named `bmi` that will compute a person's body mass index given their height and weight. Note there are two formulas, one for height in inches and weight in pounds, and another for height in centimeters and weight in kilograms; use whichever you prefer, but make sure you enter the correct units in your tests. Using the formula for U.S. units, the BMI for a person 6'4" tall and weighing 210 pounds is 25.56:

```
>>> bmi(76,210)
25.559210526315788
```

2.11. Define a function named pmt that will compute the amount of a monthly payment on a loan. The three parameters of the function should be amt, the initial loan amount; rate, the annual interest rate; and yrs, the number of years before the loan is paid off.

The algorithm for computing the payment is as follows. First, calculate a value r using the formula $r = \text{rate}/100/12$. Then calculate a value $p = 12 \times \text{yrs}$. The formula for the payment is then

$$\text{payment} = \frac{r \times \text{amt}}{1 - (1+r)^{-p}}$$

The first test asks for the monthly mortgage on a \$150,000 home with a 4.5% mortgage rate on a 30-year loan:

```
>>> pmt(150000, 4.5, 30)
760.0279647388287
```

Rounding off to the nearest penny, the loan payment would be \$760.03. The second test is a car loan of \$10,000 at 5% for 6 years:

```
>>> pmt(10000, 5, 6)
161.04932661450974
```

2.12. In the sport of baseball, a pitcher's earned run average (ERA) is defined as the number of earned runs allowed per 9 innings pitched. For example, if a pitcher plays 5 innings and allows 2 earned runs to score his ERA is 3.6 by projecting the rate of 2 runs every 5 innings out to a full 9 innings. Define a function named era that will take two arguments: runs, the number of runs allowed, and outs, the number of outs recorded. The function should return the ERA, using the formula $\text{ERA} = 27 \times \text{runs}/\text{outs}$.

```
>>> era(2, 15)
3.6
```

2.13. Define a function named repeat that will take a word s and an integer n as arguments and return n copies of s separated by a comma and a space:

```
>>> repeat('yeah', 3)
'yeah, yeah, yeah'
```

2.14. Define a function named gp that will take a string s as an argument. You can assume s is one character long and is either 'A', 'B', 'C', 'D', or 'F'. The value returned by your function should be the point value of the grade passed in s, where A is worth 4 points, B is worth 3, C is worth 2, D is worth 1, and F is 0.

```
>>> gp('A')
4
>>> gp('B')
3
```

Note: Save this function in a file named gpa.py. A programming project in the next chapter will add another function, named gpa, to the file. The new function will compute a grade point average given on a list of grades, but the gp function of this chapter just returns the number of points for a single grade.

2.15. Modify gp so it will handle + and - grades by adding or subtracting 0.3 points. For example, a B+ is worth 3.3 points, and a C- is 1.7 points.

```
>>> gp('A-')
3.7
>>> gp('B+')
3.3
```

Suggestion: You could just add a bunch of elif clauses to test each grade separately, but a simpler design is to use a call to s.startswith (similar to endswith, except it checks characters at the start of a string) to figure out the value of the letter grade, then use s.endswith to see if you should add or subtract 0.3 points.

2.16. A gerund is a part of speech where a verb becomes a noun by adding the letters "ing" to the
end. For example, "float" becomes "floating" and "hike" becomes "hiking." Using the structure
of the `plural` function in Figure 2.10 as a guide, write a function named `gerund` that will take
a word passed as an argument and return the gerund form of the word. Your function should at
least work for "float" and "hike," but see if you can figure out other cases as well.

```
>>> gerund('float')
'floating'
>>> gerund('hike')
'hiking'
>>> gerund('fly')
'flying'
```

Note: PythonLabs has a function named `randverb` that will return a random word from a list of
verbs. Import this function if you want some additional verbs to use for testing `gerund`.

```
>>> from PythonLabs.Tools import randverb

>>> randverb()
'expand'
>>> randverb()
'open'
>>> gerund(randverb())
'helping'
```

Chapter 3

The Sieve of Eratosthenes

An algorithm for finding prime numbers

Mathematicians who work in the field of number theory are interested in how numbers are related to one another. One of the key ideas in this area is how an integer can be expressed as the product of other integers. If an integer can only be written in product form as the product of 1 and the number itself it is called a **prime number**. Other numbers, known as **composite** numbers, are the product of two or more factors, for example, $15 = 3 \times 5$ and $12 = 2 \times 2 \times 3$.

Ancient mathematicians devoted considerable attention to the subject. Over 2,000 years ago Euclid investigated several relationships among prime numbers, among other things proving there are an infinite number of primes. For most of their history, prime numbers were only of theoretical interest, but today they are at the heart of a variety of important computer applications. The security of messages transmitted using public key cryptography, the most widely used method for transferring sensitive information on the Internet, relies heavily on properties of prime numbers that were discovered thousands of years ago.

One of the fascinating things about prime numbers is the fact that there is no regular pattern to help predict which numbers will be prime. The number 839 is prime, and the next higher prime is 853, a distance of 14 numbers. But the next prime right after that is 857, only 4 numbers away. In some cases they appear in pairs, such as 881 and 883, where the difference between successive primes is only 2.

Because there is no pattern or rule to help us find prime numbers we need some way to systematically search for primes in a given range. A method for finding prime numbers that dates from the time of the ancient Greek mathematicians is known as the **Sieve of Eratosthenes**. The algorithm is very easy to understand, and even simpler to implement. We don't even need to know how to do addition or multiplication: all we have to do is count.

The project for this chapter is to implement the Sieve of Eratosthenes in Python. Our goal is to write a function named sieve that will make a list of all the prime numbers up to a specified limit. For example, if we want to know all the prime numbers less than 1,000, we just have to pass that number in a call to sieve:

```
>>> sieve(1000)
[2, 3, 5, 7, 11, 13,, ... 983, 991, 997]
```

The main goal for this chapter, as it is in the other chapters in this book, is to understand the algorithm and the computation used to carry out the steps. We will start with an informal description and show how this method was used to make short lists of prime numbers for thousands of years using only paper and pencil. Working through the example raises an interesting question about the algorithm that needs to be resolved before we can get a computer to carry out the steps of the computation. Implementing the Sieve of Eratosthenes in Python and doing some experiments on our "computational workbench" also provides an opportunity to introduce some new programming language constructs and some new features of the PythonLabs software that will be used in later projects.

3.1 The Algorithm

To see why this algorithm is called a "sieve," imagine numbers are rocks, where the shape of each rock is determined by its factors. Even numbers, which are multiples of 2, have a certain distinctive shape, multiples of 3 have a slightly different shape, and so on. Now suppose we have a magic bowl with adjustable holes in the bottom. To figure out which numbers are prime, put a bunch of rocks in the bowl and adjust the holes to match the shape for multiples of 2. When we shake the bowl, all the rocks that are multiples of 2 will fall through the holes (Figure 3.1). Next, adjust the holes so they match the shape for multiples of 3, and shake again so the multiples of 3 fall out. If the goal is to have only prime numbers in the bowl we need to keep repeating the adjusting and shaking steps until all the composite numbers have fallen out.

Instead of just starting with a random collection of numbers, and sifting out values in no particular order, the steps of the algorithm tell us how to proceed in a very precise manner.

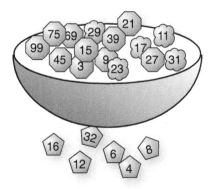

Figure 3.1: *The Sieve of Eratosthenes is a "magic bowl" that lets composite numbers fall through holes in the bottom. The first time the bowl is shaken, even numbers (multiples of 2) fall out.*

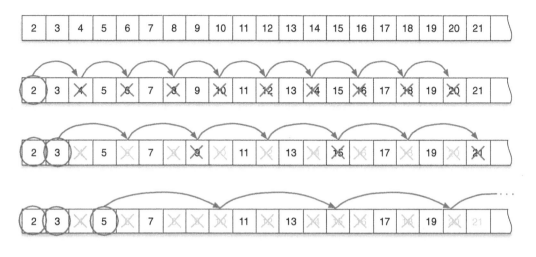

Figure 3.2: *To find prime numbers, write the integers starting from 2 on a strip of paper, then systematically cross off multiples of 2, multiples of 3, etc. until only primes are left.*

As a result of taking a more systematic approach, we are guaranteed that when we're done we will have every prime number within a specified range.

Because we don't really have a magic bowl, we need to use paper and pencil or some other real technology. Figure 3.2 shows how we can carry out the steps using a long strip of paper. Start by listing all the integers, starting with 2 and extending as far as we want. The top row of the figure shows a "paper tape" with numbers from 2 to 21, but the tape can extend arbitrarily far to the right.

To begin the process of removing composites we want to remove multiples of 2. This is easy enough: starting at 2, move to the right two spaces at a time, crossing off every number we land on, as shown on the second row in Figure 3.2. When we are done, the first unmarked number to the right of 2 is 3. We now know 3 is prime, and we step through the list again, this time starting at 3 and moving to the right three spaces at a time.

For the next round the first unmarked number is 5, since 2 and 3 were marked as prime on previous rounds, and 4 was crossed off in the first round. So on the third round we begin at 5 and move to the right in steps of size 5 to remove the multiples of 5.

At the start of each remaining round, we can make the following general statements. Let's call the first unmarked number on the paper k. Then:

- k is not a multiple of any of the numbers to its left (otherwise it would have been crossed off).
- Because all the numbers were written in order when we first created the list we now know k is prime.
- On this round we will start at k and move through the list in steps of size k.
- The first step takes us to $k + k = 2k$, the next lands on $2k + k = 3k$, and so on.
- We can cross off every number we land on because it will be a composite number of the form $i \times k$ for some value of i.

The method is fairly straightforward, and it should be clear that if we have a large piece of paper and enough patience we can make a list of all primes up to 1,000 or 1,000,000 or any value we want. If we carefully follow the steps exactly as they are specified we are guaranteed to end up with a list of all primes between 2 and the last number on the paper.

The description above leaves out a very important detail: when is the list of primes complete? If you do a few more steps of the example in Figure 3.2, where the last number on the strip of paper is 21, you'll soon notice you aren't crossing out any more numbers. For example, after you "shake the bowl" to cross out multiples of 11, the only remaining numbers will be 13, 17, and 19. You could continue and search for multiples of 13, but you can tell at a glance there aren't any. The smallest composite number that is a multiple of 13 is $2 \times 13 = 26$, and this list only goes up to 21, so clearly there aren't any more multiples of 13 in the list. By the same reasoning, there aren't any multiples of 17 or 19 either, and all the numbers remaining on the paper are prime.

Simply saying "repeat until all the numbers left are prime" might be sufficient if we're telling another person how to make a list of prime numbers, but it is not specific enough to include as part of the definition of a Python function. If we want to implement this algorithm in Python we need a more precise specification of when the algorithm is finished.

We'll return to this question later in the chapter, but the first order of business is to figure out how to create lists of numbers in Python and how to scan lists to remove composite numbers.

Tutorial Project

Take some time to make sure you understand how the Sieve of Eratosthenes works.

T1. Use the algorithm to make a list of prime numbers less than 50. Write the numbers from 2 to 49 on a piece of paper, then do rounds of "sifting" until you are left with only prime numbers.

T2. You can check your results with a version of `sieve` defined in the lab module for this chapter:.
```
>>> from PythonLabs.SieveLab import sieve
>>> sieve(50)
[2, 3, 5, 7, 11, 13, 17, 19, 23, 29, 31, 37, 41, 43, 47]
```

♦ How many passes did you make over the full list before you were left with all prime numbers? If we call the last number on the page n, can you derive a general formula in terms of n that describes a rule for when you can stop sifting?

3.2 Lists

In everyday life a **list** is a collection of items. Some lists are ordered, but many are just random collections of information. Ingredients in recipes are typically ordered because cooks like to have them presented in the order in which they are used, but shopping lists are often just random or semirandom collections of items (even if you're one of those super-organized people who arranges grocery lists by sections, the items within a section can be in any order; there's no reason to put radishes before cucumbers in the produce list).

Mathematicians also deal with collections. The idea of a **set** is one of the fundamental concepts in mathematics. Usually items in a set are not given in any particular order, but if order is important mathematicians say they have an "ordered set" or a "sequence."

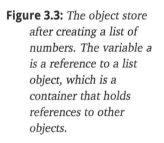

Figure 3.3: *The object store after creating a list of numbers. The variable a is a reference to a list object, which is a container that holds references to other objects.*

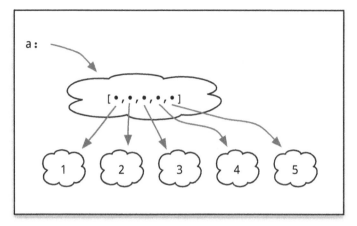

Computer scientists have a wide variety of ways of organizing collections of data, including sets, graphs, trees, and many other structures. For this project, we just need the simple linear collection in the form of a list.

The simplest way to make a list in Python is to write the items in the list between square brackets, separated by commas. To define a variable named a that refers to a list of the numbers from 1 to 5 we simply write

```
>>> a = [1, 2, 3, 4, 5]
```

The statement above is an assignment statement. Python handles this assignment statement just like it did the assignments in the previous chapter: it creates an object to represent the expression on the right side, and the name on the left becomes a reference to the object. The only difference here is the value on the right side is a list of numbers.

In programming language terminology, an object that holds a collection of data items is known as a **container**. Figure 3.3 shows what the object store looks like after the assignment statement that defines a to be a reference to a list. The new variable is a reference to a container object, and the container holds references to the items that are part of the list.

As we did in the previous chapter, we can verify the result of an assignment by asking Python to print the value of the variable:

```
>>> a
[1, 2, 3, 4, 5]
```

We can also ask Python to tell us what type of object a refers to:

```
>>> type(a)
<class 'list'>
```

Recall that "class" is the object-oriented programming terminology for "type." From this output we can see that the name of this class in Python is list.

In the previous chapter we saw Python has a function named len that counts the number of characters in a string. The same function tells us how many items are in a list:

```
>>> len(a)
5
```

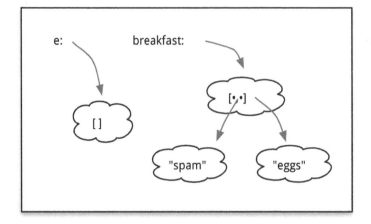

Figure 3.4: *In this example the object store has two lists. The list named e is empty; it does not contain references to any other objects. The other list, named breakfast, holds references to two strings.*

Note that it is possible to create a list that has nothing in it:

```
>>> e = []
```

This list is known as an **empty list**. It's analogous to the empty set in math, or, in real-life terms, to a three-ring binder with no paper. The binder is still a binder, even if it's empty. An empty list is an object, just like any other list object; it just doesn't contain any references to other objects. The list on the left in Figure 3.4 shows what an empty list might look like in the object store. Empty lists are very common in a wide variety of algorithms. An algorithm will often create an initially empty list, and add items to it in future steps, or start with a list of items and repeatedly delete items until the list is empty.

Python allows us to put any type of object in a list. This statement creates a list of strings:

```
>>> breakfast = ['spam', 'eggs']
```

Note what happens when we ask Python to compute the length of this list of words:

```
>>> len(breakfast)
2
```

Python did not count the number of letters in the strings. Instead, it counted the number of objects in the list, and because the list was created using two string objects the value returned by the call to len is 2 (Figure 3.4).

There are all sorts of interesting things we can do with our new lists. We can add items to the end, insert items in the middle, delete items, invert the order, and carry out dozens of other useful operations. Most of these operations are performed by calling methods defined in the list class. For example, to attach a new item to the end of a list, call a method named append:

```
>>> breakfast.append('sausage')
```

For the projects in this chapter, however, we only need to know how to create lists and access items in the list, so we'll put off the discussion of other methods until we need them in a project.

Tutorial Project

T3. Start your IDE and type this statement in the shell window to create a list containing a few even numbers:

```
>>> a = [2, 4, 6, 8]
```

T4. Ask Python to show which class the new object belongs to:

```
>>> type(a)
<class 'list'>
```

T5. Use the len function to find out how many items are in the list:

```
>>> len(a)
4
```

T6. Add a new number to the end of the list:

```
>>> a.append(10)
```

T7. Did the previous expression change the length of the list? Call the len function to check your answer.

T8. Type this expression to make a new empty list named b:

```
>>> b = []
```

T9. Use the len function to find out how many elements are in b. Did you get 0?

T10. Make a list of strings:

```
>>> colors = ['green', 'yellow', 'black']
=> ['green', 'yellow', 'black']
```

T11. Verify there are three objects in this list:

```
>>> len(colors)
3
```

T12. Use the append method to add a new string to the end of colors:

```
>>> colors.append('steel')
```

T13. Use len to figure out how many items are now in colors.

Earlier in this section we learned that len, which we first encountered as a function that returns the number of characters in a string, also works for lists. Do you suppose the same thing is true of the + and * operators? Can you predict what Python will do if we enter expressions using + or * and the operands are lists?

T14. Do an experiment to test your prediction. Create two lists named a and b:

```
>>> a = [2,4,6]
>>> b = [1,2,3,4,5]
```

T15. Ask Python to evaluate some expressions with the + and * operators:

```
>>> a + b
[2, 4, 6, 1, 2, 3, 4, 5]
>>> a * 3
[2, 4, 6, 2, 4, 6, 2, 4, 6]
>>> len((a + b) * 2)
16
```

Make sure you understand what Python did to evaluate each of the expressions above and how it evaluates similar expressions when a and b are strings instead of lists.

In Python strings and lists are special cases of a more general type of object called a **sequence**, which means operators and methods defined for lists often do a similar operation on strings, and vice versa.

3.3 Iterating Over Lists: for Statements

After a program has created a list to hold a collection of objects it often needs to carry out some calculation based on all the objects. An application might need to compute the average of a list of numbers, or search for a specific string in a list of words, or any number of similar operations.

The process of working through a list to perform some operation on every object is one form of a type of computation known as **iteration**. The word comes from the Latin word *iter*, which means "road" or "path." We often talk about iterating over a collection of items, which means we start at the front and step through the collection, one item at a time.

The easiest way to specify this sort of iteration in Python is to use a construct called a **for loop**. The word "loop" has been part of computer programming jargon since the earliest programming languages. It refers to the idea that when the computer has finished executing a set of statements it should cycle back to the beginning and repeat the same operations again. When we are iterating over a list, we want the system to execute the statements for one item, then go back to the start of the loop and execute the same steps again for the next item in the list.

In Python, a for loop consists of a **for statement**, which controls how often the loop is executed, followed by one or more statements that make up the body of the loop. Two examples are shown in Figure 3.5. The print_list function uses a for loop to print every item in a list, and total computes the sum of a list of numbers.

The statement "for x in a" says we want to do something with every item in the list a. Python will assign x to the first item in the list and execute the statements in the body. It then goes back to the top of the loop, assigns x to the next item from the list, and executes the body again. This process repeats as long as there are still items in the list. The end result is that the statement in the body is executed once for each item, in order from the front of the list to the end.

The total function in Figure 3.5 uses a for loop to compute the sum of a list of numbers. The body of the loop uses a type of assignment statement we haven't seen yet. The += operator is a combination of addition and assignment. An expression of the form n += m means "add m to n." The statement on line 4 is telling Python to add one of the numbers in the list to sum, a local variable that is keeping a running total of the values seen so far. The function starts by initializing sum to 0 and then uses the for loop to add each item in the list to sum. After the loop terminates the sum is returned as the value of the function call.

```
1    def print_list(a):
2        for x in a:
3            print(x)
```

```
1    def total(a):
2        sum = 0
3        for x in a:
4            sum += x
5        return sum
```

Figure 3.5: *Examples of for loops. (left) A function that prints every item in a list. (right) A function that computes the sum of a list of numbers. Note that when there are two or more statements in the body of a loop they must all be indented the same number of spaces.*

Tutorial Project

Use your IDE to create a new file named lists.py to use for experimenting with operations on lists.

T16. Add the definition of the print_list function of Figure 3.5 to lists.py, save the file, and then load it into an interactive shell session.

T17. Create a list to test your print_list function:
```
>>> colors = ['blue', 'yellow', 'black', 'green', 'red']
```

T18. Call your function, passing it the new list:
```
>>> print_list(colors)
blue
yellow
...
```

Did your function print every string in the test list? If not, compare your definition line by line with the function shown in Figure 3.5, paying particular attention to how the lines are indented.

The next set of experiments shows how the increment operator works.

T19. Initialize a variable and then use an assignment statement to update it to a new value:
```
>>> n = 4
>>> n = n + 1
>>> n
5
```

T20. Do another update, but this time use the increment operator:
```
>>> n += 1
>>> n
6
```

Displaying Output in the Shell Window

Python has a built-in function named print. Whenever we call print we can pass it one or more objects as arguments and Python will display the objects in the shell window.

Here is an example that prints three integers:
```
>>> print(1,2,3)
1 2 3
```

The arguments passed to print can be any type of object. In this example breakfast refers to a list of strings:
```
>>> print(breakfast)
['spam', 'eggs', 'sausage']
```

This function isn't often used in interactive sessions since we can get the value of a variable by just typing its name, but printing is going to come in very handy when we start experimenting with more complicated algorithms. At various points in an algorithm we are going to want to call print to display the values of selected variables so we can keep track of how the algorithm is progressing.

Updating a Variable

The function that computes the mean of a list of numbers illustrates a common situation in computer programming. On each iteration of the loop we want to add another item from the list to the running total. One way to write this might be

```
sum = sum + x
```

This statement always appears strange to new programmers because it looks so much like an algebraic equation that does not make any sense. However, remember that the equal sign in Python is not a comparison operator, it is the assignment operator. This statement is processed just like any other assignment: Python looks for the current value of sum in its object store, adds x to that value, and overwrites the old value of sum with the new one.

Experienced Python programmers write the statement using an "augmented assignment" which does exactly the same thing:

```
sum += x
```

T21. When updating an integer, the operand on the right side of the operator can be any integer value:

```
>>> n += 3
>>> n
9
```

T22. Add the definition of the total function (Figure 3.5) to your lists.py file. Save the file and load the function into your shell session.

T23. Create a short list of numbers to test the function:

```
>>> a = [3,4,3,2,1]
```

T24. Python already has a built-in function named sum that computes the sum of a list of numbers. Use this function to show you the result you should expect when you test your total function:

```
>>> sum(a)
13
```

T25. Call total to compute the sum of numbers in your test list:

```
>>> total(a)
13
```

If you did not get 13 as the result, go back and check every line to make sure it is exactly as shown in Figure 3.5).

T26. What do you think your function will do if you pass it an empty list? Pass an empty list to total to see if your prediction is correct:

```
>>> total( [] )
```

3.4 List Indexes

Often when we write programs that use a list we need to do more than simply access every item in order. Some algorithms use only part of a list, while others need to access items out of order. These algorithms need a way to specify the locations of the items they use, allowing a programmer to write something that tells Python "fetch the second item in the list" or "set a variable m to the item in the middle of the list."

Python and other programming languages use the idea of a **list index** to refer to a position within a list. An index for a list with n items is a number between 0 and $n - 1$. Note that the index of the first item is 0, not 1. This is a convention widely used in programming languages, but a common source of errors for new programmers who expect to use 1 as the first location.

To access an item in a list, simply write the name of the list followed by the location of the item, where the index is enclosed in "square brackets." Here is an example, using a list of five strings:

```
>>> vowels = ['a', 'e', 'i', 'o', 'u']
>>> vowels[0]
'a'
>>> vowels[3]
'o'
```

List Indexes

Computer scientists are famous for beginning at 0 instead of 1 when they start assigning labels to things. List indexes are a good example. If a list has 10 items, the locations in the list are labeled from 0 to 9.

If you type an expression with a list index operator and you see a result that doesn't look right, the first thing to check is to make sure you count starting at 0. If you type a[1] expecting to see the first string in the list shown below you'll get the wrong value. You need to remember to type a[0].

```
>>> a = ['H', 'He', 'Li', 'Be', 'B', 'C']

>>>   a[1]
'He'

>>>   a[0]
'H'

>>>   len(a)
6

>>>   a[6]
IndexError: list index out of range

>>>   a[5]
'C'
```

H	He	Li	Be	B	C
0	1	2	3	4	5

When we are reading an expression like a[i] out loud we say "a sub i." The convention is derived from mathematical notation, where a sequence x is written $x_1, x_2, \ldots x_n$. The subscript notation from math and the square bracket notation used in computer programming are two different ways of expressing the idea that we are working with an ordered collection of items. Just remember that the first object in a list in Python is at location 0 instead of location 1.

If we know an item is in a list and we want Python to tell us where it is located we can call a method named index. If a list has n items, the value returned by a call to index will be a number between 0 and $n - 1$:

```
>>> vowels
['a', 'e', 'i', 'o', 'u']
>>> vowels.index('e')
1
>>> vowels.index('u')
4
```

If the specified item is not in the list, index generates an error:

```
>>> vowels.index('x')
ValueError: 'x' is not in list
```

In Chapter 2 we saw that the word in can be used as a Boolean operator to see if a string is contained inside another one. We can do the same thing here, to see if an item is contained in a list:

```
>>> 'e' in vowels
True
>>> 'x' in vowels
False
```

Note that both the index method and the in operator will search through a list to find an item. The difference is that in will return a Boolean value, depending on whether or not the item was found, and index returns an integer that tells us *where* an item was found, and it generates an error if the item is not in the list.

Another example of a function that iterates over a list is partial_total, shown in Figure 3.6. Like the total function used in a previous example, this function computes the sum of elements in a list, but in this case it uses only the values at the front of the list. The number of values to sum is specified by the first argument, so a call of the form partial_total(n, a) means "compute the sum of the first n items in a."

```
>>> a
[3, 4, 3, 2, 1]
>>> partial_total(3,a)
10
```

Because the partial_total function is not iterating over the entire list we need to specify the part of the list we want to use. The technique shown in Figure 3.6 is to use a function named range to specify a set of index values. The expression range(i, j) means "the set of integers from i through j − 1."

```
1   def partial_total(n,a):
2       """
3       Compute the total of the first
4       n items in list a.
5       """
6       sum = 0
7       for i in range(0,n):
8           sum += a[i]
9       return sum
```

Figure 3.6: *This function computes the sum of the values at the front of a list. The number of items to include in the sum is the first argument passed to the function.* **Note:** *There is a potential bug in this code (see Exercise T39 at the end of this section).*

Download: *sieve/partialtotal.py*

If we want to access the first n items in a list the for statement is

```
for i in range(0, n):
    # ... do something with a[i] ...
```

Notice how the assignment statement on line 4 of partial_total differs from the corresponding statement in total (Figure 3.5). If the loop is written as "for x in a" then Python has already fetched a value from the list and saved it in x. But if the for statement uses a range expression to define values for an index variable we need to access the item in the list using an index expression, for example as a[i].

We don't need a list in order to use range; any situation where we need a sequence of integers is a good occasion to use a for loop with a range expression. Here is a loop that simply prints each number from 1 to 10 along with its square root:

```
for i in range(1,11):
    print(i, sqrt(i))
```

Notice that the last value of i will be 10 and not 11. This may seem awkward at first—if the function is going to make values between 1 and 10 why not just write range(1,10)? As the example with list indexes showed, location numbers start with 0 and end with n − 1, and it is convenient to write range(0,n) when the values are going to be used to index into a list.

When we implement the Sieve of Eratosthenes we are going to want to iterate over a list of numbers using different step sizes. On the first round we will cross off every second number, on the next round every third number, and so on. For situations like this, Python allows us to pass a third argument to range that specifies the distance between successive values. Here is the previous for loop, but this time it only prints the square roots of odd numbers:

```
for i in range(1,11,2):
    print(i, sqrt(i))
```

Because this loop has a step size of 2, the variable i will be assigned 1, 3, 5, *etc.*

Tutorial Project

T27. Make a small list of strings to use for testing index expressions:
```
>>> notes = ['do', 're', 'mi', 'fa', 'sol', 'la', 'ti']
```

T28. Use an index expression to get the first item in the list:
```
>>> notes[0]
'do'
```

T29. Because there are seven strings in this list we can find the last one at location 6:
```
>>> notes[6]
'ti'
```

T30. Here is another way to access the last item:
```
>>> notes[len(notes)-1]
'ti'
```

T31. Try asking Python for values at other locations, using any index between 0 and 6. Is the result what you expected?

T32. Repeat the experiment, but ask for an index that is past the end of the list; for example
```
>>> notes[12]
IndexError: list index out of range
```

T33. Next try some expressions using in, which should evaluate to True or False, depending on whether the list contains the specified item:
```
>>> 're' in notes
True
>>> 'bzzt' in notes
False
```

T34. Use Python's index method to find out where the items are located:
```
>>> notes.index('re')
1
>>> notes.index('bzzt')
ValueError: 'bzzt' is not in list
```
Do you see why the first expression above returned a value of 1?

The next set of exercises asks you to type a for loop in your shell session. After you type the first line, your prompt should change to an ellipsis (three dots) to show you that Python is expecting you to continue something you started on an earlier line. Don't forget to type spaces before each line in the body of the loop. After the last line simply hit the return key to end the loop.

T35. Use range to iterate over the notes list and print something about each note:
```
>>> for i in range(0,len(notes)):
...     print(notes[i], 'has', len(notes[i]), 'letters')
...
do has 2 letters
re has 2 letters
...
```
Make sure you understand exactly why Python prints what it does.

T36. Enter the `for` loop that prints the square roots of numbers from 1 to 10 (don't forget to import the `sqrt` function if you haven't done it already):

```
>>> from math import sqrt
>>> for i in range(1,11):
...     print(i, sqrt(i))
...
1 1.0
2 1.4142135623730951
...
```

T37. Repeat the previous exercise, but this time use `range(1,11,2)`. Did Python execute the body of the loop only for odd values of i? Do you see why?

T38. Either download the definition of `partial_total` (Figure 3.6) or add the definition to your `lists.py` file. Load the definition into your interactive session and call it a few times (using the same list you used to test `total`) to verify it works.

T39. What happens if you pass a number to `partial_total` that is larger than the number of items in the list? For example, what does Python do if a has five items and you call `partial_total(7,a)`?

♦ Add statements to your definition of `partial_total` so the function returns a correct value even if n is too large. If n is greater than the number of items in the list, simply return the sum of all the items.

3.5 The Animated Sieve

Now that we know how to create lists of integers in Python we are ready to start writing a function that uses the Sieve of Eratosthenes to create a list of prime numbers. As a first step we are going to take a closer look at how the sieve works, using a function included as part of PythonLabs.

The projects in this section use a technique known as **algorithm animation** to help visualize the steps of a computation. Animated algorithms draw 2D pictures in a graphics window called a **canvas** and then update the drawing as the algorithm proceeds. In future chapters we will use animations to illustrate how objects are moved around by different sort algorithms, show the motion of planets during a solar system simulation, and visualize many other projects throughout the book.

To use the animated version of the sieve we need to import the functions with this statement:

```
>>> from PythonLabs.SieveLab import *
```

The asterisk here means "everything," so this statement imports all the functions defined in SieveLab.

Because the lists we will experiment with are too long to fit on a single line they will be displayed on the canvas as a set of rows and columns, in a format a person might use when writing the numbers on a piece of paper, and we will refer to the list of numbers as the "worksheet." But regardless of whether the numbers are written on a single line or in a grid, the basic operation is the same: when crossing off multiples of some number k we want to iterate over the set of numbers using steps of size k.

Constructors

In Python we can use class names as functions. In the previous chapter we learned that Python's string class is named str, and this chapter introduced lists, which belong to a class named list. If we use these names as functions Python will create an object of that type. For str and list, Python creates an empty string or an empty list:

```
>>> str()
''
>>> list()
[]
```

We can also pass arguments to these functions if we want the new object to be initialized. For example, to make a list of numbers, pass a range expression to list:

```
>>> list(range(0,10))
[0, 1, 2, 3, 4, 5, 6, 7, 8, 9]
```

In object-oriented programming terminology these functions that create objects of a specified class are known as ***constructors***. We will learn more about constructors in Chapter 7, when we learn how to define our own classes.

Recall that the first step in the algorithm is to create a list of integers, starting with 2. Fortunately, Python has several features that will help us make the list without having to type a long list of numbers. One way is to use the name list as a function, passing it a range expression to tell it to make a list containing all the values in that range. For example, this is how we can make a list of numbers up to but not including 10:

```
>>> list(range(10))
[0, 1, 2, 3, 4, 5, 6, 7, 8, 9]
```

This is an example of a more general technique where we use a class name as a function in order to create an instance of that class (for more see the sidebar on "Constructors").

If we want to find prime numbers less than 100, a statement that creates a list with numbers between 2 and 99 is

```
>>> list(range(2,99))
[2, 3, 4, 5, ... 97, 98]
```

But notice that because 2 is the first item it is at location 0. The number 3 is at location 1, and in general, any number i is at location $i - 2$.

We can avoid some potential confusion if we add two "placeholder" values at the front of the list. What we want is a list that looks like this:

```
[?, ?, 2, 3, 4, ... ]
```

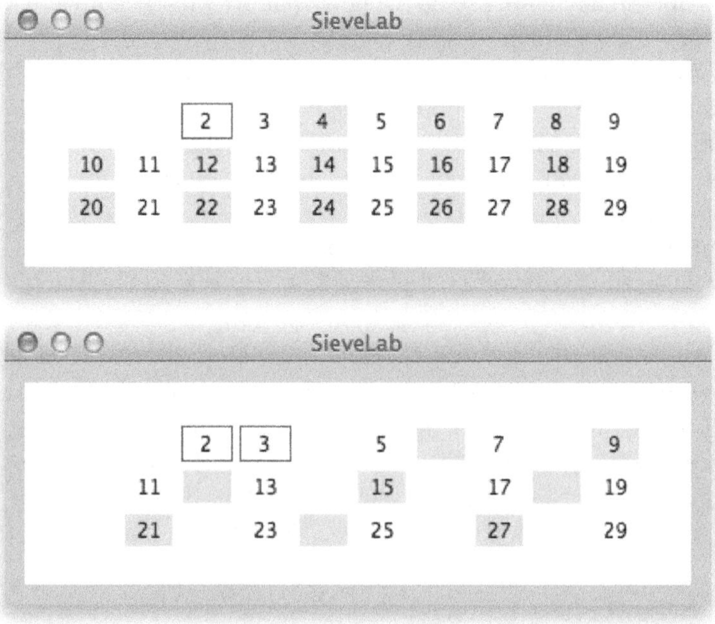

Figure 3.7: *Top:* *A snapshot of the worksheet during the first iteration of the Sieve of Eratosthenes, after 2 has been identified as a prime number and its multiples have been marked.* **Bottom:** *After multiples of 2 are erased, the second iteration marks multiples of 3.*

If we can figure out what to put in place of the question marks we will have what we want: the number 2 is in location 2, 3 is in location 3, and so on.

A natural candidate for the two placeholder values is a special Python object named None. This object is often used to introduce a variable before we know what value we want to assign it, or in situations where we need space for something.

We saw earlier that the + operator means "concatenate" when we apply it to two lists. That means we can initialize the worksheet by creating a list with two None objects and then "adding" the list of numbers beginning with 2. Putting all these pieces together, here is the statement that initializes our worksheet with numbers from 2 to 99:

```
>>> worksheet = [None, None] + list(range(2,100))
```

If we print the value of this variable we can see it's exactly what we want:

```
>>> worksheet
[None, None, 2, 3, 4, 5, ... 97, 98, 99]
```

To start the animation, pass the initial worksheet in a call to view_sieve, one of the functions imported from SieveLab:

```
>>> view_sieve(worksheet)
```

If the canvas is not already on the screen, a new window will open up and the system will draw the numbers in the list in a rectangular grid.

For this interactive session, where we're using visualization to explore the algorithm, the operations performed on each round are split into two separate functions. When we write the Python code later in this chapter all the steps will be in a single function, but in SieveLab the steps have been separated into two different functions for better control of the interactive visualizations.

A function named mark_multiples will highlight a number and all of its multiples. On the first round we need to cross off multiples of 2, so this is what we type:

```
>>> mark_multiples(2, worksheet)
```

After Python returns from this function the canvas should look something like the snapshot in the top half of Figure 3.7 (to save space only the top three rows of the worksheet are shown). Notice how the box for the number 2 is highlighted and the boxes for all the multiples of 2 have a gray background to show they will be deleted.

To continue, call a second function that removes the marked numbers from the worksheet:

```
>>> erase_multiples(2, worksheet)
```

All of the grayed-out numbers will be erased from the canvas (as seen in the snapshot in the lower half of Figure 3.7).

We can repeat the calls to these two functions as many times as we'd like. Before each call to mark_multiples, simply look for the next number following the most recently identified prime. After removing multiples of 2 we can see the next prime is 3. Note that the next prime after 3 will be 5, because 4 will be filtered out as a multiple of 2.

Let's return now to the question of how we will be able to tell when only prime numbers are left on the worksheet. The process described in the introduction said "repeat sifting until it's clear there are no composites left." The example given at the time was that we don't have to sift out multiples of 13 if the list only goes up to 21 because the lowest possible multiple of 13 is 2×13, or 26. Now that we want to automate the process we need to be more precise about the terminating condition so we can implement the algorithm in Python.

It turns out we can stop the marking and erasing process as soon as the first unmarked number is greater than $\sqrt{n}$. Every composite number less than n will have at least one prime factor p less than $\sqrt{n}$, which means the number will have been crossed off when we filter out the multiples of p (the sidebar on "Prime Factors" on page 73 explains why).

In the example used in the tutorial project for this section, where $n = 100$, we can stop filtering after we remove the multiples of 7. After removing multiples of 7 the next unmarked number will be 11, and because it's greater than $\sqrt{100} = 10$ all the numbers left on the worksheet will be prime.

◆ Prime Factors

Every composite number can be written as the product of two smaller numbers. If either of those numbers is itself composite it can also be written as a product. We can continue this process (which is known as **factoring**) until each smaller number is prime.

At least one of the prime factors of a number $k = i \times j$ must be less than or equal to $\sqrt{k}$. It's easy to see why: if both i and j are greater than $\sqrt{k}$ then their product would be larger than k because $i \times j > \sqrt{k} \times \sqrt{k} = k$.

When looking for primes less than n, every composite number between 2 and n always has one prime factor $p < \sqrt{n}$ so the number will be removed when we sift multiples of p. That means the Sieve of Eratosthenes can stop sifting when the smallest unmarked number on the worksheet is greater than $\sqrt{n}$.

As an example, suppose we are looking for primes less than 1000. The stopping point is $\sqrt{1000} \approx 31.62$. After we filter multiples of 31, the next number on the worksheet is 37, and we are done.

$65 = 5 \times 13$

$147 = 3 \times 7 \times 7$

$308 = 2 \times 2 \times 7 \times 11$

$793 = 13 \times 61$

$861 = 3 \times 7 \times 41$

$901 = 17 \times 53$

$961 = 31 \times 31$

When n = 1000 every composite number has at least one prime factor smaller than 31.62

Tutorial Project

T40. Import the functions defined in SieveLab:
```
>>> from PythonLabs.SieveLab import *
```

T41. Pass a range to the list constructor to see how it can be used to create a list of numbers:
```
>>> list(range(0,10))
[0, 1, 2, 3, 4, 5, 6, 7, 8, 9]
```

T42. Note how the last number in the list is one less than the upper limit passed to range. Try making a few more lists with different starting and ending values to make sure you understand how the constructor works.

T43. Make a list of odd integers:
```
>>> list(range(1,10,2))
[1, 3, 5, 7, 9]
```
Do you see why this list contains every other number starting with 1?

T44. Do some experiments with the + operator to verify it concatenates two lists:
```
>>> a = ['do', 're', 'mi']
>>> b = ['fa', 'sol', 'la']
>>> a + b
['do', 're', 'mi', 'fa', 'sol', 'la']
```

T45. Initialize the worksheet with integers less than 100:
```
>>> worksheet = [None, None] + list(range(2,100))
```

T46. Display the worksheet on the PythonLabs canvas. You should see a 10 x 10 grid:
```
>>> view_sieve(worksheet)
```

T47. Call mark_multiples to highlight the multiples of 2:
```
>>> mark_multiples(2, worksheet)
```
Is the number 2 circled? Are all the multiples of 2 grayed out?

T48. Erase the multiples of 2. All the grayed-out numbers should disappear:
>>> *erase_multiples(2, worksheet)*

T49. Mark the multiples of 3:
>>> *mark_multiples(3, worksheet)*

If your IDE supports "command line editing" this is a good place to use it. You can save yourself some typing if you hit the up-arrow key twice, change the 2 to a 3, and hit the return key.

T50. Notice how every third square is gray, including locations that contained numbers that were erased in the previous step.

T51. Erase the multiples of 3:
>>> *erase_multiples(3, worksheet)*

T52. Do you see how 5 is next number to sift out? It's the lowest unmarked number because 4, which is a multiple of 2, was removed in the first round.

T53. Keep repeating the marking and erasing steps until all the numbers left on the worksheet are prime numbers. Note that one way to tell there are no composite numbers left is when all the cells grayed out by a call to mark_multiples have already been erased on a previous round.

T54. How many rounds of marking and erasing did you do before you decided there were no more composite numbers left?

T55. Call sieve(100) to get the list of primes and compare this list with the worksheet on your canvas. Do they agree?

The SieveLab implementation of sieve will draw the worksheet and show the steps of the algorithm on the canvas if we pass it some additional arguments.

T56. The following command shows how to call sieve and have it display the worksheet as a 25 × 40 grid:
>>> *sieve(1000, vis=True, nrows=25, ncols=40, delay=1.0)*

The last argument determines the speed of the animation. If you want the program to pause for 2 seconds after each marking and erasing step use delay=2.0.

T57. Is the final display of the list of primes less than 1000 what you expected? Can you see why the last number circled on the canvas is 31?

3.6 A Helper Function

The animations in the previous section show what we need to do when we write our own Python program to implement the Sieve of Eratosthenes. We've already seen how to initialize the worksheet; now we need to figure out how to remove composite numbers.

We're going to take an approach commonly used by programmers and break the problem into two smaller sub-problems. In this section we will write a function named sift that will implement a single round of finding and removing multiples. For example, if we call sift(5, worksheet) we will expect it to scan through the list of numbers to remove all the multiples of 5. When we are satisfied this function is working correctly, we will turn our attention to the main part of the algorithm, which will have a loop that repeatedly calls sift to systematically remove all the composite numbers.

The sift function has a very specific purpose, and it is unlikely we would ever want to use it for anything other than as part of a project that implements the Sieve of Eratosthenes.

Programmers write these sorts of specialized functions as a way of focusing on a single aspect of a complex problem, allowing them to design and test the code for each piece separately. A function like `sift` that is intended only to be called for one special purpose is called a **helper function**. After we implement and test our helper function we will use it in the next section as we implement our own version of `sieve`.

The Python code for the function will start with a `def` statement that looks like this:

```
def sift(k, a):
```

When the function is called, the value of a newly discovered prime number will be in k, and a will be a reference to the main worksheet list.

The first question we need to consider is what it means to "cross off" a number. When we used paper and pencil it was sufficient to draw a line through a number or erase it. Now that we are using a Python list to hold the numbers on the worksheet we need to figure out how to record the fact that a number is composite.

It's tempting to use a statement that removes numbers from the list. Python has a statement named `del` (short for "delete") that does just that. Here is an example that shows a list and what it looks like after using `del` to remove the middle item:

```
>>> greeks = ['alpha', 'beta', 'gamma', 'delta', 'epsilon']
>>> del greeks[2]
>>> greeks
['alpha', 'beta', 'delta', 'epsilon']
```

The problem with this approach is that `del` shortens the list. In the example above we can see `greeks` is a list of four strings after the `del` statement is executed. This poses a big problem for the `sift` function because the algorithm depends on removing every k^{th} item when it is working its way through the list to remove multiples of k. If we start removing items from the worksheet successive multiples will no longer be k numbers apart.

So instead of actually removing a number we should figure out a way to leave it there but mark it as composite. A straightforward way of doing this is to simply overwrite the number with a placeholder. We've already decided to use two `None` objects at the front of the list as placeholders, so a natural choice for crossing off a composite number is to simply store a `None` in the corresponding location in the list.

From the original presentation of the algorithm in Section 3.1 and the animations in the previous section it should be clear that to sift out the multiples of some number k we want to store a `None` object at regular intervals in the worksheet. The first multiple is $2 \times k$. In Python terms, we want to store a `None` in a[2*k] and then skip ahead k places to get to the next multiple. This suggests we should use a `range` expression that has an initial value of 2*k, an upper limit that is the length of the list, and a step size of k:

```
range(2*k, len(a), k)
```

A Python implementation of `sift` that uses this strategy is shown in Figure 3.8. As an example of how it works, consider what the `for` loop does when k is 5. The locations set to None are a[10], a[15], etc., which are all the locations in the worksheet that contain multiples of 5.

This version of the `sift` function is an accurate implementation of the steps illustrated in the previous section, but it actually is doing more work than necessary. An optional programming exercise at the end of this chapter gives some hints about how one might write a more efficient `for` loop that requires fewer iterations.

```
1   def sift(k, a):
2       "Remove multiples of k from list a"
3       for i in range(2*k, len(a), k):
4           a[i] = None
```

Figure 3.8: *The* sift *helper function iterates through the worksheet in steps of size k, replacing numbers with placeholders (None objects).*

Download: sieve/sift.py

Tutorial Project

The goal for this section is to create an initial version of a file named sieve.py that will eventually contain the complete definition of sieve and its helper functions. This first version will contain only the definition of sift; the remaining code will be added in the next section.

To get started, you can either download the function shown in Figure 3.8 and rename the file to sieve.py or use your IDE to create new file named sieve.py and enter the code shown in the figure.

T58. Load your sift function into an interactive shell session in your IDE.

T59. Make a short worksheet to test sift:
>>> worksheet = [None, None] + list(range(2,30))

T60. Call sift to remove multiples of 2:
>>> sift(2,worksheet)

T61. Take a look at the worksheet:
>>> worksheet
[None, None, 2, 3, None, 5, None, 7, None, ... 27, None, 29]
Do you see how all the even numbers have been replaced by None?

T62. Call sift(3, worksheet) and verify the multiples of 3 are removed.

T63. What do you suppose would happen if you try to remove multiples of 4 at this point? Check your answer by calling sift(4, worksheet).

T64. Call sift one more time to remove multiples of 5, then ask Python to display the worksheet:
>>> sift(5,worksheet)
>>> worksheet
[None, None, 2, 3, None, 5, None, 7, None, ... None, None, 29]

T65. Are there any more composite numbers in the worksheet?

3.7 The sieve Function

With our new helper function implemented and tested we are ready to start working on the sieve function itself. In programming jargon, sieve is the **top-level** function. It is the function that can be called, either by a user in an interactive session or from another program, whenever a list of prime numbers is needed.

As we saw in earlier examples, sieve will have a single argument, and it should return a list of prime numbers less than that value, for example

>>> sieve(50)
[2, 3, 5, 7, 11, 13, 17, 19, 23, 29, 31, 37, 41, 43, 47]

Because we are going to be passing sieve a parameter that specifies the upper limit, the def statement at the beginning of the definition will specify a single argument, which we will call n:

```
def sieve(n):
```

The first thing the function needs to do is create the worksheet. All we need to do is modify the expression we used earlier so the upper limit is n, the argument passed to sieve:

```
worksheet = [None, None] + list(range(2,n))
```

Note that if we call sieve(100) the initial list will go up to 99.

The heart of the implementation will be a loop that calls the sift helper to remove composites. We clearly want to use some sort of for loop to generate integers to pass to sift, and it's also clear that the range of values should start with 2. But what should we use for the upper limit? What is the last number we need to pass to sift before we know all the remaining values on the worksheet are prime?

Recall from Section 3.5 that whenever we are looking for primes up to some number n we can stop sifting as soon as we reach a value greater than $\sqrt{n}$. Unfortunately, if we try to specify $\sqrt{n}$ as the upper limit of a range expression we will get an error message. The square root function in Python's math library returns floating point values, but range expects us to pass it integers.

Another function in Python's math module will help us convert the upper limit to an integer to use in the range expression. The function is called ceil, which is short for "ceiling." A call ceil(x) returns the smallest integer greater than x, that is, it "rounds up" to the next whole number.

Putting all these pieces together, here is a for loop that will call sift using all the integers from 2 up to the desired limit:

```
for k in range(2, ceil(sqrt(n))):
    sift(k, worksheet)
```

This loop would work, but it does too much work. It will call sift to remove multiples of 2, then 3, and then 4. Exercise T63 showed there is no harm in sifting out multiples of 4: the helper function will just store None in the worksheet at locations 8, 12, 16, etc. But those locations are already set to None because they are also multiples of 2, so we would be wasting time if we called sift with k set to 4.

An improvement is to test if the current value has already been crossed off. We can add an if statement to the body of the loop so that we call sift only if k has not already been crossed off. When we are checking to see if something is None the idiom in Python is to use a Boolean operator named is:

```
if worksheet[k] is not None:
    sift(k, worksheet)
```

The complete code for our Python implementation of the sieve function is shown in Figure 3.9. It includes the definitions of the two helper functions, the one described earlier that sifts out multiples and a new helper that makes the final list of numbers. The new helper, named non_nulls, condenses the worksheet by removing all the None objects to create the list that will be returned by the call to sieve. The logic used by non_nulls is straightforward. It creates an empty list to hold the final result and then passes through the worksheet to collect all the numbers, appending them to the result list.

```
1    from math import sqrt, ceil
2
3    def sieve(n):
4        "Return a list of all prime numbers less than n"
5
6        worksheet = [None, None] + list(range(2,n))
7
8        for k in range(2, ceil(sqrt(n))):
9            if worksheet[k] is not None:
10               sift(k, worksheet)
11
12       return non_nulls(worksheet)
13
14   def sift(k, a):
15       "Remove multiples of k from list a"
16       for i in range(2*k, len(a), k):
17           a[i] = None
18
19   def non_nulls(a):
20       "Return a copy of list a with None objects removed"
21       res = []
22       for x in a:
23           if x is not None:
24               res.append(x)
25       return res
```

Download: sieve/sieve.py

Figure 3.9: *A Python implementation of the Sieve of Eratosthenes.*

Tutorial Project

The tutorial for this section will complete the implementation of the sieve function, using the iterative and incremental development style described in the "Art of Programming" sidebar on page 79. If you completed the tutorial exercise from the previous section you have a file named sieve.py with the definition of the sift helper function. The exercises in this section will add the top-level sieve function and the second helper, non_nulls.

T66. Start a new shell session with your IDE.

T67. Type this import statement in your shell window so you can try some experiments with the ceil and sqrt functions:

     from math import sqrt, ceil

T68. Call ceil a few times to make sure you understand what it does:

     >>> ceil(10.7)
     11
     >>> ceil(10.2)
     11
     >>> ceil(10)
     10

The Art of Programming (Part I)

At times programming is as much a craft as it is a profession. This sidebar describes some "tricks of the trade" programmers use when developing new programs.

Iterative and Incremental Development

Many Python programmers have adopted a process from software engineering known as ***iterative and incremental development***. The idea is to make very small changes to a program and to continually test the program, performing a set of tests after each change.

This approach to programming also encourages programmers to reuse and adapt code that is already known to work. Python is very well suited to this style of programming. We can type statements in an interactive shell window, for example to see how a function s is called, and then copy and paste from the shell into the text of our program.

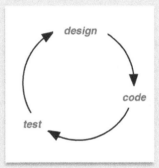

Here are some suggested milestones if you want to try this style of programming with your sieve function. Once each version is working to your satisfaction move on the the next one:

- The first version of sieve should do nothing more than create the worksheet and return the entire list.
- Add a `for` loop that prints "filter out" followed by the value of k.
- Comment out the print statement and add a call to `sift`, passing it k and the worksheet.
- Type in and test the `non_nulls` helper.
- The final version of sieve will call `non_nulls` and return the resulting list.

An important benefit of this style of programming is that we find mistakes as soon as we make them. There are dozens of things that can go wrong with even the simplest program: forget a colon at the end of the `def` statement? Leave off the closing parenthesis at the end of the list of parameters? Use the wrong number of spaces in front of statements in the body of a function? By testing a program each time we make a small change we will find these errors as soon as they are made, and when part of a program passes its tests we can move on with confidence to the next phase of the project.

Print Statements as Scaffolding

A straightforward technique for debugging a program is to add `print` statements so you can follow the flow of execution. An example is putting a `print` statement at the beginning of a `for` loop so you can see the value of variables at the start of each iteration.

```
for k in range(2, ceil(sqrt(n))):
    # print('sifting', k)
    if worksheet[k]:
        sift(k, worksheet)
```

After verifying this loop works as expected "comment out" the print statement but leave it in the file.

When the program is working, rather than delete the print statements, programmers often just insert a comment symbol at the start of the line. If it turns out the program needs further debugging it's easy to remove the comment symbol to reactivate the statement.

These sorts of statements are known as "scaffolding" — they are created and used when the program is under construction, but not intended to be active in the final version.

T69. Compute the upper limit of the range of values to sift for different values of n:
```
>>> ceil(sqrt(30))
6
>>> ceil(sqrt(100))
10
>>> ceil(sqrt(1000))
32
```

T70. Add the `import` statement to your `sieve.py` file so `ceil` and `sqrt` will be available for your `sieve` function.

T71. Add the definition of the `non_nulls` helper function to your file and load it into your shell window.

T72. Make a small list to test `non_nulls`:
```
>>> colors = ['red', None, 'green', None, 'blue']
```

T73. Test the function:
```
>>> non_nulls(colors)
['red', 'green', 'blue']
```

T74. Make up a few other lists and pass them to `non_nulls`:
```
>>> non_nulls([None,5,7,9,None])
[5, 7, 9]
```
Is it doing what you expect?

You're now ready to add the code for the top-level function. The incremental development philosophy described in the sidebar suggests you do this in two steps. First write and test the code that creates the worksheet, and when that is working, add the for loop that calls the `sift` helper.

T75. Type in a version of the `sieve` function that simply creates the worksheet and returns it as the value of the function:
```
def sieve(n):
    worksheet = [None, None] + list(range(2,n))
    return worksheet
```

T76. Save the file and load the outline into your interactive session. Test the function:
```
>>> sieve(30)
[None, None, 2, 3, 4, 5, 6, 7, ... 27, 28, 29]
```
Try calling the function with different values of n. Do you always get back a list that starts with two `None` objects followed by a list of integers from 2 to $n - 1$?

T77. Edit the `return` statement so the worksheet is passed to `non_nulls`:
```
    return non_nulls(worksheet)
```

T78. Save the file and reload the function. Repeat the tests you made earlier.
```
>>> sieve(30)
[2, 3, 4, 5, 6, 7, ... 27, 28, 29]
```
This time are you getting back lists that have the `None` objects removed?

T79. Add the `for` loop that iterates over the worksheet:
```
    for k in range(2, ceil(sqrt(n))):
        if worksheet[k] is not Null:
            sift(k, worksheet)
```

At this point your `sieve` function should look like the one in Figure 3.9. You should be able to load it into an interactive session and create a complete list of prime numbers up to any specified upper limit.

3.8 ◆ Running a Program from the Command Line

Once we have written a set of functions and tested them with the IDE we are ready to start using them to solve problems. This is where a command line interface will be useful. If we want to see a list of prime numbers, it would be convenient if we could just type a single command in a terminal emulator window instead of starting the IDE, loading the code into an interactive session, and calling the function.

The simplest way to execute a Python program from the command line is to type the program's name as part of the shell command that starts Python. As explained in the Lab Manual, if we run Python without specifying any file names the system starts an interactive session in the terminal emulator window. In this example, the % symbol is the system prompt and the three greater than symbols are Python's interactive prompt:[1]

```
% python3
>>>
```

If we also type the name of a file containing Python source code Python will load that file and execute all the commands in the file. This example shows what we would type to use our sieve function to generate a list of prime numbers up to 50 and what will be printed in the terminal window as a result:

```
% python3 sieve.py 50
[2, 3, 5, 7, 11, 13, 17, 19, 23, 29, 31, 37, 41, 43, 47]
```

However if we type the command shown above using the version of sieve.py in Figure 3.9 it will look like nothing happens. The problem is that the file will be loaded, and the function will be defined, but Python doesn't know what we want to do with it. We need to tell Python to get the upper limit from the command line and pass it in a call to sieve.

In order to introduce the technique for creating a command line application we will use the celsius function from Chapter 2. The program shown in Figure 3.10 reads an argument from the command line and passes it to the celsius function. This command shows how to run the program to convert 77°F to Celsius:

```
% python3 celsius.py 77
25.0
```

The first thing to notice is the program uses a standard library named sys. This library includes a variety of items used to connect a program to the operating system. In this case, the statement on line 1 imports argv, which is a list containing the collection of arguments the user typed on the command line. The name stands for **argument vector** ("vector" is another term for a list, and the name argv has been a standard part of Unix since the 1970s).

The if statement on line 7 is the key part of this program. Think of it as a "magic incantation" that effectively says "if this file is being run as a command line program execute the following statements." We'll dig into the parts of the Boolean expression later in the book when we look at modules and the set of variables Python automatically creates when it loads a module. For now, however, it's sufficient to know that any time you want to turn

---

[1]On Microsoft Windows the command is simply python, without the "3".

```
1    from sys import argv
2
3    def celsius(f):
4        "Convert temperature f to Celsius."
5        return (f - 32) * 5 / 9
6
7    if __name__ == "__main__" :
8        # print(argv)
9        t = int(argv[1])
10       print(celsius(t))
```

Figure 3.10: *This version of* celsius.py *will read a string from the command line, convert it to an integer, pass that number to the* celsius *function, and print the result.*

Download: CLI/celsius.py

a collection of functions into a command line application you should add an if statement that looks exactly like the one on line 7 of this file.

An important detail about argv is is that it is a list of strings. Even though we type a number on the command line, it is passed to Python as a string of digits. If we uncomment the print statement on line 8, so the first thing the program does is print the contents of argv, this is what we would see when we run the celsius program:

```
% python3 celsius.py 77
['celsius.py', '77']
25.0
```

Notice how the first item in argv is the name of the program and the temperature string is in argv[1]. The statement on line 9 converts the argument into an integer and saves it in a variable named t.

Finally, line 10 has the statement that passes the input temperature to our celsius function and prints the result returned by the function. With all these new statements added, this is what we will see when we run the program in a terminal emulator window:

```
% python3 celsius.py 80
26.666666666666668
```

The reason we went to all the trouble of adding the if statement on line 7 is that it gives us a choice of running interactively or from the command line. When want to load the file into an interactive session we will type this statement in the shell window:

```
>>> from celsius import *
```

In this case the program is not being run from the command line so the Boolean expression in line 7 is False. Python will just load the function but it won't call it yet. To convert temperatures we just call the function interactively, like we did previously:

```
>>> celsius(80)
26.666666666666668
```

♦ Tutorial Project

To work the exercises in this section you can use the editor built in to your IDE to add statements to your program but you won't be using the interactive shell to test the changes. Instead, start a separate terminal emulator window where you can type the commands that will launch your program.

♦ Add the statement on line 1 of Figure 3.10 to the front of your `celsius.py` file so it imports `argv`.

♦ Add the "magic incantation" `if` statement from line 7 of Figure 3.10 to the end of your file. But for the first version of your command line program put these two statements in the body of the `if` statement:

```
print(argv[1])
print(type(argv[1]))
```

♦ Save the file, start a terminal window, and use the `cd` command to navigate to the directory where you saved `celsius.py`.[2]

♦ Using the terminal window, run this first version of your program (the percent sign here is the system prompt, which you don't type):

```
% python3 celsius.py hello
hello
<class 'str'>
```

From this little experiment it should be clear that `argv[1]` is a string.

♦ Repeat the experiment but this time type a number for the argument:

```
% python3 celsius.py 100
100
<class 'str'>
```

Python is still interpreting the argument as a string, in this case the three-character string with the digits "1", "0", and "0".

♦ If you want to see how the `int` function converts strings to integers, start an interactive shell in your IDE and try some tests:

```
>>> s = "42"
>>> type(s)
<class 'str'>
>>> n = int(s)
>>> type(n)
<class 'int'>
>>> n / 7
6.0
```

♦ Replace the two `print` statements in your program with the statements shown on lines 9 and 10 of Figure 3.10.

♦ Your program should now convert the string in `argv[1]` to an integer, pass that number to `celsius`, and print the result. Test your program with an argument of 100:

```
% python3 celsius.py 100
37.77777777777778
```

♦ Add the `import` statement and "magic incantation" lines to your `sieve.py` file to turn it into a command line application. Can you run the program from the command line to make a list of primes up to 100? Can you still load the file into an interactive session and call `sieve` interactively?

---

[2]See your Lab Manual if you need information about how to use the `cd` command.

3.9 Summary

This chapter explored the Sieve of Eratosthenes, an algorithm that has been used for thousands of years to make lists of prime numbers. The algorithm finds all primes up to some maximum value n by making a list of all numbers from 2 to n and then systematically "sifting out" the composite numbers.

We saw how to implement the algorithm in Python in the form of a function named `sieve`. After we load the function into an interactive session, we can call it to create primes up to a specified value:

```
>>> sieve(50)
[2, 3, 5, 7, 11, 13, 17, 19, 23, 29, 31, 37, 41, 43, 47]
```

The project introduced an important type of object called a container, which is an object that holds references to other objects. One of the most widely used kinds of containers in Python is a list. We saw how to create lists of numbers and lists of strings, and how to write statements that carry out some important operations on lists:

- We can access an item in a list using an index expression
- Several built-in functions operate on lists, for example we can use `len` to count the number of items in a list
- Methods defined in the list class perform other common operations; for example, if a is a reference to a list object, a call to a.append(x) will add x to the end of the list

Another important idea in computing introduced in this chapter is iteration, which generally means "repetition." We saw how to use a "for loop" to iterate over a list:

```
for x in a:
    ...
```

This statement tells Python to set the variable x to the first element in a and then execute the statements in the body of the loop. After the last statement has been executed, Python gets the next item from the list, stores that in x, and again executes all the statements in the body. This process repeats until every item in the list has been processed.

The implementation of the `sieve` function was also our first introduction to a very important programming skill: the ability to look at a complex algorithm and break it into smaller pieces that can be worked on independently. If we can implement the smaller pieces as helper functions, we can separately test and debug the operations they perform. When all the pieces are working properly we can assemble them into the final program. This approach to designing solutions to complex problems will be used often throughout the book.

Prime numbers play an important role in modern cryptography. Algorithms that encrypt messages, for example credit card numbers you submit to a secure website when you order something online, use "encryption keys" that are created by multiplying together two large prime numbers. Encrypted messages are safe as long as an intruder cannot break the key into its prime factors. The prime numbers used to make keys are huge. To make it very difficult for an intruder, keys should be around 600 digits long.

The programs at secure websites that choose prime numbers to make keys can't simply use the Sieve of Eratosthenes to generate their primes because they would have to make an initial worksheet with every number from 1 to 10^{600} (to put this in perspective, there are

---

Concepts and Terminology Introduced in This Chapter

prime number	An integer that is not evenly divisible by any numbers except 1 and itself
composite number	An integer that can be expressed as the product to two or more other integers
list	An ordered collection of objects
index	A number that specifies a location in a list; if a list has n items the indexes range from 0 to $n - 1$
iteration	A technique for solving a problem by repeating a set of steps
for **loop**	A programming construct often used to apply an operation to each item in a collection, e.g., to iterate over a list
top-level function	A function that implements a complete solution to a problem
helper function	A function designed to solve a small part of a larger problem
range	A built-in function in Python used to generate a sequence of values in a specified range
ceil	A function from Python's math library that "rounds up" to the nearest integer
None	A special name that means "no object," used as a place-holder in lists or for variables that will be given a value later

---

only 10^{80} atoms in the entire universe). But the sieve is still used by these websites. One of the steps in the algorithm that makes keys relies on a list of small primes, less than 10,000, and Eratosthenes' sieve is still used today to make these shorter lists.

Exercises

3.1. Using a paper and pencil, follow the steps prescribed by the Sieve of Eratosthenes to make a list of prime numbers up to 30. How many rounds of sifting did you have to do?

3.2. The last number on the tape in Figure 3.2 is 21. Is the last round of sifting, which removes multiples of 5, necessary if the tape ends at 21? Explain why or why not.

3.3. In the example shown in Figure 3.2 the number 12 is marked twice, once on the first round and then again on the second round. Explain the situations where a number is crossed off in more than one round and why it doesn't affect the final result.

3.4. What is the value of k when the for statement on line 5 of Figure 3.9 terminates when $n = 500$? When $n = 5,000$?

3.5. Suppose a list is defined with this assignment statement:

```
>>> names = ['pete', 'john', 'paul', 'george']
```

Explain what Python will print if we ask it to evaluate each of these expressions:

   a) `names[0]`

   b) `names[2]`

   c) `names[4]`

   d) `names[2] == 'george'`

3.6. Using the same list of names, explain what Python would do if we ask it to evaluate each of the following expressions.

   a) `len(names)`

   b) `len(names[0])`

   c) `len(names[0] + names[1])`

   d) `len(names[0]) + len(names[1])`

3.7. Do you think Python will let you change an item in a list using an assignment statement? What do you think will happen if you type this expression (assuming `names` has been defined as above)?

```
>>> names[0] = 'ringo'
```

Use a shell session in your IDE to check your answer.

3.8. Suppose a list is defined with this statement:

```
>>> gas = ['He', 'Ne', 'Ar', 'Kr', 'Xe', 'Rn']
```

What are the values of the following Python expressions?

   a) `'Ne' in gas`

   b) `'Fe' in gas`

   c) `gas.index('Ne')`

   d) `gas.index('Xe')`

   e) `gas.index('Rb')`

3.9. Suppose we define two lists of numbers as follows:

```
>>> a = [1,1,2,3,5,8]
>>> b = [13, 21, 34]
```

Explain what Python will do if we ask it to evaluate each of these expressions:

```
a[0] + b[0]

a + b
```

3.10. Using the definitions of a and b from the previous problem, does the following statement make any sense?

```
>>> a += b
```

Explain what Python would do if a and b were both integers. Does it do something analogous if they are both lists of integers?

3.11. Briefly summarize, in one or two sentences, how Python would execute the following for loop:

```
for i in range(0,10):
    print(i, i * i)
```

3.12. What would Python print if you entered the following call to `total` (the example function shown on page 62)?

```
>>> total([4,2,9,6,5,3,6])
```

Programming Projects

3.13. Define a function named `prime_before` that returns the largest prime number less than a specific number:

```
>>> prime_before(100)
97
>>> prime_before(1000)
997
```

Hint: Use `sieve` as a helper function.

3.14. Write a function named `pi` that prints a table that shows how many primes are less than n for various values of n. The argument to your function should be a list of numbers to use for n, and the output should be a set of lines that shows the value of n and the number of primes less than n:

```
>>> pi( [10, 100, 1000, 10000, 100000] )
10 4
100 25
1000 168
10000 1229
100000 9592
```

3.15. A straightforward solution to the previous problem is to call `sieve` using each value of n and print the length of each list. Can you modify your program so it only calls `sieve` once?

3.16. A **Mersenne prime** is a prime number that is one less than a power of 2. For example, $131,071$ is Mersenne prime because $131,071 + 1 = 2^{17}$. Write a function named `mersennes` that returns a list of all Mersenne primes less than a specified value. This is how you would call the function to make a list of Mersenne primes less than 1,000:

```
>>> mersennes(1000)
[3, 7, 31, 127]
```

3.17. A **prime pair** is a set of two numbers that are both prime and that differ by 2. For example, 599 and 601 are a prime pair. Write a function named `prime_pairs` that prints all the prime pairs less than a specified value:

```
>>> prime_pairs(50)
3 5
5 7
11 13
17 19
29 31
41 43
```

3.18. ♦ The specification for the `sift` function says it should remove multiples of k from the worksheet by removing $2 \times k$, $3 \times k$, and so on. But $2 \times k$ has already been removed because it is a multiple of 2. Similarly, $3 \times k$ should have been removed by an earlier call to sift out multiples of 3. Can you think of a more effective starting point for removing multiples of k? Derive a formula for the location of the first place in the worksheet that contains a multiple of k and edit your definition of `sift` so it starts at this location.

3.19. Write a function named `mean` that will return the mean (arithmetic average) of a list of numbers:

```
>>> mean([4,6,3,8,1])
4.4
```

Hint: Copy the definition of `total`, use the same loop to compute the sum of the numbers, and return the sum divided by the length of the list.

3.20. Write a function named `palindrome` that returns True if a string reads the same backward and forward:

```
>>> palindrome('madam')
True
>>> palindrome('madman')
False
>>> palindrome('maam')
True
```

Hint: Write a for loop with an index variable i that goes only half-way through the string. In the body of the loop, compute an index j that is i characters from the end, and compare the characters and locations i and j.

3.21. Add a function named gpa to the file named `gpa.py` you created in the previous chapter (this is the file with the definition gp, a function that translates a single letter grade into a number). Your function can use a for loop to iterate over a list of grades passed as an argument and return the grade point average:

```
>>> gpa(['A', 'B', 'B'])
3.3333333333333335
```

If you implemented the checks for + and - grades:

```
>>> gpa(['A', 'A+', 'B+'])
3.866666666666667
```

3.22. ♦ What does your gpa function do if one of the strings is not a valid letter grade? Here is the result from a program where gp returns 0 if the grade is not valid, and gpa computes the average by summing points and dividing by the length of the list:

```
>>> gpa(['A', 'Q', 'B'])
2.3333333333333335
```

Note how the letter 'Q' was treated as an 'F'. A better design might be to print a message when a string is not a valid letter grade and to ignore it in the calculation:

```
>>> gpa(['A', 'Q', 'B'])
Unknown grade: Q
3.5
```

Modify your gpa function so it prints a warning message and computes the average using only valid letter grades.

3.23. A for loop can also be used to iterate over a string. For example, this loop prints every character in a string:

```
for ch in s:
    print(ch)
```

Write a function named vcount that will count the number of vowels in a string s:

```
>>> vcount("banana")
3
>>> vcount("uno, dos, tres")
4
```

Hint: Instead of writing a complicated if statement like

```
if ch == 'a' or ch == 'e' or ...
```

define a string named vowels before the loop:

```
vowels = 'aeiou'
```

Then, inside the loop, check to see if a character is in the string:

```
if ch in vowels:
```

Lists of Lists

In this chapter we saw lists of numbers and lists of strings. We can also make a list of lists:

```
>>>   a = [ [8,10], [13,12], [6,8] ]
```

If we ask Python how long this list is it should tell us how many lists are contained in a:

```
>>>   len(a)
3
```

If we ask for the first item in a we should get back the first list in a:

```
>>>   a[0]
[8, 10]
```

3.24. Each item in this list is a pair of numbers that represents the dimensions of rooms in a house:
```
>>> h = [ [18,12], [14,11], [8,10], [8,10] ]
```
Write a function named area that computes the total area of all rooms:
```
>>> area(h)
530
```

3.25. Write a function named avg_bmi that computes the average body mass index of a group of people, using your bmi function from the previous chapter (Project 2.10) as a helper function. The input to avg_bmi will be a list of pairs of numbers, where each pair represents the height (in inches) and weight (in pounds) of one person:
```
>>> pop = [ [76, 200], [70, 180], [66, 150] ]
```
If that list is passed to avg_bmi this should be the result:
```
>>> avg_bmi(pop)
24.791528013264173
```

3.26. Write a new version of the function prime_pairs (see Problem 3.17 above) that returns a list of prime pairs instead of printing the pairs in the terminal window:
```
>>> prime_pairs(50)
[[3, 5], [5, 7], [11, 13], [17, 19], [29, 31], [41, 43]]
```

3.27. ♦ Add statements to your BMI program (Problem 2.10) so it can be used as a command line application:
```
% python3 bmi.py 65 140
23.294674556213018
```

3.28. ♦ Add statements to your loan rate calculator program (Problem 2.11) so it can be used as a command line application:
```
% python3 pmt.py 150000 4.5 30
760.0279647388287
```
Note that the second argument is a floating point number, so you will have to figure out how to convert argv[2] to a float instead of an int.

Chapter 4

A Journey of a Thousand Miles

Iteration as a strategy for solving computational problems

One of the things computers do best is manage information. Computers store everything from small collections of personal data, like address books, photo catalogs, and financial records, to huge databases containing scientific data, medical records, and corporate finances. People constantly use their computers to search for information, whether it's in a music library on a personal computer, at a commercial music site, or even across the entire Internet. Computers also spend a lot of time rearranging data to make it easier for users to find what they need, for example sorting a list of contacts by name, or sorting music by album name or artist name.

Two fundamental operations performed by almost every information management application are *searching* and *sorting*. In their most basic forms, search algorithms scan through a collection to locate a particular item of interest, and sort algorithms reorganize collections so they are put in a particular order. In this chapter we will study two of the most common searching and sorting algorithms. The search algorithm, called *linear search*, scans a collection from beginning to end to see if it contains a specific item. The sort algorithm, *insertion sort*, systematically works through a collection, putting each successive item in its proper place.

After we see how the linear search and insertion sort algorithms can be implemented in Python we will do a set of experiments to watch the algorithms in action. These experiments will also introduce one of the most important concepts in computer science: the idea of **scalability**. These two algorithms are efficient and work well for small collections of data, but as a collection grows they become less and less effective. Computer scientists have developed a notation based on a precise definition of what it means for an algorithm to be scalable, and we will use this notation to analyze the scalability of linear search and insertion sort.

"A Journey of a Thousand Miles
   Begins with a Single Step"

The title of this chapter is based on a quote from
Tao Te Ching, by Lao-Tzu, a Chinese philosopher
who lived in the 6th century BC.

The quote succinctly captures the nature of
iteration as a strategy for solving computational
problems. By repeating the same basic steps
over and over, one can eventually solve some very
large problems.

In this chapter we will see how a simple iteration
strategy can be used to find an item in an array or to
rearrange the array so it is sorted.

The algorithms described in this chapter both use a straightforward strategy based on
iteration. Iteration was used in the Sieve of Eratosthenes to step through a list to remove
multiples of a number, and we saw how to implement iteration in Python with different
types of for statements. A similar approach will be taken in this chapter as we systematically
compare items in a search algorithm or move items to new locations in a sorting algorithm.

In Chapter 5 we will learn about a more sophisticated strategy known as "divide and
conquer." Through the use of a more sophisticated problem-solving strategy, these other
algorithms are much more efficient on large data sets. But simplicity is a virtue, and for
small collections the two algorithms we will study in this chapter are still widely used.

4.1 The Linear Search Algorithm

As the name implies, a linear search is a simple process that starts at the front of a list and
"walks through" one item at a time until it finds what it is looking for (Figure 4.1).

In the last chapter we saw how to use a method named index to search for an item in a
list. When we call a.index(x) to look for the object x in a list named a Python will compare
the items one by one, starting at the front of the list. To gain some experience with the
algorithm, we will pretend the index method is not already part of Python and that we
have to write our own search function.

In this section we will run some experiments using a function named isearch that has
been defined in PythonLabs. The function takes two arguments, a list to search and the
item to look for, and returns the location of the item. Here is an example, using the list in
Figure 4.1, that shows the number 7 is at location 4 in the list:

```
>>> a = [8,1,9,2,7,5,3]
>>> isearch(a, 7)
4
```

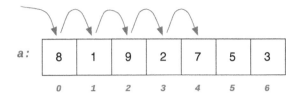

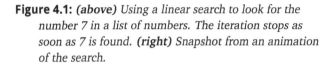

Figure 4.1: *(above) Using a linear search to look for the number 7 in a list of numbers. The iteration stops as soon as 7 is found.* **(right)** *Snapshot from an animation of the search.*

The functions in IterationLab use algorithm animation to highlight objects in a list as they are being compared. Animation is a bit of overkill for a simple algorithm like linear search, but introducing the displays here will set the stage for future animations of searching and sorting algorithms. Animations are particularly effective for showing how sort algorithms work, and later in the chapter we will use the animations to show how objects are moved around during the execution of the insertion sort algorithm.

To draw a list on the canvas call a function named view_list. The window on the right side of Figure 4.1 is a snapshot from the visualization of isearch after calling view_list to display the list shown in the figure. A list of integers is shown as a series of bars, where the height of a bar corresponds to the value of the corresponding integer.

As the search progresses the bars change colors, with the item currently being compared shown in a darker highlight color. During the animation, the highlight will move to the right as the next item in the list is compared. Beneath the row of bars is another bar, displayed horizontally. This bar represents the part of the list that has yet to be searched. At the start, the bar will be below the entire list, and as the search progresses, the region remaining to be searched will get shorter and shorter.

In our experiments on searching and sorting algorithms we are going to want to try the algorithms on some larger lists with many thousands of numbers or strings. PythonLabs has a class that will generate lists of random data to use in these experiments. The class is known as RandomList, and it is included automatically whenever you import PythonLabs modules with searching and sorting algorithms. This example shows how to make a list of five random integers:

```
>>> RandomList(5)
[12, 47, 37, 13, 50]
```

If we want a list of random strings we can pass a second argument to specify what type of string we want, for example, a list of color names:

```
>>> RandomList(5, 'colors')
['turquoise', 'spring green', 'ivory', 'yellow green', 'sienna']
```

As we run these experiments we will want to control whether or not a search will succeed. Lists made by a call to RandomList have a method named random that will give us a value to search for, and we can specify whether or not we want an item that is in the list. Passing 'success' to random tells it to return a value that will lead to a successful search, and passing 'fail' tells random to pick a value that is guaranteed *not* to be in the list:

```
>>> a = RandomList(5, 'cars')
>>> a
['mini', 'opel', 'mg', 'infiniti', 'mercury']
>>> a.random('success')
'infiniti'
>>> a.random('fail')
'isuzu'
```

It may seem strange to call a method to choose a random value in a list and then turn around and call a function that searches for that same value. Here's one way to think of the situation. When a magician does a card trick, he asks a member of the audience to pick a card at random, look at it, and insert it back in the deck. He then does his "magic" to find the chosen card. We're doing something similar here. A call to random is like picking a card from the deck. Having chosen an object, we then ask Python to do its magic by applying the linear search algorithm to find where the object is located.

Python list vs. RandomList

The RandomList objects used to test searching and sorting algorithms are implemented in PythonLabs and are not a standard part of Python. They do have a lot in common with regular lists, however. Every operation defined for lists also works for RandomLists:

```
>>> r = RandomList(10)
>>> r
[33, 45, 77, 86, 39, 84, 12, 18, 46, 92]
>>> r.append(42)
>>> r
[33, 45, 77, 86, 39, 84, 12, 18, 46, 92, 42]
>>> len(r)
11
```

The converse is not true, however. The random method is defined as part of the RandomList class, so it will not work on regular lists:

```
>>> a = list(range(0,10))
>>> a.random('fail')
AttributeError: 'list' object has no attribute 'random'
```

Tutorial Project

Each time you start a new shell session in your IDE to work on the projects in this chapter type the following statement to import IterationLab, the lab module for this chapter:

```
>>> from PythonLabs.IterationLab import *
```

T1. Creates a small list of strings:

```
>>> notes = ['do', 're', 'mi', 'fa', 'sol', 'la', 'ti']
```

T2. Python's index method will tell us where an object can be found in a list:

```
>>> notes.index('re')
1
```

Recall that list indices start with 0, so the result of this call is telling us 're' is the second item in the list.

T3. Repeat the previous experiment, this time using IterationLab's isearch function to find the location of the string:

```
>>> isearch(notes, 're')
1
```

T4. Try looking for a string that is not in the list:

```
>>> isearch(notes, 'bzzt')
```

It looks like nothing happened, but in fact the method returned None. When the value of an expression is None the Python shell doesn't print anything.

T5. To find out what sorts of strings you can request from the RandomList function call a method named sources:

```
>>> RandomList.sources()
('cars', 'colors', 'elements', 'fruits', 'fish', 'languages',
    'words', 'nouns', 'verbs')
```

T6. Make a list of five fish names (or any other type of string from the list above) and ask Python to display the list in your shell window:

```
>>> fish = RandomList(5, 'fish')
>>> fish
['flounder', 'angelfish', 'tetra', 'sturgeon', 'halibut']
```

T7. Ask Python to tell you what sort of object your new list is:

```
>>> type(fish)
<class 'PythonLabs.Tools.RandomList'>
```

This name shows that RandomList is defined in the Tools module of the PythonLabs package.

T8. Ask for a random string in the list:

```
>>> fish.random('success')
'sturgeon'
```

Repeat that expression a few times. Does the method return a string from the list each time?

T9. Ask for a string that is *not* in the list:

```
>>> fish.random('fail')
'guppy'
```

Repeat this new expression a few times. Are you getting back fish names that are not in your list?

T10. Make the list of numbers shown in Figure 4.1:

```
>>> a = [8,1,9,2,7,5,3]
```

T11. Call `view_list` to display the list on the canvas:

>>> *view_list(a)*

You should see a new window, similar to the one in Figure 4.1, with seven vertical bars, one for each item in a (if not, refer to the Trouble Shooting section of your Lab Manual).

T12. Compare the bars on the canvas with the numbers printed in your shell window. Do the smaller numbers correspond to shorter bars, and larger numbers to taller bars?

T13. Call `isearch`, and watch the canvas window to see how the animation is updated during the search:

>>> *isearch(a, 7)*
4

Did you see how the dark bar, which indicates the item currently being compared, moves steadily to the right? And how the horizontal bar at the bottom of the canvas, which shows the portion of the list that has not yet been searched, grows shorter?

T14. Make a list of random integers:

>>> *a = RandomList(50)*

T15. Ask Python to show you the list:

>>> *a*
[477, 68, 271, 321, ... 299, 82, 179]

Of course your list won't be the same as the one above, but you should see 50 random values between 0 and 500.

T16. Display the new list on the canvas:

>>> *view_list(a)*

T17. You can speed up the animation by setting a value named `delay` that is defined in the `Canvas` module. Type this assignment to set the animation delay to 0.1 seconds:

>>> *Canvas.delay = 0.1*

T18. This statement will get a random number and then tell `isearch` to find it:

>>> *isearch(a, a.random('success'))*
33

Repeat that call to `isearch` a few times. Each search should stop somewhere in the list, and the number printed as the result of the call will be the location where the search stopped, that is, the location where the random item was found.

T19. This statement will start a search that is guaranteed to fail:

>>> *isearch(a, a.random('fail'))*

Repeat that statement a few times. Notice how each search goes all the way to the end of the list before it returns. Also notice that no number is printed as the result of the search because `isearch` is returning `None` as the result of each call.

4.2 Implementing Linear Search

If we want to write our own version of `isearch` an obvious choice is a `for` loop. To search a list a we could use a `range` expression to generate all index values from 0 up to one less than the length of a. The body of the loop would test each location to see if it contains the item we're looking for.

The first function definition in Figure 4.2 has a `for` loop with a `range` expression. The key step is the statement in the body of the loop that tells Python to return the value of the index variable i as soon as a[i] matches the search target x.

The return statement on line 4 is an alternative way of terminating the loop: if Python executes that statement it will immediately return to where the function was called. Even if the iteration is only part of the way through the range of values, the return statement will force Python out of the loop. If the program makes it all the way through the list without executing the return statement inside the for loop it means the item x was not found. The for loop will terminate and the program will continue at line 5, and the special value None will be returned to indicate the search failed.

Note carefully how the statement on line 5 is indented. It is not part of the for loop because it is not indented as far as the other statements in the loop body (lines 3–4). It is, however, part of the body of the function, and it will be executed after the for loop terminates.

The second implementation of isearch shown in Figure 4.2 introduces an alternative looping construct called a **while loop.** The loop consists of a while statement followed by one or more statements that make up the body of the loop. The while statement at the head of the loop controls the iteration. The statement consists of the keyword while followed by a Boolean expression (an expression that evaluates to either True or False). Python evaluates the expression, and if it is True, the statements in the body are executed. Python then goes back to the top of the loop to evaluate the expression again, and if it is still True the statements in the body are executed again. The loop terminates when the while statement tests the Boolean expression and finds it is False.

At this point you might be wondering why Python has two different looping constructs. The answer is that for loops are very flexible and can solve a variety of problems, but there are iterative algorithms that do not examine an easily described sequence of locations. We'll see one in the next chapter: a binary search begins in the middle of a list and might go in either direction from there, and it would be very difficult to specify a sequence of items to examine with a for loop.

```python
1  def isearch(a, x):
2      "Use a for loop to search for x"
3      for i in range(0, len(a)):
4          if a[i] == x:
5              return i
6      return None
```

```python
1  def isearch(a, x):
2      "Use a while loop to search for x"
3      i = 0
4      while i < len(a):
5          if a[i] == x:
6              return i
7          i += 1
8      return None
```

Figure 4.2: *Two implementations of the linear search algorithm, one using a for loop and the other a while loop.*

The for statement with its range expression does three important jobs automatically: it determines the first value of the index variable, i; at the end of each iteration it figures out the next value of i; and it specifies the final value so the loop will terminate properly.

With a while loop it is our responsibility to write statements that do each of these jobs separately. Notice how there is a statement on line 2 of the second program that initializes i, a statement on line 6 that updates i for the next iteration, and a Boolean expression in the while statement on line 3 that checks the terminating condition.

All three pieces are necessary, but arguably the most important is the code that increments the index variable. If we forget to include this statement the program will be trapped in an infinite loop because the Boolean expression will always be true. The general rule is that somewhere in the body of a while loop there must be a statement that changes a value that is part of the Boolean expression in the while statement.

As a matter of programming style the first version of isearch, the one that uses the for loop, is probably the best choice. It is shorter, and because Python takes care of managing the index variable for us it is less error prone. The version with the while statement is simply a good way to see how this alternative looping construct works so we can put it to use when we need it later in this chapter.

Tutorial Project

Note: In the exercise for this section you will be defining your own version of the isearch function. When you load your function into an interactive session it will replace the one from IterationLab. If you want to revert to using the function from IterationLab (e.g. to do some more visualizations) you can retype the import statement to reload the module.

T20. Use your IDE to create a new file named isearch.py for your implementation of the isearch function.

T21. Type in the second definition of isearch (the one that uses a while loop in Figure 4.2). Save the file and load the function into an interactive shell session.

T22. Create a list to test isearch:
```
>>> alkali = ['Li', 'Na', 'K', 'Rb', 'Cs', 'Fr']
```

T23. Test isearch by looking for some strings in the list:
```
>>> isearch(alkali, 'K')
2
>>> isearch(alkali, 'Cs')
4
>>> isearch(alkali, 'Fr')
5
```

If you are not getting the results you expect, check each line in your definition and make sure it looks exactly like the one in Figure 4.2.

T24. Add the following print statement immediately after the while statement in your definition of isearch. It should be on line 4, and the if statement should move down to line 5:
```
print('looking in location', i)
```

T25. Save your file, reload the definition, and repeat the tests. For the first search this is what you should see:

```
>>> isearch(alkali, 'K')
looking in location 0
looking in location 1
looking in location 2
2
```

Do you see how the `return` statement caused the `while` loop to exit as soon as it found the item it was looking for?

♦ If you want to verify that a `return` statement will also cause an early exit from a `for` loop, add the version of isearch shown at the top of Figure 4.2 `isearch` to your file. Give this version of the function a new name like `isearch2`, and add a `print` statement to the loop. Load this function into your shell session and repeat the experiments, this time calling `isearch2`.

If we ask one of our versions of isearch to find an item that is not in the list it fails silently, unlike Python's index method, which prints an error message. The next few exercises should convince you that `isearch` is working correctly and in fact returning a `None` object when the item is not found.

T26. Try a search that fails:

```
>>> isearch(alkali, 'U')
```

It should look like nothing happened; Python won't print anything, and all you will see is the prompt for the next expression.

T27. Ask Python to show you the value of a variable that has not yet been defined in your current shell session:

```
>>> z
NameError: name 'z' is not defined
```

T28. Save the result of a failed search in z:

```
>>> z = isearch(alkali, 'U')
```

T29. Repeat the statement that asks for the value of z:

```
>>> z
```

There is no error message this time, so z must have been defined as a result of that assignment statement.

T30. Ask Python to print the type of the object referred to by z:

```
>>> type(z)
<class 'NoneType'>
```

It might seem at first that z has no value. Nothing was printed when you asked Python to show you its value, but in fact z has the value `None`. What is happening here is that Python doesn't print anything when an expression (including a function call) has `None` as its value.

4.3 The Insertion Sort Algorithm

Sorting is one of the most studied problems in computer science. There are hundreds of different algorithms for sorting information. Insertion sort, the algorithm we will explore in this chapter, is a simple technique that is easy to understand and easy to implement. The computation carried out by this algorithm is a simple extension of the iterative solution to the search problem, and it is another example of how a seemingly complex problem can be solved by repeatedly executing very simple steps.

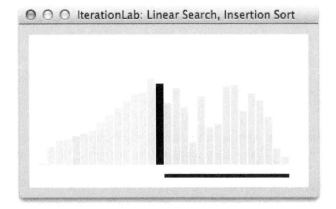

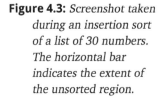

Figure 4.3: *Screenshot taken during an insertion sort of a list of 30 numbers. The horizontal bar indicates the extent of the unsorted region.*

To get an idea of how insertion sort works, imagine you are playing a card game. When a hand is dealt, you want to sort the cards in your hand, putting them in order defined by their rank (we'll ignore suits for now). One way to do this systematically is to first make sure the two cards on the left are in the right order. Then pick up the third card and place it in the proper location among the first two. Then pick up the fourth card and place it among the first three cards. Repeat this operation of picking out each successive card and finding a location for it until all cards have been placed.

If we step back from this problem and think about what is going on each time we pick up a card, an important feature of this algorithm becomes apparent. At any point during this process there are two parts to the hand. The part on the left is already sorted, and the part on the right is still unordered. A succinct statement of the main step is that we remove the first card from the unsorted part and insert it into its proper location in the sorted part (thus the name of the algorithm, "insertion sort"). Note also that after each round of insertions the sorted part grows longer, and eventually the algorithm terminates when the final card has been placed where it belongs and the entire hand is sorted.

When we implement this algorithm in Python it will be in the form of a function that will sort a list of objects. The tutorial project includes exercises that sort lists of numbers and lists of strings, but in fact any objects that can be compared with one another can be sorted.

Our first experiments will use a function from IterationLab named `isort` that rearranges items in a list into increasing order. We can observe how the algorithm works by watching an animation that uses the same techniques introduced in Section 4.1. If a list of numbers has been displayed on the canvas after a call to `view_list`, `isort` will move the bars around as the list is sorted. Figure 4.3 shows a snaphsot taken during a sort of a list of 30 numbers.

As was the case for the linear search animation, the horizontal bar at the bottom of the display is the "to-do" list. In this case it is the unsorted portion of the list. Notice how the bars to the left of the unsorted region are all in order from lowest to highest. As you watch the algorithm in progress, the first item in the unsorted region will be highlighted, and then it will move steadily left until the function finds the right place for the object.

Tutorial Project

T31. Make a small list of strings to use in the first tests with isort:

```
>>> a = RandomList(7, 'elements')
```

T32. Ask Python to show you the list:

```
>>> a
['Er', 'Li', 'Cf', 'Rb', 'B', 'Ir', 'Ag']
```

Your list will be different, but you should see a list of seven short strings.

T33. Use isort (defined in IterationLab) to sort the list:

```
>>> isort(a)
```

T34. Look at the list again. The strings should be sorted:

```
>>> a
['Ag', 'B', 'Cf', 'Er', 'Ir', 'Li', 'Rb']
```

T35. The output shows lists of strings are sorted in increasing alphabetical order. Try a few more experiments with other lists of strings or lists of numbers.

T36. Make a small list of numbers and display it on the canvas:

```
>>> a = RandomList(10)
>>> view_list(a)
```

T37. Call isort and watch how the animation is updated as the sort progresses:

```
>>> isort(a)
```

Do you see how the sorted region in the left part of the list grows steadily longer while the unsorted region shrinks? And that after the last step the unsorted region has shrunk to 0 bars and the entire list is sorted?

T38. If you want to experiment with larger lists, set the animation delay to a smaller value:

```
>>> Canvas.delay = 0.1
```

T39. The default screen size will show up to 75 bars, but you can resize the window if you want to see more.

```
>>> a = RandomList(50)
>>> view_list(a)
>>> isort(a)
```

As you watch the insertion sort algorithm working on longer lists, notice how the first few rounds go by very quickly. But as the sort progresses it takes longer and longer for the algorithm to find the final location of items from the end of the list. Some of those items have to move all the way from the right side to the left as the algorithm compares the current item to every other item in the list. Insertion sort works quite well for fewer than a dozen items, but the amount of time required to sort a list increases quite rapidly as lists get longer, a topic we will return to in Section 4.6.

4.4 A Helper Function for Insertion Sort

The key step in the insertion sort algorithm is the process that moves an item toward the front of the list, stopping when the item has reached the correct location. This operation is itself defined by an iterative algorithm. If i is the current location, we want to scan left from $i - 1$ and work back toward the beginning of the list. The scan will stop when we find an item smaller than the one we are moving or when we reach the front of the list.

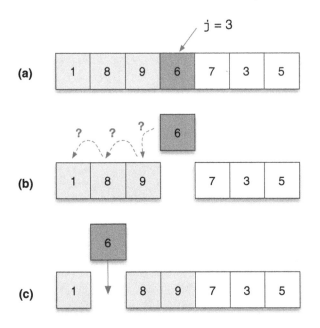

Figure 4.4: *Operations performed by the move_left helper function during one round of the insertion sort algorithm.* ***(a)*** *Light gray squares are the part of the list that is already sorted. The dark gray square is the item to be moved.* ***(b)*** *The item is compared with all the items in the sorted region until its correct location is found.* ***(c)*** *The item is inserted back into the list.*

The strategy we are going to use for our own implementation of insertion sort is similar to what we did for the Sieve of Eratosthenes in the previous chapter. We are first going to implement a helper function that will perform the key step—in this case removing a single item and finding where it belongs—and then we will write a top-level function that repeatedly calls the helper to carry out the complete sort.

An example that shows what we expect our move_left helper to do is shown in Figure 4.4. The first step is to remove the item at the beginning of the unsorted region from the list. Then we compare this item with the ones to the left, checking all the way to the front of the list if necessary. Finally, having found a spot for the current item, it is inserted back into the list.

The two arguments passed to move_left will be a, a reference to the list being sorted, and j, the location of the item we want to move. Removing the item at a specified location is straightforward: we simply have to call a method named pop defined in the list class. The method has an unusual name, but like a lot of other terminology in Python, it has a long history, and the designers of the language decided to keep a term that was familiar to most programmers. This statement removes the object at location j and saves it in x:

```
x = a.pop(j)
```

After executing this statement, the variable x refers to the object that used to be in the list at location j and the object is no longer in the list.

Here is the general outline of a while loop that implements our leftward scan:

```
while a[j-1] > x:
    j -= 1
```

Because j is initially the location where x used to be, the first time Python executes the while statement it will compare x to the item that was immediately to its left in the list. If that item is larger it means we need to check further to the left. The statement in the body

```
1    def move_left(a, j):
2        "isort helper function that moves a[j]"
3        x = a.pop(j)
4        while j > 0 and a[j-1] > x:
5            j -= 1
6        a.insert(j, x)
```

Figure 4.5: *A helper function that removes an object from a list and reinserts it in its correct location to the left.*

Download: `journey/moveleft.py`

of the loop subtracts one from j, using Python's **decrement** operator, which is similar to +=
except it subtracts instead of adding. Python then goes back to the top of the loop where it
will compare x with the object two places to the left of where it used to be. This process of
comparing and decrementing continues until x is larger than one of the items in the list, at
which point we have found the place to reinsert x.

There is a problem with the loop shown above, however. What will happen if x is smaller
than any object in the sorted region? The algorithm makes it clear what *should* happen: x
should be inserted at the front of the list, at location 0. But if x is smaller than the object in
a[0] the loop will continue, and the next iteration of the while statement is going to try to
compare x with a[-1].

In some languages, trying to make a reference to a[-1] would lead to a "list index out of
range" error. In Python, however, a negative index is legal, and it is used to index from the
end of the list instead of the beginning: a[-1] refers to the last item in the list, a[-2] is the
second to the last, and so on. As a result, we won't see an "out of range" error when j is -1,
but the item will not be placed back in the right location (see Exercise 4.19 at the end of the
chapter if you want to find out exactly how the program will fail eventually).

We can solve this problem by modifying the Boolean expression in the while statement
so it checks two conditions. The decrement statement should be executed only if the index
is still greater than 0 *and* the item to the left is greater than x:

```
while j > 0 and a[j-1] > x:
    j -= 1
```

The word and is a Boolean operator that evaluates to True only if both conditions are True.
We saw a similar operator named or back in Chapter 2 that was True if either condition was
True. The sidebar on "Boolean Operators" on page 104 explains how these operators work
in more detail.

With this new terminating condition, the loop will end either when j falls all the way to
0 or as soon as the program finds a place where x is larger than some item in the list.

When the while loop terminates, the variable j will hold the location where we want to
put the x back into the list. Here again we can use a method defined for lists:

```
a.insert(j, x)
```

This statement tells Python to insert the object x into the list before the object at location j.

The complete definition of the helper function is shown in Figure 4.5.

Tutorial Project

T40. Create a list to use in experiments with the pop and `insert` methods:
```
>>> a = ['spam', 'eggs', 'sausage']
```

T41. Remove the string at a[1] and save it in a variable named x:
```
>>> x = a.pop(1)
```

T42. Ask Python to show you the value of x and the new value of the list:
```
>>> x
'eggs'
>>> a
['spam', 'sausage']
```

Recall that `a.insert(i,x)` means "insert x before location i in a."

T43. Insert x back into the list before a[0], which places it at the front of the list:
```
>>> a.insert(0,x)
>>> a
['eggs', 'spam', 'sausage']
```

T44. We can insert an object at the end of the list by specifying the last location in the list:
```
>>> a.insert(3,'spam')
>>> a
['eggs', 'spam', 'sausage', 'spam']
```

Boolean Operators

Python allows us to combine individual Boolean operations into a single Boolean expression using operators shown in the table at right.

p **and** q	True only if p and q are both True
p **or** q	True if either p or q is True
not p	True if p is False

Complex expressions can also have parentheses to form groups, as in

```
p and not (q or r)
```

An important detail about Boolean expressions is that Python does as little work as possible to figure out if an expression is True. Suppose we ask Python to evaluate p **and** q. If it finds out p is False, Python won't even bother trying to evaluate q because p **and** q can only be True if p is True.

This type of "lazy evaluation" is what allows us to write an expression like

```
while j > 0 and a[j-1] > x:
```

If Python always evaluated both parts of this expression it would try to access a[-1] when j is 0, which is the situation we are trying to avoid. Instead, when j is 0 the left side of this expression is False and Python knows it doesn't have to evaluate the right side.

Python also does lazy evaluation with the **or** operator. When evaluating p **or** q, if Python learns p is True it won't bother evaluating q because it already knows the expression will be True.

T45. Insert a new word into the middle of the list:
```
>>> a.insert(3, 'lobster')
>>> a
['eggs', 'spam', 'sausage', 'lobster', 'spam']
```

The next part of this project is to create an initial version of a file named isort.py that will eventually contain the complete definition of isort and its helper function. This first version will contain only the definition of move_left; the remaining code will be added in the next section.

Either download the function shown in Figure 4.5 and rename the file to isort.py or use your IDE to create new file named isort.py and enter the code shown in the figure. Load the function into your interactive shell session.

T46. Type this expression to create a list of strings to test move_left:
```
>>> gas = ['H', 'O', 'He', 'Ar', 'Xe']
```

T47. Note that the first two strings are in order, but the third is not. This call to move_left should move the string at a[2] so it is between the 'H' and the 'O':
```
>>> move_left(gas, 2)
>>> gas
['H', 'He', 'O', 'Ar', 'Xe']
```

T48. Now the first three items are sorted. The next call will move the fourth string ('Ar'). It will also test the case where the item being moved is smaller than anything to the left, so this string should be at the front of the list after the call:
```
>>> move_left(gas, 3)
>>> gas
['Ar', 'H', 'He', 'O', 'Xe']
```

T49. If we call move_left to move the last string Python will put it right back where it was because it is already in the correct location:
```
>>> move_left(gas, 4)
>>> gas
['Ar', 'H', 'He', 'O', 'Xe']
```

If you're still not sure what move_left does make a few tests of your own. You can also add print statements to your code so the function prints messages as it works its way through the list. When you are confident in your understanding of move_left you are ready to move on to the next section.

♦ You may have noticed that the variable used in the decrement operation in the body of the while loop is one of the arguments passed to move_left. In Python j acts like a local variable. When the function is called, j is initialized with the integer passed in the call, but we are free to update this variable however we see fit.

4.5 Implementing Insertion Sort

Now that we have a helper function to implement the process of moving a single item to its correct location, writing a Python program to implement insertion sort is very simple. The code is shown in Figure 4.6. The idea is to use the index variable i as the boundary between the sorted and unsorted regions. At any point in the algorithm, all locations from a[0] to a[i-1] will be sorted. Because a[i] is the first item in the unsorted region all we need to do on each iteration is call move_left to move the object in a[i] to the proper location.

```
1    def isort(a):
2        "Sort list a using insertion sort"
3        for i in range(1, len(a)):
4            move_left(a,i)
5
6    def move_left(a, j):
7        "isort helper function that moves a[j]"
8        x = a.pop(j)
9        while j > 0 and a[j-1] > x:
10            j -= 1
11        a.insert(j, x)
```

Figure 4.6: *The final version of* isort, *a Python implementation of the insertion sort algorithm.*

Download: *journey/isort.py*

The for loop in the isort function is almost identical to the loop in our implementation of linear search. The only difference is the starting point. Before the first iteration we can consider the single item in a[0] to be the "sorted region" because a list of size 1 is already sorted. That means the iteration can start at a[1] instead of a[0].

As you are experimenting with your own version of this function you may want to include a print statement to show the progress of the sort. An effective technique is to print two lists, one showing the sorted region and one showing the unsorted region.

Python makes it very easy to print parts of a list. We can specify a portion of a list using a **slice** expression. If a is a list, the notation a[i:j] means "the items from a[i] through a[j-1]." Notice the similarity with the range expression: range(i,j) means "the numbers from i up to but not including j" and a[i:j] means "the list of objects from a[i] up to but not including a[j]."

When we write a slice expression we can leave out one of the indices. The notation a[:i] means "the list from a[0] to a[i-1]" and a[i:] is "the list from a[i] through the end a."

To trace the progress of your isort function add a single print statement that prints two list slices, one for the sorted region and one for the unsorted region. Because the index variable i marks the border between the two regions, and in fact is the location of the first object in the unsorted region, this statement is all we need to print the two regions:

```
print(a[:i], a[i:])
```

After adding this print statement to the for loop, just before the call to move_left, this is what the shell window will show when sorting a small list of strings:

```
>>> a = RandomList(5,'elements')
>>> a
['U', 'Cs', 'I', 'Md', 'Rn']

>>> isort(a)
['U'] ['Cs', 'I', 'Md', 'Rn']
['Cs', 'U'] ['I', 'Md', 'Rn']
['Cs', 'I', 'U'] ['Md', 'Rn']
['Cs', 'I', 'Md', 'U'] ['Rn']

>>> a
['Cs', 'I', 'Md', 'Rn', 'U']
```

Tutorial Project

T50. Add the definition of the isort function shown on lines 1 through 4 of Figure 4.6 to your isort.py file.

T51. Add this print statement to the body of the for loop, just before the call to move_left on line 4:

```
print(a[:i], a[i:])
```

Save your file and load the function definition into an interactive shell session.

T52. Create a list of strings to test your function:

```
>>> a = RandomList(7, 'elements')
```

T53. Ask Python to show you the list it just made:

```
>>> a
['Ti', 'Ir', 'Ar', 'Cu', 'I', 'Lu', 'Mn']
```

The list you see will be different, but you should see a list of seven short strings.

T54. Pass your list to isort. You should see something like this in your shell window:

```
>>> isort(a)
['Ti'] ['Ir', 'Ar', 'Cu', 'I', 'Lu', 'Mn']
['Ir', 'Ti'] ['Ar', 'Cu', 'I', 'Lu', 'Mn']
['Ar', 'Ir', 'Ti'] ['Cu', 'I', 'Lu', 'Mn']
...
```

Can you see how the sorted region starts out being one item long? And that on each round, the first string in the unsorted region is removed and placed somewhere in the sorted region on the next line? And that as the sort progresses, the sorted region grows longer by one string per iteration, so that when the algorithm terminates the entire list is sorted?

T55. Make some more test lists of varying sizes and pass them to isort.

T56. Notice how many lines are printed on your console: if you make a list with n items your function should print $n - 1$ lines as the list is sorted. Can you explain why there are always $n - 1$ lines?

4.6 Scalability

The projects in the previous sections are similar to the tests programmers use when they are developing software. By using a small set of test data (in this case a few short lists) programmers can see how the algorithms work and verify the implementations are correct. But these experiments don't give any insight into how the methods will perform when they are used on real data, where the lists might be much longer than the test data.

Unfortunately, it is often the case that software will work well for short test data, but it takes far too long when given larger inputs. It would be a disaster for a company if the software developers use a dozen songs to test an application that manages music libraries, only to find the program crashes when users try to import thousands of tunes, or that it takes the program several minutes to sort lists with more than a few hundred songs.

The ability of an algorithm to solve increasingly larger problems is an attribute known as **scalability**. The term refers to the fact that we want algorithms to continue to work well as the size of the input "scales up."

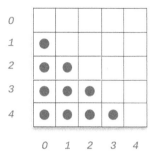

Figure 4.7: *The cells in this matrix represent potential comparisons made during a call to* isort *as it sorts a list of five items. Each row shows the comparisons that might be made by a single call to* move_left, *the helper function that does the comparisons. The label next to a row is the value of* i *passed to* move_left.

The experiments in this section will give us some insight into how well the algorithms used in the isearch and isort methods work on longer lists. Because the key step in each method is the one that compares two objects from a list we will measure performance by counting the number of comparisons made when a function is called.

Counting comparisons in a linear search is straightforward. We can use the value returned by a call to isearch to figure out the number of comparisons made. For example, if the return value is 2, we can infer the function made three comparisons: at a[0], a[1], and finally a[2], where it found what it was looking for. So in general if the function returns a number i we know it made $i + 1$ comparisons. If the search fails, we know there were n comparisons, where n is the size of the list, because the algorithm looked at all of them before returning None.

Counting the number of comparisons required to sort a list is not as simple. The output from the experiments in the previous section show one line for each iteration of the for loop, but the problem is the algorithm does not always do the same number of comparisons on each iteration. Each line printed by the function shows the result of a call to move_left, the helper function that moves each item to its correct location, but this function might make a different number of comparisons each time it is called.

Taking a closer look at what happens inside move_left, recall that each time the function is called, it will remove an item from the list and then scan to the left to find a location to reinsert it. The key observation is that a call to move_left(a,i) will make as many as i comparisons to find a home for item a[i]. For example, when i is 3, move_left might have to compare a[3] with a[2], a[1], and a[0].

The highest total number of comparisons that isort would ever have to make to sort a list with n items is roughly $n^2/2$. The picture in Figure 4.7 shows why. The drawing shows the comparisons that might need to be made when sorting a list of five items. Each row represents one iteration of the for loop; the value of i for each iteration is shown to the left of the row. The dots in a row indicate the potential comparisons made by move_left on that iteration. For example, the row labeled 3 has dots in columns 0, 1, and 2 because move_left might have to compare a[3] with a[2], a[1], and a[0].

The matrix in Figure 4.7 is for a list with five objects, but it should be apparent that for a list of any length n we could draw an $n \times n$ matrix and fill in dots using the same logic. If we were to draw a line on the diagonal of the matrix, from upper left to lower right, it would cut the matrix in half, and all the cells below the line would have dots in them. Because roughly half the n^2 cells in the $n \times n$ matrix would contain dots the algorithm would make approximately $n^2/2$ comparisons in the worst case.

We now have formulas that estimate the number of comparisons made by our two algorithms as they operate on lists of size n. Linear search will make up to n comparisons, and insertion sort up to $n^2/2$ comparisons. The equations are simply estimates, because the actual number depends on the contents of the lists. Linear search will do fewer comparisons when the item it is looking for is found near the front of the list. The helper method used in insertion sort will only have to do one comparison if an item is already in the right location.

When computer scientists write a formula to characterize the number of comparison steps made by linear search they use the notation $\mathcal{O}(n)$. The $\mathcal{O}$ comes from the phrase "on the order of." When read out loud, $\mathcal{O}$ is pronounced "oh" or sometimes "big oh," so we say the linear search algorithm is "oh of n."

When we characterize the scalability of an algorithm we usually ignore constants because we are interested in the overall growth trend. From this perspective, algorithms that grow as $2n^2$ or $3n^2$ are similar. They both grow quadratically because the dominant term in the equation is the n^2. So when we use the $\mathcal{O}$ notation to describe insertion sort we simplify $n^2/2$ to n^2 and write the equation as $\mathcal{O}(n^2)$, which is pronounced "oh of n squared."

These equations describe the *computational complexity* of an algorithm. They are intended to capture the basic trend of how the number of steps will grow as the size of the problem increases. There is a very precise and formal definition for what it means to describe an algorithm as $\mathcal{O}(n)$ or $\mathcal{O}(n^2)$, but in this book we will just use the notation informally to give a rough sense of how the number of steps executed by an algorithm depends on the size of the input.

The "big $\mathcal{O}$" notation can be used to estimate scalability since it gives us an idea of what to expect as n, the size of the input, grows larger. For example, if we double the size of the input list passed to `isort`, we can expect it to make roughly four times as many comparisons to sort the longer list. When we increase the size of the input list by a factor of 10, the execution time should increase by a factor of 100. The reason the number of steps increases so dramatically is because the equation that predicts the number of comparisons includes the term n^2.

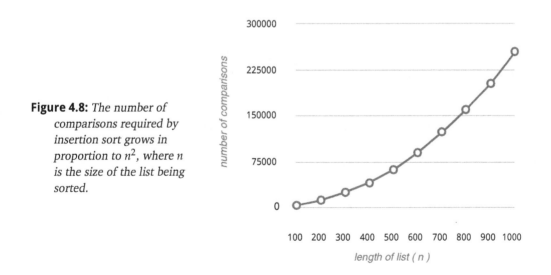

Figure 4.8: *The number of comparisons required by insertion sort grows in proportion to n^2, where n is the size of the list being sorted.*

If we actually measure the number of comparisons made by our programs we will see the number does in fact grow according to the predictions made by the scalability equations. Figure 4.8 shows the number of comparisons made by isort as it sorted lists of varying sizes. Each circle represents the average number of comparisons made when sorting a list, where the size of the list is indicated along the x-axis.

The versions of isearch and isort that are implemented in IterationLab automatically keep track of the number of comparisons they make. After calling one of these functions we can ask the system to print the number of comparisons. Here is an example of how to access the number of comparisons after calling isearch:

```
>>> a = RandomList(10, 'elements')
>>> isort(a)
>>> Counter.value('comparisons')
29
```

The first two lines above create a random list of 10 strings and then use isort to sort the list. The method named Counter.value will return the value of the specified counter. In this case we can see the sort algorithm made 29 comparisons to sort the test list.

The Counter class is defined in the Tools module. If you would like to use counters in your Python programs, the Lab Manual explains how to create your own counters and how to have your functions update them.

This project also introduces a new construct in Python that will make it easier to repeat experiments. We can combine two or more statements on a single line if we separate the statements by a semicolon. For example, this statement will make a test list and sort it:

```
>>> a = RandomList(10, 'elements'); isort(a)
```

If we want to run this small experiment with a new list we can use our IDE's command line editing feature to repeat the previous line. With many IDEs this can be done with only two keystrokes: hit the up-arrow key to display the previous command, then hit the enter key to execute the command.

Tutorial Project

Note: These experiments are based on the versions of isearch and isort from IterationLab. If you have been working with your own versions you need to import the functions again:

```
>>> from PythonLabs.IterationLab import *
```

T57. Type the statement that creates the short list of strings used in earlier experiments:
```
>>> gas = ['H', 'O', 'He', 'Ar', 'Xe']
```

T58. Use isearch to find the location of the string He:
```
>>> isearch(gas, 'He')
2
```
As expected, the function says this string is at gas[2].

T59. Find out how many comparisons were made in this search:
```
>>> Counter.value('comparisons')
3
```
Is this the right value?

T60. Do a search for a string that is not in the list and check the number of comparisons:
```
>>> isearch(gas, 'C')
>>> Counter.value('comparisons')
5
```
The output confirms that every item in the list was compared before the search failed.

T61. The following line combines two operations into a single Python statement that makes a random list and searches for a random element:
```
>>> a = RandomList(10, 'elements'); isearch(a, a.random('success'))
3
```
Your output will be different, but it should be a number between 0 and 9.

T62. Repeat the previous expression several more times. Are all your results numbers between 0 and 9?

T63. Edit the previous command so the call to isearch fails:
```
>>> a = RandomList(10, 'elements'); isearch(a, a.random('fail'))
```
As expected, nothing is printed because a failed search returns None.

T64. Ask Python to show the number of comparisons made by the failed search:
```
>>> Counter.value('comparisons')
10
```

The next set of experiments will create test lists of varying sizes and count the number of comparisons required to sort them.

T65. Type the following three Python statements all on a single line, separated by semicolons:
```
>>> a = RandomList(10); isort(a); Counter.value('comparisons')
32
```
Your output will likely be different, because the number of comparisons depends on the contents of the list. Do you see how Python executed all three commands? It made a random list, sorted it, and displayed the number of comparisons made in the call to isort.

T66. Repeat the previous command line a couple more times. Because n, the length of the list, is 10 in this experiment, the number of comparisons should always be less than $10^2/2 = 50$. Is that the case?

T67. Edit the command to change the 10 to 100 and execute the commands again:
```
>>> a = RandomList(100); isort(a); Counter.value('comparisons')
3007
```
Repeat this line a few times. Is the number of comparisons always less than $100^2/2 = 5,000$? What is the average number of comparisons?

T68. Repeat the experiment, sorting lists of 200, 400, and 1,000 items a few times each.

T69. Earlier the claim was made that doubling the size of the input list should quadruple the number of comparisons. Is the average count for $n = 200$ approximately four times the average count for $n = 100$? Is the average for $n = 800$ approximately four times the average for $n = 200$?

T70. When you increase the size of the list by 10 the number of comparisons should increase by a factor of 100. Is the average count for $n = 1,000$ around 100 times greater than the average count for $n = 100$?

♦ If you have a spreadsheet application, make a table with one column for list size and another column for the number of comparisons used to sort a list of that size. Make a graph that shows how the number of comparisons grows as a function of list size. Do you see a curve that grows roughly like n^2 (Figure 4.8)?

4.7 Summary

The main theme of this chapter was how a strategy of repeating a few very simple steps can be used to solve computational problems. For searching and sorting, the individual steps of an algorithm compare objects in a list or move objects around in a list. What we learned is that repeated execution of these simple steps will eventually find an item in a list or rearrange the order of the items.

The two algorithms we looked at were linear search and insertion sort. As its name implies, a program doing a linear search simply looks at each object in order, from front to back, in effect "walking through" the list.

The insertion sort algorithm has the same basic structure, iterating over each position in a list, but on each iteration the program removes an item and reinserts it someplace closer to the beginning of the list. The main thing to understand about insertion sort is that at the start of each iteration an index variable points to the location of the element that will be moved. The region to the left of this location is already sorted, and the algorithm just has to find the correct location for the item somewhere in this sorted region.

Our experiments with these algorithms introduced several new constructs in Python. The previous chapter introduced for statements to control iteration, and here we learned about a new statement called a while statement. Either of these statements can be used to control the repetition of a set of statements in the loop body. We also saw how to combine Boolean expressions into more complex expressions through the use of Boolean operators (and, or, and not) and how to use the special object called None as a signal that no value was found during a search.

Finally, an important idea in computing introduced in this chapter is the concept of scalability. Linear search and insertion sort are easy to understand and easy to implement, so they are widely used for small data sets. However, as the lengths of the input lists grow longer the algorithms become less and less efficient.

"Big-oh" notation, used by computer scientists to describe the computational complexity of an algorithm, is a quick way to estimate how many steps an algorithm will perform for a

---

Concepts and Terminology Introduced in This Chapter

linear search	An algorithm that searches a list one item at a time, starting at the first location and working toward the end
insertion sort	An algorithm that sorts a list by repeatedly removing items and reinserting them in a continually growing sorted region at the front of the list
while **loop**	A Python statement that repeatedly executes a set of statements as long as a certain condition is true
scalability	An attribute of an algorithm that determines how well it will perform on larger data sets
$\mathcal{O}$ **notation**	A formula that summarizes the number of steps an algorithm will execute as a function of the problem size

---

The Art of Programming (Part II)

Here is more practical advice to help you learn the craft of programming.

for Statement or while Statement?

Programmers often have a choice between using a for statement or a while statement to implement a loop. Often either one can be used, so the choice is a matter of personal preference, but here is some general advice.

```
for x in a:
    # do something with x
```

A for statement is preferred if you are going to iterate over a collection or a range of values, for example if the loop is going to do something with every object in a list.

```
i = 0
while i < len(a):
    # do something with a[i]
    i = i + 1
```

A while statement might be a better choice if there is a complex termination condition (such as the while statement in the move_left function). The binary search algorithm described later in the book uses a while loop because the loop index starts in the middle of the list and hops around instead of progressing steadily through all values in a range.

In most situations if you can use a for statement it is probably the best choice because it takes only one line to describe the loop. You don't have to worry about initializing a loop index or causing an infinite loop by forgetting to update the variables used in the Boolean expression that controls the loop.

Helper Functions *vs.* Nested Loops

A common way of writing the insertion sort function in Python is shown below. This version uses ***nested loops*** (one loop inside the body of another loop) instead of calling a helper function to move an item to the left. Deciding whether to write one large function or to split the code into smaller pieces is again often a matter of personal preference.

One very strong reason to create a helper function is size: if a function definition grows to be a dozen lines or more it is time to start considering whether part of it can be moved into a helper function. If a function is too long to be viewed all at once in the editing window of your IDE it is definitely time to divide it into smaller pieces.

Another reason to create helper functions is the ability to work on and test pieces independently. This is the approach we took with the move_left helper function. Even though the steps take only a few lines of code and fit nicely inside the for loop, it was much easier to test and debug this operation in a function of its own and then call that function from the main loop.

A version of isort *written using nested loops (a* while *loop inside a* for *loop) instead of a call to a helper function*

```
def isort(a):
    for i in range(1, len(a)):
        j = i
        x = a.pop(j)
        while j > 0 and a[j-1] > x:
            j -= 1
        a.insert(j, x)
```

given problem size. Linear search is $\mathcal{O}(n)$, and the number of comparisons is proportional to the size of the list. Insertion sort is $\mathcal{O}(n^2)$, so the number of comparisons grows quadratically with the size of the list.

Scalability is an attribute of an algorithm, not a program. We did our experiments with Python functions that are based on the algorithms, but the same general results would have been seen no matter which language was used to write the methods that do a linear search or an insertion sort. Whether they are implemented in Java, C++, or any other language, these two algorithms will not give the best performance for large inputs. If an application is going to be searching or sorting very large lists we need to consider other algorithms, like the ones presented in the next chapter.

Exercises

4.1. Use the insertion sort algorithm to sort a set of playing cards. Choose seven cards from the same suit, shuffle them, and lay them out face up in a row on a table in front of you. As you work your way through the list, slide the cards around on the table, using the process illustrated in Figure 4.3.

4.2. Suppose a list is defined with this statement:

```
>>> languages = ['perl', 'python', 'ruby', 'java', 'c++']
```

Describe what would be printed in your shell window if you were to type these expressions:

a) `isearch(languages, 'perl')`

b) `isearch(languages, 'java')`

c) `isearch(languages, 'fortran')`

4.3. Suppose a list a is defined with this statement:

```
>>> a = [11, 0, 6, 12, 7, 8, 3, 15, 4, 10]
```

How many comparisons will be made by the following searches using the linear search method?

a) `isearch(a, 0)`

b) `isearch(a, 3)`

c) `isearch(a, 9)`

d) `isearch(a, 10)`

e) `isearch(a, 12)`

4.4. What would happen if `isearch` is asked to search through a list that contains duplicate entries? Explain what the function would do if it searches for 7 in this list:

```
[6, 2, 7, 8, 2, 9, 7, 4, 5]
```

4.5. Suppose you make a list of 100 random integers and perform a set of experiments where you count the number of comparisons made during successful searches of the list. On average how many comparisons do you expect `isearch` to make? Why?

4.6. Assume a list is defined with this statement:

```
>>> halogens = ['F', 'Cl', 'Br', 'I', 'At']
```

Explain how the list would be sorted by a call to `isort`. The easiest way to do this is to show the lines that would be displayed by the `print` statement that displays the sorted and unsorted regions. Here are the first two lines to get you started:

```
>>> isort(halogens)
['F'] ['Cl', 'Br', 'I', 'At']
['Cl', 'F'] ['Br', 'I', 'At']
```

How many lines in all will be printed before the list is sorted?

4.7. Repeat the previous exercise, using this list:

>>> *heavy = ['U', 'Np', 'Pu', 'Am', 'Cm', 'Bk', 'Cf']*

4.8. We already know Python makes a distinction between upper and lowercase letters in variable names and function names. What do you think it will do with string objects? If Python doesn't care about upper and lowercase, the following comparison would result in True:

>>> *'MPEG' == 'mpeg'*

How do you think Python will evaluate this expression? Use your IDE to check your answer.

4.9. Use your IDE to ask Python to compare a string that starts with an uppercase letter with a string that starts with a lowercase letter:

>>> *'Fred' < 'fred'*

4.10. Using what you learned from the previous exercise, how do you think Python would sort this list?

>>> *names = ['Mendeleev', 'Pasteur', 'Pascal', 'da Vinci', 'Darwin',*
           'von Neumann', 'Galileo', 'Turing']

4.11. ♦ How many comparisons would isort make if you passed it a list that was already sorted? Would it always be the fewest possible number of comparisons? Explain.

4.12. ♦ Explain why isort would make the maximum number of comparisons if it is passed a list that is already sorted, but in the opposite order, from largest to smallest.

4.13. ♦ Figure 4.7 was used to derive an estimate of n^2 as the highest number of comparisons that would be needed to sort a list of n objects. Can you derive an equation that gives an exact number of comparisons? Hint: The first iteration of the for statement starts with i = 0, so the top row of the matrix in Figure 4.7 can be ignored when counting the number of dots.

4.14. ♦ Figure out the Python commands that would create a random list and arrange it in order from largest to smallest (Hint: Use the sort and reverse methods of the list class). Pass the list to isort. Does the number of comparisons match the equation you derived for Problem 4.13?

Programming Projects

4.15. Write a new version of isort called isort_less that will sort items in reverse order, from largest to smallest.

4.16. Write a function named scrabble_sort that will sort a list of strings according to the length of the string, so that shortest strings appear at the front of the list (but words that all have the same number of letters can be in any order). To test your function make a list of random words by passing 'words' to RandomList:

>>> *a = RandomList(20, 'words')*
>>> *scrabble_sort(a)*
>>> *a*
['mum', 'gawk', 'wist', 'forgo', 'caring', ... 'unquestioned']

4.17. ♦ Modify scrabble_sort so it arranges all the words of the same length in alphabetical order.

4.18. Write a function named test_search that will compute the average number of comparisons made by a successful linear search. Your function should take two arguments, size and ntests, which determine how long the test list should be and how many tests to run. For example, to search a list of 100 numbers 25 times, the call to test_search would be

    test_search(100,25)

Your function should create an empty list to hold the results of the tests. Use a loop to call isearch the specified number of times, making sure it searches for a random value known to be in the list. Figure out how many comparisons were made by each test and append the count to the list of results. Finally, use your mean function to compute the average number of comparisons.

Here is an example for 250 tests on a list of 1,000 numbers:

```
>>> test_search(1000,250)
512.384
```

The results show that on average it took 512 comparisons to search a list of 1,000 items.

4.19. ♦ Suppose the test for the beginning of the list is not part of the Boolean expression in the while statement in move_left but instead is simply

```
while a[j-1] > x:
```

Explain what would happen if you try to sort a list using this version of the while statement. Confirm your explanation by changing your version of move_left and adding print statements to show the value of j inside the loop. Call isort on a small test list. Did this version of move_left behave the way you expected it to?

4.20. The non_nulls helper function from the previous chapter (Figure 3.9, page 78) returns a copy of the list it is passed. Write a new function named remove_nulls that modifies the list instead of returning a copy. Your code should use a while loop to iterate through the list. If a is the name of the list, the while statement would be:

```
while i < len(a):
```

On each iteration, either delete the element a[i] or add 1 to i.

4.21. Define a function named gcd that implements Euclid's algorithm, another early algorithm that is over 2,000 years old. The function should compute the greatest common divisor of two integers a and b. In modern terminology, the algorithm uses a while loop. The statements in the body of the loop compare a to b and subtract the smaller value from the larger one. The loop terminates when $a = b$, and the value of a can be returned as the result of the call. Here are some examples to test your implementation:

```
>>> gcd(100,25)
25
>>> gcd(100,15)
5
>>> gcd(100,33)
1
```

4.22. Define a new function named collatz that will create a sequence of numbers that starts with a specified value. Initialize a list so it contains only the value passed as a parameter. Then use a while loop to extend the list using the following process. Let n be the value currently at the end of the list. If n is even, extend the list with $2 \div n$, but if n is odd, extend the list with $3 \times n + 1$. An unproven conjecture from number theory is that eventually the number 1 will be appended to the list, which is when the process terminates. Here are some examples:

```
>>> collatz(5)
[5, 16, 8, 4, 2, 1]
>>> collatz(6)
[6, 3, 10, 5, 16, 8, 4, 2, 1]
```

Some useful hints to consider when translating this specification into Python:

• To test if n is odd check to see if n % 2 == 1 (the expression n % 2 means "the remainder after dividing n by 2").

• Remember to use integer division when you divide by 2, that is, append n // 2 when you extend the list.

Chapter 5

Divide and Conquer

A new strategy: Breaking large problems into smaller subproblems

The previous two chapters introduced important ideas about algorithms and how they can be implemented in a programming language like Python. The algorithms presented—the Sieve of Eratosthenes, linear search, and insertion sort—all used a very straightforward approach to solving their respective problems. At the heart of those algorithms is a simple operation, such as storing a placeholder in a list or comparing two objects, and the problem was solved by repeating these operations until the the final result was produced.

The problem-solving strategy used by these algorithms is often characterized as "brute-force." There is nothing very sophisticated or clever going on. The algorithms are effective because a computer can perform list updates and comparisons very quickly, and it was a simple matter of repeating these operations over and over until the problem was solved.

Brute force works well for searching through small collections or sorting short lists, but as the problem size grows it is going to be necessary to use a better approach. The important new concept introduced in this chapter is a problem-solving strategy called **divide and conquer**, which breaks a large problem into smaller subproblems and then works on each subproblem independently.

A search algorithm known as **binary search** uses divide and conquer to look for an item in a list. The word "binary" in the name of the algorithm comes from the fact that when the list is divided it is cut into two equal pieces. The strategy is similar to what you would use to look for a word in a dictionary or for a book on a library shelf.

As an example of how this type of search works, suppose the goal is to find a recipe that starts with the word "eggplant" in a box of recipe cards sorted by title. First pick a card near the middle of the box and compare the title on that card with "eggplant." Suppose the title on the card is "minestrone." Because the cards are sorted, any recipes starting with "eggplant" must be somewhere closer to the front of the box. Pick another

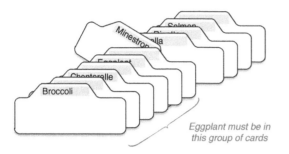

Figure 5.1: *If a set of cards is in alphabetical order we can use a binary search to find a card. Start in the middle of the collection and then search the cards before or after the middle card.*

card somewhere between the "minestrone" card and the front of the box and check again (Figure 5.1). Repeating this process of picking a card in the middle of the current group of cards, and then focusing on cards before or after the chosen card, will eventually zero in on the eggplant recipe. This is an example of divide and conquer because at each step the process splits the recipes into two groups, one before the card it just selected and one after, and in the next round the search can be limited to one group and the other group can be ignored.

Several different sorting algorithms also use a divide-and-conquer strategy. One famous algorithm, known as **quicksort**, works by breaking a list to be sorted into smaller lists and then sorting each sublist.

Suppose the recipe box is dropped on the floor, and the cards must be sorted again so they can be placed back in the box. To use the quicksort algorithm, divide the cards into two piles, the first containing all the recipes with titles that start *A* through *M*, and the second containing titles *N* through *Z* (Figure 5.2). There is no need to organize these piles when the cards are divided; the goal at this point is to just look at the title on a card and then place it on top of one of the piles. Next, push the *N* through *Z* pile to the side of the table, saving it for later, and repeat this process with the *A* to *M* pile. Divide the *A* to *M* group into smaller piles, one for *A* through *F* and the other for *G* through *M*. Then push *G* through *M* to the side, and break *A* through *F* into two more smaller groups, *A* to *C* and *D* to *F*. Keep repeating this process of making two piles, saving one, and continuing with the other. Eventually there will only be three or four cards in a pile, and it can be sorted quickly and put back in the box. At this point, go pick up the most recently saved pile and start the dividing process again with this pile.

The sorting algorithm we are going to look at in this chapter is known as **merge sort**. Quicksort can be characterized as a "top-down" approach, because it starts with a big problem and breaks it into smaller and smaller pieces. Merge sort takes the opposite approach. It works "bottom up" by starting with small groups and repeatedly merging them into bigger and bigger lists until it finally has just one complete sorted list. Both approaches have advantages, and depending on the situation, either one might be the best algorithm to solve a particular problem. We are going to look at merge sort simply because it is easier to follow the state of the sort, watching how smaller pieces are combined into bigger pieces, and easier to count the number of comparisons that are made.

The obvious question at this point is whether these algorithms are really that much more efficient than the simple iterative techniques presented in the previous chapter. Does the divide-and-conquer strategy lead to fewer steps, or are these new algorithms going to end up doing the same number of comparisons and take the same amount of time? What we will see is that for large lists, binary search is far more efficient than linear search, and merge sort is far more efficient than insertion sort. In this chapter we will do a set of experiments that will count the number of comparisons made by the algorithms, and the results of the experiments will supply empirical evidence that the divide and conquer strategy pays off. We will also develop equations that characterize the scalability of the new algorithms and show that divide-and-conquer is more effective from a theoretical point of view as well.

"Divide and conquer" is a general term that can be applied to many different situations. Binary search and quicksort are examples of a special type of divide and conquer, where each subproblem is a smaller version of the main problem. When a problem can be broken into two or more smaller problems, where each subproblem is exactly the same type of problem as the original, we say the problem is **recursive** or **self-similar**.

When a programmer uses the term "recursive" it refers to a particular way of writing a function to solve a recursive problem. If a function is recursive, it means that somewhere in

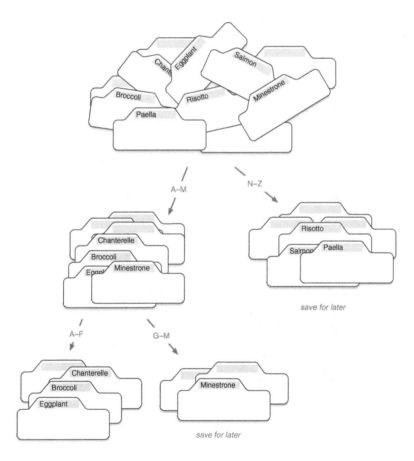

Figure 5.2: *To use the quicksort strategy to sort a set of cards, first divide them into two piles. Save the N–Z pile for later and use quicksort to sort the A–M pile. Then save G–M and use quicksort on the A–F pile. When a pile has only a few cards transfer them in order to the box, then go back and use quicksort to sort the most recently saved pile.*

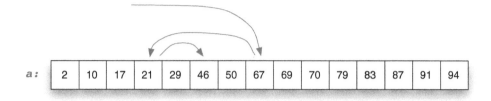

a :

| 2 | 10 | 17 | 21 | 29 | 46 | 50 | 67 | 69 | 70 | 79 | 83 | 87 | 91 | 94 |

Figure 5.3: *The arrows show how a binary search looks for the number 46 in a sorted list of numbers. The first comparison is in the middle of the list; after that the search continually moves backward or forward, eventually settling on the item it is looking for.*

the body there is a statement where the function calls itself. For example, one way to write a program to do a binary search would be to write a function named rsearch. When the problem is reduced to searching one of two sublists, the new search can be done simply by passing the sublist in a call rsearch. In other words, the function calls itself.

The use of recursion can lead to some very elegant pieces of code that are easy to analyze and easy to understand. But learning to write recursive function takes a bit of practice, so the Python code we will use for the experiments in this chapter has been developed without recursive calls. The important concept of using divide and conquer to break a problem into smaller, more manageable pieces, will still be there, and it's accurate to say that the problems themselves have a recursive nature; it's just that to keep the programs simple we will use familiar iterative statements in our Python programs. For readers who would like to see how recursion works in practice an optional section at the end of the chapter will show how binary search and quicksort can be written as a recursive function in Python.

5.1 The Binary Search Algorithm

The basic idea for a binary search is shown in Figure 5.3. The first comparison is made in the middle of the list. In this example, the number we're looking for is less than the one in the middle, so the second comparison is to the item halfway back toward the front of the list. The number there is less than the one we're looking for, so the algorithm moves forward, to a point halfway between the current location and the middle of the list.

In abstract terms, we can describe how the process works in terms of a "search region." The initial region consists of the entire list, but on each round the region shrinks so the new region is half as big as the previous region.

There is an important caveat, however: for this type of search to work, the collection must be arranged in order. If the items in the list are not sorted, the most reliable and efficient search is a linear search that checks each item, one after the other, starting at the front of the list. When we do experiments that apply binary search to lists, we have to make sure the lists are sorted or the method won't work.

The screenshot in Figure 5.4 shows how the same list is displayed by a call to view_list, the function that displays a list on the PythonLabs Canvas. The list is represented by a series

Figure 5.4: *When a list is displayed on the canvas the horizontal bar shows the region being searched. Items in the list are represented by vertical bars. The brightest bar, in the middle of the region, corresponds to the item compared on the current iteration.*

of vertical bars, where the height of a bar corresponds to the value at that location. The search region is indicated by a horizontal bar below the items in the region. When the list is first drawn, the horizontal bar is below the entire list. As a binary search is carried out the visualization will show how the region is cut in half on each round.

To help users keep track of which items have been compared, and where the boundaries of the current region are, the highlight colors in the visualization gradually fade away after an item is compared. When the algorithm checks a value at a location the corresponding bar is set to a bright color. At the same time, the bars for any items checked previously are dimmed slightly. That means on any round of the binary search visualization the brightest bar is the one in the middle of the current region. The upper and lower boundaries of the region should also be highlighted, but in a dimmer color because they were examined on previous rounds.

Tutorial Project

The experiments in this section use a function named bsearch, the PythonLabs implementation of binary search.. Make sure you import bsearch and the other functions in RecursionLab when you start an interactive session with your IDE:

```
>>> from PythonLabs.RecursionLab import *
```

T1. Make a small list of strings to make sure you know how to call bsearch:
```
>>> consonants = ['b', 'c', 'd', 'f', 'g', 'h', 'j']
```

T2. Use bsearch to find the locations of some of these strings:
```
>>> bsearch(consonants, 'd')
2
>>> bsearch(consonants, 'h')
5
```

T3. Look for a letter that is not in the list:
```
>>> bsearch(consonants, 'e')
```
It looks like nothing happened, but in fact the function returned None.

T4. Make a longer list to use in visualizations:

```
>>> a = RandomList(63, sorted=True)
>>> view_list(a)
```

Note the optional argument (sorted=True) passed to RandomList: if we're going to do a binary search the list needs to be sorted.

T5. Notice how the bars on the canvas increase in size. They should all be in order, like the ones in Figure 5.4.

T6. Start a successful search:

```
>>> bsearch(a, a.random('success'))
```

Repeat the previous statements a few more times. Do you see how the horizontal bar, which represents the region that needs to be searched, is initially below the entire list and then decreases by a factor of two on each iteration until the item is found? The result printed in the console window will always be a number between 0 and 62 (one less than the length of the list) that indicates where the search succeeded.

T7. Search for an item that is not in the list:

```
>>> bsearch(a, a.random('fail'))
```

Repeat that statement a few times. Does the active region always shrink to 0 items? Is the result always None?

5.2 Implementing Binary Search

Given the general description of the binary search algorithm from the previous section, we are now going to take a closer look at the details so we can write our own version of bsearch. We need to figure out how to represent the search region and how to cut the region in half for the next round of the search. We also need to address a detail that has been ignored in the discussion so far: how do we know to stop searching if the item is not in the list?

One way to implement regions is to use two index variables, named lower and upper, to mark the boundaries of the region. At any point in the algorithm, lower will be one less than the index of the first item in the region, and upper will be one more than the index of the last item in the region. The first row in Figure 5.5 shows the initial region for a search of the list with 15 items that was shown earlier in Figure 5.3. Because the list indexes run from 0 to 14, the initial value of lower is −1 and upper is initialized to 15.

The first step on each round of the search is to compute the location in the middle of the region. If the item we're looking for (let's call it x) is at that location the search terminates successfully. Otherwise we need to find the middle of the region. It's simply the index value that is halfway between lower and upper:

```
mid = (lower + upper) // 2
```

If the list item at this location is equal to x, the value we're searching for, we return mid as the result of the function call.

If the item is not at the midpoint we have to continue the search. To reduce the size of the region we simply have to move one of the boundaries so it is equal to the current midpoint. If x is less than the value at a[mid], we can restrict the region to the lower half of the list by

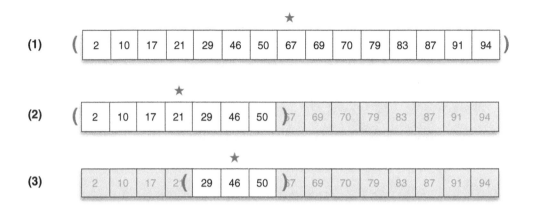

Figure 5.5: *The binary search algorithm uses three variables to represent the current state of the search: the lower boundary of the current region (shown as an open parenthesis), the upper boundary (a closing parenthesis), and the index of the item in the middle of the region (indicated by the star). The figure shows the values of the variables during the three rounds of the search for the number 46.*

setting upper to mid. On the other hand, if x is greater than a[mid] we want to set lower to mid in order to restrict the search to the upper half of the region.

In the example search shown in Figure 5.5 the item we're searching for is 46. Because that value is less than 67 the first round cuts the region in half by moving the upper boundary to the midpoint. Notice how on the second line in the figure the value of lower is the same but upper is where mid used to be.

In the second round of this search, the function again computes the midpoint. This time 46 is greater than the value in the middle of the current region. The third row of Figure 5.5 shows that the region has again been cut in half, this time by setting lower to the midpoint of the previous round but leaving upper where it was.

The important thing to keep in mind about this algorithm is that lower and upper mark locations that are *outside* the region: lower is just to the left of the first item in the current region, and upper is just to the right. Each time the region shrinks, either lower or upper will move, and its new location is the place that was the middle of the current region. Because mid is half way between lower and upper, the region shrinks to half its size on each iteration of the algorithm.

To answer the question about how to terminate the algorithm during an unsuccessful search, when the item we're looking for is not in the list we will eventually reach a point where lower will be at some location i and upper will be at the adjacent location i + 1. That means there is nothing in the region between the two boundary values. Thus the loop that controls the iteration should terminate when lower is one less than upper.

The Python definition of bsearch is shown in Figure 5.6. After initializing lower and upper so they are next to the first and last elements in the list, the function goes into a while loop. The while statement that controls the iteration checks to see whether the current range has shrunk all the way to zero items. In the body of the loop we compute the

```
1    from PythonLabs.RecursionLab import print_bsearch_brackets
2
3    def bsearch(a, x):
4        "Use a binary search to find x in list a"
5        lower = -1
6        upper = len(a)
7        while upper > lower + 1:
8            mid = (lower + upper) // 2
9    #            print_bsearch_brackets(a, lower, mid, upper)
10           if a[mid] == x:
11               return mid
12           if x < a[mid]:
13               upper = mid
14           else:
15               lower = mid
16       return None
```

Download: `divide/bsearch.py`

Figure 5.6: *Python implementation of binary search. The arguments are a list to search and the item to look for. The function called on line 9 (commented out) is a helper function implemented in RecursionLab.*

midpoint, then see if the item there is the one being searched for. If not, the range will be cut in half, either by moving the lower or upper boundary to the midpoint of the current range, and the execution continues on to the next iteration.

Here is a call to the function, using the example list (Figure 5.5):

```
>>> bsearch(a, 46)
5
```

Line 9 in Figure 5.6 has a call to a helper function that is defined in RecursionLab. If this line is uncommented the function will print a line of text at the start of each iteration. Each line will show the contents of the list with a pair of brackets around the region that needs to be searched, and an asterisk will be printed in front of the item in the middle of the region:

```
>>> bsearch(a, 46)
[2   10   17   21   29   46   50 *67   69   70   79   83   87   91   94]
[2   10   17 *21   29   46   50] 67   69   70   79   83   87   91   94
 2   10   17   21 [29 *46   50] 67   69   70   79   83   87   91   94
5
```

The three lines printed between the command on the first line and the result on the last line were printed by the helper function. Because there are three lines we can tell it took three iterations to find the number 46. On first line we can see the search region was initially the entire list, with the number 67 at the midpoint. 46 is less than 67, so the region was narrowed to the left half of the list, as shown on the second line. On the third line the region was cut in half again, and the search succeeded on this iteration.

Tutorial Project

The projects in this section work with the version of bsearch in Figure 5.6. Use your IDE to create a new file named bsearch.py and type in the program, or download the file from the book website.

T8. To see what the print_bsearch_brackets helper does, import it into your interactive session:

```
>>> from PythonLabs.RecursionLab import print_bsearch_brackets
```

T9. Define the same list of strings used in tests in the previous section:

```
>>> consonants = ['b', 'c', 'd', 'f', 'g', 'h', 'j']
```

T10. Call the helper, telling it the lower boundary is 1, the midpoint is 3, and the upper boundary is 5:

```
>>> print_bsearch_brackets(consonants, 1, 3, 5)
 b   c [d *f  g] h   j
```

Because the lower boundary is outside the region, the opening bracket is printed between locations 1 and 2. Similarly, the upper boundary is outside the region, so the closing bracket is between locations 4 and 5.

T11. Call print_bsearch_brackets again, this time passing it the actual values of the boundary points that will be used on the first iteration of the search:

```
>>> print_bsearch_brackets(consonants, -1, 3, 7)
[b   c   d *f  g   h   j]
```

As expected, the brackets enclose the entire list, and the midpoint is half-way between.

T12. Load the function definition from your version of bsearch.py into the interactive session, and then call it a few times to make sure it is working:

```
>>> bsearch(consonants, 'd')
2
>>> bsearch(consonants, 'h')
5
```

The expressions above are the ones used to test the RecursionLab version of bsearch, and the results should be the same.

T13. Edit the function definition to uncomment the call to the helper function and reload the function.

T14. Repeat the searches above and notice how the search region changes on each iteration:

```
>>> bsearch(consonants, 'd')
[b   c   d *f  g   h   j]
[b *c   d] f   g   h   j
 b   c [d] f   g   h   j
2
>>> bsearch(consonants, 'h')
[b   c   d *f  g   h   j]
 b   c   d   f [g *h   j]
5
```

T15. Try the search that fails:

```
>>> bsearch(consonants, 'e')
[b   c   d *f  g   h   j]
[b *c   d] f   g   h   j
 b   c [d] f   g   h   j
```

Do you see how the search region will be empty after that last round, so the while loop terminated and the function returned None?

♦ Create some large test lists and try some successful searches:

```
>>> a = RandomList(31, sorted=True)
>>> bsearch(a, a.random('success'))
```

♦ The number of iterations (which you can figure out by counting the number of lines printed) should vary from 1 to 5. Is that what you see?

♦ Do some unsuccessful searches:

```
>>> bsearch(a, 99)
```

♦ These searches should always make five iterations. Can you explain why?

5.3 Binary Search Experiments

It is not uncommon to see a newspaper or magazine article describe something as "growing exponentially." In everyday usage the phrase simply means something is increasing very rapidly. To a mathematician or scientist, however, the term has a very specific meaning: if something has exponential growth, the equation that explains the rate of growth has a term with one of the variables in an exponent. One of the simplest exponential equations is $y = 2^x$. The y in this equation grows very quickly indeed. Each time x increases by 1 the value of y doubles because by definition $2^{x+1} = 2 \times 2^x$.

Exponential growth is relevant to this chapter because it gives us a way to appreciate how truly efficient a binary search can be. To see why, let's first turn the problem around and ask

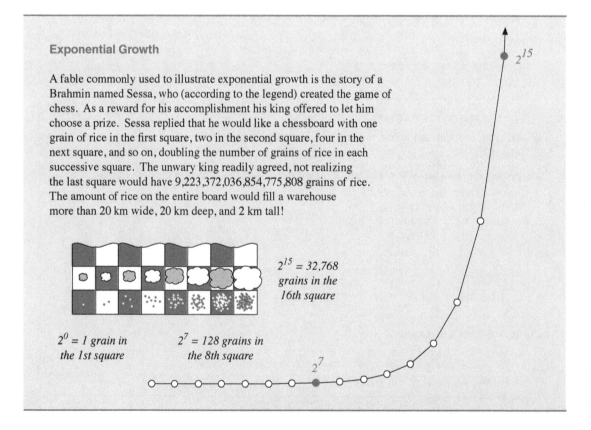

Exponential Growth

A fable commonly used to illustrate exponential growth is the story of a Brahmin named Sessa, who (according to the legend) created the game of chess. As a reward for his accomplishment his king offered to let him choose a prize. Sessa replied that he would like a chessboard with one grain of rice in the first square, two in the second square, four in the next square, and so on, doubling the number of grains of rice in each successive square. The unwary king readily agreed, not realizing the last square would have 9,223,372,036,854,775,808 grains of rice. The amount of rice on the entire board would fill a warehouse more than 20 km wide, 20 km deep, and 2 km tall!

$2^{15} = 32{,}768$
grains in the
16th square

$2^0 = 1$ *grain in*
the 1st square

$2^7 = 128$ *grains in*
the 8th square

2^{15}

2^7

Table 5.1: *To estimate the number of comparisons required to search a list with n items, calculate* $\log_2 n$ *and round up to the nearest integer. A binary search through a list of 1,000,000 items will need at most 20 comparisons.*

n	$\lceil \log_2 n \rceil$
2	1
4	2
8	3
16	4
1,000	10
2,000	11
4,000	12
1,000,000	20
1,000,000,000	30
1,000,000,000,000	40

how big a list we can search if we are given a "budget" of a certain number of comparisons. Suppose we know already, perhaps by running a binary search and counting the number of comparisons, that it takes c comparisons to search for something in a list of length n. With one more comparison, the algorithm would be able to search a list of twice the size, that is, with $c + 1$ comparisons we can search a list of $2n$ items. That's because the comparison in the first iteration would cut the list from $2n$ to n items, and we already know that finding the item in a list of size n takes only c comparisons. Because the list size n doubles each time c increases by one, the equation that defines the size of the list we can search with c comparisons is $n = 2^c$.

Now let's go back to the original question, which is how many comparisons it will take to search a list of a given size. To answer this question, simply invert the equation we just derived. If $n = 2^c$, that means $c = \log_2 n$, using the definition of a logarithm. In other words, if we have a sorted list of n items, we can determine in $\log_2 n$ steps whether or not a particular item is in the list.

Unless n is a power of 2, $\log_2 n$ will not be an integer; for example, $\log_2 100$ is 6.64. In these cases we simply "round up" to the next integer, so our estimate is that 7 comparisons will be required to search a list with 100 items. The more formal way to express the relationship between c, the number of comparisons, and n, the length of the list, is

$$c = \lceil \log_2 n \rceil$$

where the $\lceil n \rceil$ is the "ceiling" function we saw earlier in Chapter 3 (the ceiling of a number n is the smallest integer greater than n). This function grows very, very slowly. The list size n can be quite large—much larger than the number of items we can fit in a computer's memory—and a function based on the binary search algorithm will take only a few dozen comparisons to find any item in the list (Table 5.1).

In Chapter 4 the notation $\mathcal{O}(n)$ was used to describe algorithms where the number of steps grows linearly with the length of an input list. Because the binary search requires at most $\log_2 n$ comparisons to search a list with n items, this algorithm is characterized as $\mathcal{O}(\log n)$.

Binary search could, like the linear search algorithm, be lucky and find what it is looking for on the first comparison. But what is the worst case? What situations would cause an algorithm to make the most comparisons? The answer, for both algorithms, is that the highest number of comparisons will be made when an item is not in the list. In Chapter 4 we didn't bother with experiments involving unsuccessful searches: it was obvious that if a list has n items, a linear search function will make n comparisons before returning None. What is so impressive about binary search is that, even for an unsuccessful search, there will be $\log_2 n$ comparisons before the algorithm determines the item it is looking for is not in the list.

Tutorial Project

T16. Start a new shell session and import the functions in RecursionLab (the version of bsearch implemented in the lab module keeps track of the number of comparisons):

```
>>> from PythonLabs.RecursionLab import *
```

T17. Import two functions from the math library: the ceiling function (ceil) we used in Chapter 3 and a function named log2 that computes $\log_2$ of a number:

```
>>> from math import ceil, log2
```

T18. Use log2 to find $\log_2 8$:

```
>>> log2(8)
3.0
```

The result is 3 because $8 = 2^3$ and thus, by definition, $\log_2 8 = 3$.

T19. Call log2 to compute the logarithms of some numbers that are not powers of 2:

```
>>> log2(10)
3.321928094887362
>>> log2(50)
5.643856189774724
```

T20. Use ceil to find the next highest integer of the logarithms above:

```
>>> ceil(log2(10))
4
>>> ceil(log2(50))
6
```

T21. Use log2 and ceil to verify some of the values shown in Table 5.1.

T22. Make a list of 50 random integers and use bsearch to find a value known to be in the list:

```
>>> a = RandomList(50, sorted = True)
>>> bsearch(a, a.random('success'))
36
```

If you repeat the search you should always see a location between 0 and 49.

T23. To find out how many comparisons were made, ask for the value of the comparison counter:

```
>>> Counter.value('comparisons')
3
```

In this example bsearch found the item it was looking for on the third round.

T24. This statement will combine the previous two steps into a single command:

```
>>> loc = bsearch(a, a.random('success')); Counter.value('comparisons')
5
```

The number printed by Python is the value of the second expression on the line, which is asking for the counter value.

T25. Repeat the previous expression a few more times. Is the result always less than $\lceil \log_2 50 \rceil$?

T26. Change 'success' to 'fail' and repeat the expression a few times. Is the result ever greater than $\lceil \log_2 63 \rceil$?

Try some tests using larger lists of random numbers. Python will easily search lists of millions of numbers ($\lceil \log_2 1,000,000 \rceil$ is only 20) but it may take a few seconds for the RandomList class to make a list that big as it needs to make sure there are no duplicates.

5.4 The Merge Sort Algorithm

The sorting algorithm we are exploring in this chapter uses the same general divide-and-conquer strategy that makes binary search so efficient. As explained in the introduction, a merge sort works "bottom up" by first dividing the input list into several small chunks and then combining the chunks into bigger and bigger groups until the final merged group includes the full list.

The algorithm makes a series of passes through the list. On the first pass, it combines groups of size 1 into a sorted group of size 2; in other words, it simply examines each pair of values and makes sure the smallest one is on the left. On the next pass, the algorithm combines groups of size 2 into sorted groups of size 4. The algorithm keps repeating this step of combining small groups into sorted regions twice as big until the complete list is one large sorted region.

The key step in the algorithm is an operation that merges two small groups that are already in order into a single bigger group. Figure 5.7 shows how a merge is done, using stacks of cards with numbers on them to illustrate the process. In this example, there are two stacks of four cards, and each stack is in order, with the smallest card on top. The result of the merge will be a row of eight cards, in order from smallest to largest. The idea is to

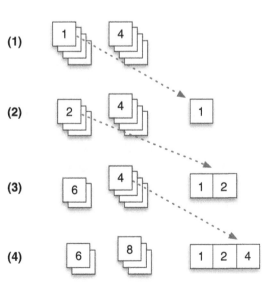

Figure 5.7: *Example of a merge operation that combines two stacks of cards into a single output list. If each stack is sorted, it is only necessary to compare the two cards on top of each stack to decide which card to move to the output list. In steps 1 and 2, the card on top of the left stack is smaller, so it is moved to the output. In step 3, the card on top of the right stack is copied.*

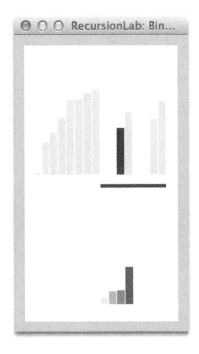

Figure 5.8: *A snapshot of the PythonLabs Canvas taken while* msort *was sorting a list of sixteen numbers. The horizontal bar representing the active region is below two groups of four numbers at the end of the list. As these groups are merged the bars are moved in order into a sorted group of eight numbers at the bottom of the screen.*

repeatedly take a card from one of the stacks and move it to the output area, stopping when all the cards have been moved. Because the two stacks are sorted, we only have to look at the top card in each stack when choosing the next item to place in the output. At each step, the merge process just moves the smaller of the two cards to the end of the output list—it doesn't have to look at any of the other cards.

The easiest way to understand how merge sort uses this list-merging process to rearrange the items in a list is to watch it in action. The PythonLabs implementation of merge sort is a function named msort. If a list has been displayed on the canvas, msort will use algorithm animation to show the progress of the sort.

Figure 5.8 shows what the canvas looks like during a merge sort. The snapshot was taken near the end of the call to msort, when the algorithm was merging two groups of size 4 into a group of size 8. The horizontal progress bar is drawn below the two groups in the main list. The four smallest items from these groups have already been moved to a temporary area at the bottom of the screen, and the next smallest item (which is highlighted) is about to join them. When the algorithm has finished combining two groups the merged list is moved from the temporary area back to the main list directly above.

Tutorial Project

If you have started a new shell session, import the functions in RecursionLab so you can call msort:

```
>>> from PythonLabs.RecursionLab import *
```

T27. Make a list of 16 numbers and draw it on the canvas:

```
>>> a = RandomList(16)
>>> view_list(a)
```

T28. Pass the list to `msort`:

```
>>> msort(a)
```

T29. Repeat the previous three commands (make a list, view it, sort it) until you understand the general concept of the merge sort algorithm: small groups are merged into larger groups until the entire list is sorted.

T30. Try some experiments with larger lists. You can speed up the animation by specifying a shorter amount of time to pause after each step (the default is 0.5 seconds):

```
>>> Canvas.delay = 0.01
>>> a = RandomList(64); view_list(a); msort(a)
```

5.5 Implementing Merge Sort

The approach we will take to writing our own version of `msort` will be similar to the one we took in the previous chapter when implementing insertion sort. First we will develop a helper function that carries out the operations in a single round of the algorithm, and then we will write the top-level function that repeatedly calls the helper until the entire list is sorted.

The helper function is named `merge_groups`. It will take two arguments: a reference to the list to work on and an integer that specifies the size of the groups to merge. Starting at the front of the list, it will merge two small groups and replace them with a larger group twice the size and then move on to the next pair of groups.

The key step is the operation that merges two small parts of the list. It turns out Python already has a function that performs this operation. It is named `merge`, and it is defined in a standard library module called `heapq`.

To experiment with `merge` in an interactive session we just need to fetch it from the library:

```
>>> from heapq import merge
```

The `merge` function is an example of a special type of function called a **generator**. Unlike other functions, generators are "lazy." They do not actually produce any results until we force them to give us the values we need. Generators will be explained in more detail in Chapter 12. For this project, it is sufficient to know that we can get a merged list by using `merge` along with the `list` constructor. This example shows how to create a new list by merging two existing lists:

```
>>> a = [1,4,7,8]
>>> b = [3,5,6,9]
>>> list(merge(a,b))
[1, 3, 4, 5, 6, 7, 8, 9]
```

Figure 5.9 shows how we can use `merge` to combine two regions in the input list. We will use three index variables to specify boundaries:

- i will refer to the first location in the first group.
- j will be the first location in the second group.
- k will mark the end of the second group.

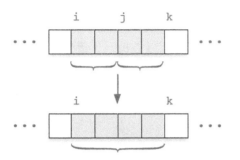

Figure 5.9: *The helper function that merges smaller groups into larger groups uses three index variables to keep track of group boundaries.*

Note that, following the Python convention, k actually refers to the location one past the end of the second list. That allows us to use the slice notation to create the arguments passed to merge. To merge all the items from location i up to j with all the items from j up to k we simply have to write

```
list(merge(a[i:j], a[j:k]))
```

The next question is, where do we put the result of the call to merge? We want the new merged list to replace the two groups. We can do this by using the slice notation on the left side of an assignment statement to put the merged items right back into the list:

```
a[i:k] = list(merge(a[i:j], a[j:k]))
```

With the merge function doing all the hard work, writing merge_groups is straightforward. We simply use a for loop that starts with i set to 0 and have it iterate through the list in steps of size 2*gs, where gs is the size of a group. Inside the loop we can compute the values of j and k as offsets from the current value of i. The final code is shown in Figure 5.10.

The top-level msort function simply has to call its helper once for each round of the algorithm. On the first iteration the helper should merge groups of size 1 into groups of size 2. In order to double the group size we just need to multiply gs by 2. The assignment statement on line 10 in Figure 5.10 uses the *= operator to implement the doubling operation. This is similar to the increment operator += we've used in previous projects, except in this case putting *= 2 on the right side means the variable on the left side is multiplied by 2. After merging groups of size 1, the while loop keeps iterating, merging groups of size 2, 4, 8, etc. until there is one group that encompasses the entire list.

If the call to the helper function on line 8 in Figure 5.10 is uncommented merge will print the state of the list at the start of each iteration. The helper prints brackets around each group in the list. Here is an example that shows how a list of eight items is sorted:

```
>>> a = [8, 5, 1, 7, 9, 2, 6, 3]
>>> msort(a)
[8] [5] [1] [7] [9] [2] [6] [3]
[5 8] [1 7] [2 9] [3 6]
[1 5 7 8] [2 3 6 9]
```

```
1    from PythonLabs.RecursionLab import print_msort_brackets
2    from heapq import merge
3
4    def msort(a):
5        "Sort list a using the merge sort algorithm"
6        groupsize = 1
7        while groupsize < len(a):
8    #        print_msort_brackets(a, groupsize)
9            merge_groups(a, groupsize)
10           groupsize *= 2
11
12   def merge_groups(a, gs):
13       for i in range(0, len(a), 2*gs):
14           j = i + gs
15           k = j + gs
16           a[i:k] = list(merge(a[i:j], a[j:k]))
```

Download: divide/msort.py

Figure 5.10: *The Python implementation of merge sort consists of a top-level function called* msort *and a helper function named* merge_groups. *Line 8 has a call to a second helper, defined in RecursionLab, that shows the progress of the sort after each round of merges.*

The first line shows how there is initially one item in each group. The next line shows what happened when these groups were merged to form groups of size two. Notice how [8] and [5] were merged to form [5 8], and how [1] and [7] were merged to make [1 7]. The following line shows how [5 8] and [1 7] were merged into a group of four items, namely [1 5 7 8]. To verify that the last step merged two four-item groups into the final sorted list with all eight numbers in order we can ask Python to print the list:

```
>>> a
[1, 2, 3, 5, 6, 7, 8, 9]
```

Tutorial Project

T31. Import the merge function so you can test it in an interactive session:
```
>>> from heapq import merge
```

T32. Make two small lists to test merge (note the lists must be sorted):
```
>>> a = RandomList(4, 'elements', sorted=True)
>>> b = RandomList(4, 'elements', sorted=True)
```

T33. Ask Python to show you the two lists. Your lists will be different, but you should see two lists with four short strings in each list:
```
>>> a
['Cr', 'Hf', 'Rb', 'Te']
>>> b
['Bi', 'I', 'S', 'Ti']
```

T34. Call merge to combine the lists and verify it works as expected:

```
>>> list(merge(a,b))
['Bi', 'Cr', 'Hf', 'I', 'Rb', 'S', 'Te', 'Ti']
```

T35. Enter a small list of eight numbers, typing the numbers in the order shown:

```
>>> a = [5, 8, 1, 7, 2, 9, 3, 6]
```

T36. Notice that each pair of numbers is in order. Define the group boundaries for a call to merge to make a group of size 4 from the first two groups:

```
>>> i = 0
>>> j = 2
>>> k = 4
```

T37. Ask Python to print the first two groups:

```
>>> a[i:j]
[5, 8]
>>> a[j:k]
[1, 7]
```

T38. Call merge to create a group of size 4 by combining these two groups:

```
>>> list( merge(a[i:j], a[j:k]) )
[1, 5, 7, 8]
```

T39. The call to merge just created the combined group but it didn't save it anywhere. Type this statement to tell Python to replace the first 4 items in the list with the output of merge:

```
>>> a[i:k] = list( merge(a[i:j], a[j:k]) )
```

T40. Now look at a again:

```
>>> a
[1, 5, 7, 8, 2, 9, 3, 6]
```

Do you see how the two sorted groups of size 2 have been replaced by a single sorted group of size 4?

T41. Repeat the previous exercises for the second half of the test list. Define values for i, j, and k so they mark the boundaries of the groups of 2 at the end of a, then see how the groups are merged and how the list is updated.

The rest of projects in this section work with the version of msort and its helper shown in Figure 5.10. Use your IDE to create a new file named msort.py and enter the two functions, exactly as shown, or you can download the file from the book website.

T42. Uncomment the call to print_msort_brackets on line 8 of the file, then load the file into an interactive session.

T43. Make a list of strings and ask Python to show you the list:

```
>>> a = RandomList(8, 'elements')
>>> a
['Rf', 'Sn', 'Au', 'Ge', 'Bh', 'Sr', 'Cn', 'Y']
```

Your list will be different, of course, but you should see a list of eight short strings.

T44. Sort the list:

```
>>> msort(a)
[Rf] [Sn] [Au] [Ge] [Bh] [Sr] [Cn] [Y]
[Rf Sn] [Au Ge] [Bh Sr] [Cn Y]
[Au Ge Rf Sn] [Bh Cn Sr Y]
```

Do you see how items within a group are always in order, and that groups in each line are formed by merging two adjacent groups from the line above it?

Notice how the helper function printed three lines at the start of each round in the example above. Because the group size doubles on each iteration, it only took three rounds to have the group size go from 1, to 2, and then to 4. After merging the two groups of size 4 on the last round the entire list of eight items was sorted.

T45. Repeat the experiment, using a list of 16 numbers. We can combine the call that creates the list with the call to msort so it all fits on one line:

```
>>> msort(RandomList(16))
```

T46. Run the experiment again, using lists of 32 and 64 numbers. How many iterations are required for these sorts?

T47. It's easy to follow the "divide-and-conquer" strategy if the size of the list is a power of 2, but msort will work for any size list. Watch what happens when we ask it to sort a list of size 10:

```
>>> msort(RandomList(10))
[43] [66] [22] [4] [36] [48] [90] [24] [21] [16]
[43 66] [4 22] [36 48] [24 90] [16 21]
[4 22 43 66] [24 36 48 90] [16 21]
[4 22 24 36 43 48 66 90] [16 21]
```

The "extra" items form a small group of 2 that is put in order on the first round. The small group hangs around until all the items to the left are merged, and finally on the last step this small group is merged with the main group to make the final result.

The examples with $n = 10$ bring up a detail in the code for merge_groups that we didn't cover before. In the call to merge(a[i:j],a[j:k]) what happens if k is past the end of the list? Or if j and k are both greater than len(a)?

♦ Suppose we are sorting a list of 10 items, as shown above. On the third round we will be merging groups of size 4. For the first call to merge we will have i = 0, j = 4, and k = 8. What are the values for the second call?

♦ Make a list to see what happens if we try to make a slice using indexes past the end of the list:

```
>>> a = RandomList(10, 'elements')
```

♦ The last item is at a[9]. Notice what happens if we ask for a slice of four items beginning with a[8]:

```
>>> a[8:12]
['Nd', 'Th']
```

Python just gave us all the remaining items by making a list using a[8] and a[9].

♦ What will happen if the start and end of the slice are both past the end of the list?

```
>>> a[12:16]
[]
```

Python is basically telling us "nothing is there."

♦ Is there a time during the call to msort with a list of 10 items where j or k or both are greater than len(a)?

So it turns out Python handles slices where index values are out of range exactly the way we needed for this project.

5.6 Merge Sort Experiments

The lines printed by calls to msort in the previous section followed a pattern that should be familiar by now: when the list size is some value n, the number of iterations of the outer loop is $\log_2 n$. This should not be surprising. Each time through the loop the group size doubles, a hallmark of exponential growth. Because the while loop is controlled by the group size, the loop makes only $\log_2 n$ iterations before the group size reaches n.

As was the case for the other searching and sorting algorithms, we would like to develop an equation that predicts the amount of work required to sort a list of n items. The most straightforward estimate is to simply assume the inner loop does n steps, because each item in the current list is part of a group, and every group in the list is merged with one other group. If there are $\log_2 n$ iterations, and each iteration does n steps, the equation for the number of operations is $n \times \log_2 n$.

Using the $\mathcal{O}$ notation introduced in Chapter 4, we can characterize the scalability of merge sort as $\mathcal{O}(n \times \log n)$. This is a significant improvement over the $\mathcal{O}(n^2)$ for insertion sort. Table 5.2 shows the maximum number of comparisons that will be made by both algorithms for some different list sizes. In each case, merge sort should make fewer comparisons. For lists of 1,000 or more items the difference is dramatic.

In the tutorial for this section we are going to do a series of experiments that count the number of comparisons made duing a call to msort. The results of these experiments will be similar to what we saw in the experiments with insertion sort in the previous chapter: the equation $n \times \log_2 n$ is an upper bound on the total number of steps, but for random lists of data, the actual number of comparisons is lower than this upper limit.

The version of msort included in RecursionLab keeps track of the number of times the function compares two items. As was the case before, to find out how many comparisons were made, we just have to ask for the value of a counter. This expression combines three operations into a single Python statement: a call to RandomList to create a list of numbers, passing the list to msort, and asking Python to display the value of the counter:

```
>>> msort(RandomList(100)); Counter.value('comparisons')
560
```

Because $100 \times \log_2 100 \approx 664$ this number is in the right ballpark.

n	$n^2/2$	$n \times \log_2 n$
8	32	24
16	128	64
32	512	160
1,000	500,000	10,000
5,000	12,500,000	65,000
10,000	50,000,000	140,000

Table 5.2: *The maximum number of comparisons required to sort a list of n items using insertion sort (roughly $n^2/2$ comparisons) and merge sort (roughly $n \times \log_2 n$ comparisons).*

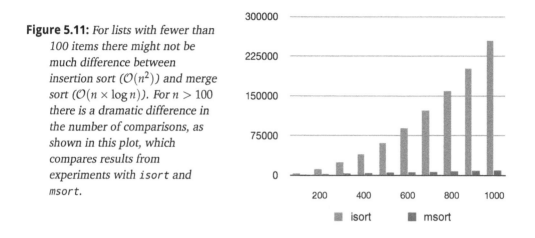

Figure 5.11: *For lists with fewer than 100 items there might not be much difference between insertion sort ($\mathcal{O}(n^2)$) and merge sort ($\mathcal{O}(n \times \log n)$). For $n > 100$ there is a dramatic difference in the number of comparisons, as shown in this plot, which compares results from experiments with* isort *and* msort.

Tutorial Project

T48. Start a new shell session and import the functions in RecursionLab so you will be able to use the version of msort defined there:

```
>>> from PythonLabs.RecursionLab import *
```

T49. Import the log2 function from the math module:

```
>>> from math import log2
```

T50. Enter the command that creates a list, sorts it, and prints the number of comparisons:

```
>>> msort(RandomList(100)); Counter.value('comparisons')
557
```

The number you see will probably be different because the actual number of comparisons depends on the values in the list.

T51. Use the log2 function to compute the upper bound for $n = 100$:

```
>>> 100 * log2(100)
664.3856189774725
```

The worst case for the number of comparisons to sort a list of 100 items is approximately 664.

T52. Compute the upper bound for sorting the same list using insertion sort:

```
>>> 100 * 100 / 2
5000.0
```

We can expect merge sort to do roughly ⅟₁₀ as many comparisons as insertion sort for a list of 100 items.

T53. Repeat the commands that create and sort a list and print the number of comparisons, using several different size lists. Is the result always just below $n \times log_2 n$?

If you did the optional exercises at the end of Section 4.6 and recorded the number of comparisons made by the insertion sort function (isort) in a spreadsheet, you can repeat the experiments here, except call msort instead of isort. Add a new column to the spreadsheet, and make a chart that displays the number of comparisons made by the two functions.

♦ Record the number of comparisons for $n = 100, 200, \ldots, 1000$.

♦ Plot your results. Do you get a general trend like the one shown in Figure 5.11?

5.7 ◆ Recursive Functions

In programming, a recursive function is a function that calls itself. Not only is it hard to imagine what such a thing might mean, it also seems like a recipe for disaster, as it would appear to lead to an infinite loop. For example, here is the general outline of a recursive function named f:

```
def f(x):
    ...
      f( ... )
    ...
```

When f is first called, it executes statements in the body of the function. Eventually it will reach the line that contains the recursive call. When the function calls itself, Python starts over again at the beginning of the function, where it again executes the first statements in the body, followed by another recursive call. Each time Python reaches the recursive call, it goes back to the beginning, and it appears the computer has been caught in an endless cycle.

Infinite recursion is something to guard against, but there are two simple rules to make sure it does not happen. First, the parameter value passed in the recursive call must be different from the one passed in the original call. In cases where recursion is used to implement a divide-and-conquer strategy, the recursive calls are made with a smaller piece of the original problem. Somewhere in the body of the function there are statements that divide the problem into smaller pieces that are passed as arguments in the recursive call.

Second, the recursive call needs to be inside an if statement that tests whether or not to make the call. Typically the if statement checks to see if the problem has shrunk all the way to the smallest possible problem. If so, the problem is solved without a recursive call, otherwise it is further divided and the pieces handed off to the recursive calls.

```
1   def bsearch(a, x):
2       "Recursive binary search to find x in list a"
3       return rsearch(a, x, -1, len(a))
4
5   def rsearch(a, x, lower, upper):
6       "Search a between lower and upper for item x"
7       if upper == lower + 1:
8           return None
9       mid = (lower + upper) // 2
10      if a[mid] == x:
11          return mid
12      if x < a[mid]:
13          return rsearch(a, x, lower, mid)
14      else:
15          return rsearch(a, x, mid, upper)
```

Figure 5.12: *A recursive implementation of binary search. The top-level function named bsearch sets the initial boundaries of the search region and passes them to a helper function named rsearch. Notice how boundaries passed in the two recursive calls in rsearch differ from those specified in the function header.*

Download: divide/rsearch.py

```
1   from PythonLabs.RecursionLab import partition
2
3   def qsort(a):
4       "Sort list a using the Quicksort algorithm"
5       qs(a, 0, len(a)-1)
6
7   def qs(a, p, r):
8       if p < r:
9           q = partition(a, p, r)
10          qs(a, p, q-1)
11          qs(a, q+1, r)
```

Figure 5.13: *This implementation of quicksort divides a list into two smaller regions by calling a helper function named partition and then making recursive calls to sort the two regions.*

Download: `divide/qsort.py`

Binary search obeys the two rules because (a) the recursive call is alway told to search a region that is one half as big, and (b) the `if` statement checks to see if the region has shrunk all the way down to the empty list, in which case the recursion ends.

The Python code for a recursive version of `bsearch` is shown in Figure 5.12. This implementation has two pieces: a top-level function, which has the same name and same set of parameters as the iterative version we saw earlier, plus a helper function named `rsearch` that uses recursion to do the search.

Notice that the helper function has two additional arguments. These are used to specify the part of the list that remains to be searched. Instead of using local variables named `lower` and `upper` to hold the region boundaries, the recursive version expects the boundaries to be passed as arguments.

When we make the top-level call to `bsearch` it figures out what the initial boundary values should be and starts the search by calling the helper function with these boundaries. As we saw earlier in the iterative version, the initial search has −1 for the lower boundary and the length of the list as the upper boundary.

The key steps in this implementation are the recursive calls on lines 13 and 15. Notice how each call specifies the values to use for the new region boundaries in such a way that the recursive call works on only the upper or lower half of the current region.

If you would like to experiment with the quicksort algorithm you can find a function named qsort in RecursionLab. One way to implement qsort is shown in Figure 5.13. This implementation also features a top-level function that sets the boundaries of the region to work on and a helper function that uses recursion to solve the problem.

Each time the helper function qs is called it will be passed two locations, which are given the names p and r, and the goal is to sort the portion of the list between these two boundaries. If the two variables have the same value it means the region to sort has shrunk all the way to 0 items, so there is nothing left to sort. Otherwise the function needs to divide the region into two parts and make recursive calls to sort each part.

A helper function named `partition`, imported from RecursionLab, does all the hard work. It uses the first item in the region being sorted as a special value called the "pivot," and then it rearranges all the items in the region so values less than the pivot are in the first part of the region and values greater than the pivot are in the second half.

Here is an example, using the following list of eight strings:

`['I', 'Cu', 'B', 'U', 'Si', 'P', 'Au', 'Na']`

If the region being sorted includes the entire list, the lower boundary is 0 and the upper boundary is 7. The pivot is the string `'I'`, so we expect `partition` to move items around so every string that comes before `'I'` alphabetically is at the front of the list and every string that comes after `'I'` is at the end:

`['Au', 'Cu', 'B', 'I', 'Si', 'P', 'U', 'Na']`

Notice how the `'I'` is now in the middle, and everything to the left is less than `'I'` and everything to its right is greater than `'I'`.

The important detail about `partition` (besides the fact that it moves items around) is that it returns the location where it placed the pivot. This location is now the dividing line between the two parts of the list that remain to be sorted. In the example above, because `'I'` was placed at location 3, the two regions to sort are between 0 and 2 (strings less than `'I'`) and between 4 and 7 (strings greater than `'I'`). The two regions are sorted by the recursive calls on lines 10 and 11. The value returned by the call to `partition` is the place where it put the pivot value, and thus it is the dividing line between the two subregions.

Tutorial Project

Either create a file named `rsearch.py` and enter the function definition in Figure 5.12 or download the file from the book website.

♦ The initial tests for the recursive version of `bsearch` will be the same as before. Make a sorted list of random numbers and try searches that succeed or fail:
```
>>> a = RandomList(15, sorted=True)
>>> bsearch(a, a.random('success'))
```

♦ Repeat the command above, except search for an item not in the list.

♦ If you want to watch the progress of the search as it runs, add a line to the front of the file that imports `print_bsearch_brackets` from PythonLabs.RecursionLab and call it right after `rsearch` has computed the value of `mid`:
```
print_bsearch_brackets(a,lower,mid,upper)
```

The following experiments use the PythonLabs implementation of `qsort` and its helper functions because they take advantage of the PythonLabs canvas. You can also download the code from Figure 5.13 or create your own version if you want to experiment with your own code.

♦ Import the functions from PythonLabs:
```
>>> from PythonLabs.RecursionLab import *
```

♦ Create a list a with the following numbers:
```
>>> a = [30, 49, 65, 20, 11, 6, 4, 19]
```

♦ Since 30 is the first item in the list it will be used as the pivot. A call to `partition` will move the 30 somewhere to the middle, put numbers less than 30 at the front of the list, and put numbers greater than 30 at the end:
```
>>> partition(a, 0, 7)
5
```

The value returned by `partition` is the location where it placed the pivot.

♦ Print the updated list to verify that the pivot value 30 is now in a[5] and that everything in a[0] to a[4] is less than 30 and everything in a[6] to a[7] is greater than 30:

```
>>> a
[19, 20, 11, 6, 4, 30, 65, 49]
```

♦ The first recursive call will sort the sublist from 0 to 4. Call partition again, but this time pass it the boundaries for the sublist:

```
>>> partition(a, 0, 4)
3
```

♦ Print the list again:

```
>>> a
[4, 11, 6, 19, 20, 30, 65, 49]
```

Notice how the number 19 was moved to location 3, that a[0] through a[2] are less than 19, and a[4] is greater than 19.

We'll skip the remaining recursive calls and move on to watching a visualization. The animation for partition is harder to follow than the other sorting algorithms, but it should give you a sense of how the partitioning process works.

♦ Retype the command that creates the test list:

```
>>> a = [30, 49, 65, 20, 11, 6, 4, 19]
```

♦ Display the list on the canvas:

```
>>> view_list(a)
```

♦ Slow down the animation by setting a 1-second delay between steps:

```
>>> Canvas.delay = 1
```

♦ Call partition, using the entire list as the region to sort:

```
>>> partition(a, 0, 7)
5
```

As you watch the animation, notice how the first item in the list is given a new highlight color to identify it as the pivot. At each step, values larger than the pivot are being pushed to the right side of the list, and smaller values are gathered on the left side. As the final step, the pivot is moved to the dividing line between the smaller and larger values.

5.8 Summary

This chapter introduced two new algorithms, binary search and merge sort. These algorithms solve the same searching and sorting problems that were introduced in the previous chapter. The difference between the new algorithms and the algorithms from the previous chapter is the problem-solving strategy: linear search and insertion sort use "brute force" iteration to repeatedly compare items and move them around in a list, but binary search and merge sort are based on a more sophisticated divide-and-conquer strategy.

Binary search is very similar to what people do when looking for an item in a sorted list, such as a dictionary or phone book. The search starts in the middle of the list. If the item being searched for happens to be at that location, the search is done. Otherwise, because the collection is sorted, the problem has now been divided in two. The search will continue in one half of the list, and the other half can be ignored. By repeatedly checking the middle of a region and continuing with a new region that is half the size the problem will be solved

---

Concepts and Terminology Introduced in This Chapter

divide and conquer	A problem-solving strategy that breaks a problem into smaller pieces and addresses each subproblem separately
binary search	A divide-and-conquer search algorithm; it divides a list into smaller regions so it can search one region and ignore the other
merge sort	An algorithm that sorts a list by combining small groups into larger groups, using a bottom-up application of the divide-and-conquer strategy
quicksort	An algorithm that sorts a list through a top-down application of the divide-and-conquer strategy
recursive problem	A problem that can be broken into one or more subproblems that are each smaller instances of the main problem
recursive function	A function, written in a programming language like Python, where a statement in the body of the function is a call to the same function

---

with far fewer comparisons. In the worst case, only $\lceil \log_2 n \rceil$ comparisons are needed to search a sorted list of n items.

Merge sort starts by repeatedly merging small groups, each of which is already sorted, into larger groups. The fact that the small groups are sorted makes the merge an efficient operation, as the comparisons only need to be made to the items at the start of each group. By the time merge sort has finished, it will have made at most $n \times \log_2 n$ comparisons to sort a list with n items.

Scalability, a concept introduced in the previous chapter, is the motivation for studying binary search and merge sort. For small lists containing only a few dozen items, linear search and insertion sort are perfectly adequate. In fact, because they are simpler to implement, they may even run faster. But as lists become longer, the more sophisticated algorithms will be much more efficient and do far fewer comparisons.

Binary search is a natural way to solve the problem of finding an item in an ordered collection, and it is something people have been doing intuitively for many years, probably since the first dictionaries were published. But merge sort and quicksort were invented by computer scientists who analyzed the problem of sorting lists of numbers and were motivated to find more efficient algorithms that would work on large lists. Merge sort was first described in 1945, by John von Neumann, and quicksort in 1960, by C.A.R. Hoare.

This raises an interesting question: if merge sort is such an effective algorithm for computers, wouldn't it also be an efficient way for people to sort real objects? If you drop your box of recipes, and you need to sort several hundred cards, would a merge sort be more effective than a more intuitive insertion sort as a way to reorganize the cards so you can put them back in the box? If you would like to pursue this question further, Exercises 5.13 and 5.14 below have some suggestions for how to organize a merge sort using real cards.

Exercises

5.1. Do a web search for newspaper or magazine articles with the phrase "exponential growth." Is the term being used in a mathematical sense, that is, can the situation be described by an exponential equation, or is it being used in the colloquial sense, of "growing very rapidly"?

5.2. Here is a list with fifteen numbers:

1 3 9 25 26 27 29 32 48 53 64 82 88 94 95

What sequence of values is compared by a binary search algorithm when it searches for the following numbers? You can either print the list with brackets and a star, the way Python would if bsearch calls print_bsearch_brackets, or draw a picture with arrows, as in Figure 5.3, or just list the sequence of numbers used in each comparison.

a) 53

b) 25

c) 26

d) 95

e) 42

5.3. Show how a binary search would look for values in the following list of element names:

Ce Dy Er Eu Gd Ho La Lu Nd Pm Pr Sm Tb Tm Yb

a) Pm

b) Nd

c) Dy

d) Rb

e) Ce

5.4. Here is an unordered list of numbers:

23 53 39 71 11 92 88 65 16 56 79 95 18 68 86

The following numbers are in the list, but binary search won't find them. In each case, show which items are compared to explain why bsearch gets lost:

a) 88

b) 18

c) 39

5.5. Are there any items in the list for the previous problem that would be found, by luck, even though the list is not sorted?

5.6. Does bsearch always make fewer comparisons than isearch (the linear search function)? Given the list shown in Problem 5.2, is there any value that isearch would find with fewer comparisons than bsearch?

5.7. Below are several test lists with sixteen numbers. Show how the lists would be sorted by a call to msort. The easiest way to do this is to show the groups before each round of merges, the way Python prints them when it calls print_msort_brackets.

1 99 3 47 50 37 79 71 15 51 87 28 19 93 91 70

25 69 64 92 10 7 27 51 54 12 71 65 59 74 79 46

66 38 79 70 45 20 16 69 52 67 72 13 5 28 39 82

37 61 40 53 89 10 72 68 99 17 67 74 47 36 3 23

14 27 32 57 34 9 56 79 44 89 35 90 84 43 59 41

5.8. About how many comparison or copy operations do you think msort will make when it sorts a list of 2,000 numbers? Check your answer by making a few test lists, sorting them, and checking the value of the comparison counter after each search.

5.9. Repeat the previous problem, but with a list of 5,000 numbers, then with a list of 10,000 numbers.

5.10. ◆ What will happen if merge sort is asked to sort a list that is already sorted? Will it make fewer comparisons than for a random test list?

5.11. ◆ Which function will make fewer comparisons when it is passed a list that is already sorted, isort or msort? Explain why.

5.12. ◆ Can you develop a formula for the number of comparisons msort will make if it is passed a list that is already sorted?

5.13. If you want to try your hand at using merge sort to organize a real-world collection of objects, here is one way to do the sort. You can try this method with any group of objects that can be ordered: a deck of playing cards, recipe cards, homework papers that have been graded and need to be entered into a gradebook, etc.

    a) In the first round, pick up two items, put the smaller one on top, and set them face down on a table. Pick up two more items, put the smaller one on top, and put this group face down on top of the first group, but turn them so they are at a right angle to the first group. Keep picking up pairs of items, sorting them, and placing them at right angles on top of the pile until all items are in a group of two (unless there is an odd number, in which case just put the last item on the pile in its own group of one).

    b) Turn the pile over, so it is face up.

    c) Take the top two groups off the pile and place them side by side on the table. Remove the smaller of the two items showing on the top of a group and place it face down on the table. Repeat this step until the two groups have been merged into a new group, which will be face down.

    d) Repeat the previous step, but this time the new group should be at a right angle to the first one. Keep removing two groups, merging them, and building up the output pile with each new group at right angles to the previous one.

    e) When the last groups have been merged, begin a new round of merges by repeating this process from step (b). Eventually the output pile will be a single group with all the objects in order, at which point you are done.

5.14. ◆ If a person uses insertion sort to sort a group of real world objects (cards, papers, etc.) do you think they would do a linear search to find the place to insert each new object? Or could they use a binary search?

5.15. ◆ In Chapter 4, the process used by the helper function named insert_left was described as an iteration that required up to i steps, where i marks the current location in the sort. Because all the items in the list from 0 up to $i-1$ are already sorted, would it be possible to use a binary search for this operation? Would the new version of insertion sort then be $\mathcal{O}(n \times \log n)$?

Programming Projects

5.16. Modify your copy of bsearch so it prints a[mid] on each iteration. Load the complete list of words from the PythonLabs dictionary and use bsearch to look up some words:

```
>>> words = RandomList('all', 'words')
>>> bsearch(words, 'recursion')
lemon
rhinology
palatialness
...
```

Try several more searches, including words in the middle of the dictionary and some close to each end. Does it seem like binary search is looking through the dictionary the way a person would? What are some similarities or differences?

5.17. Write a program that will print a table similar to Table 5.2. You can choose any set of values you like for n. This for statement will generate the values from 1,000 to 10,000 in increments of 1,000:

```
for n in range(1000, 10000, 1000):
```

5.18. Modify the test_search function from Problem 4.18 so it calls both search and bsearch and prints the average number of comparisons made by each function. To make accurate comparisons, both functions should search for the same word on each test.

5.19. Modify test_search so every search will fail. Do the search algorithms consistently return the same number of comparisons (n comparisons for search, $\lceil log_2 n \rceil$ comparisons for bsearch)?

5.20. ♦ Add statements to test_search.py so it can be run as a command line application. The name of a search function and the search parameters should be specified on the command line. For example, to run bsearch on lists of size 1,000 the command would be

```
$ python3 test_search.py bsearch 1000
```

To run search on lists of 5000 and to have the search fail every time the command would be

```
$ python3 test_search.py search 5000 fail
```

5.21. Do a systematic test of the number of comparisons made during successful searches by the two algorithms, using lists of size 1,000, 2,000, etc. up to 10,000. Copy and paste the number of comparisons to a spreadsheet and plot the results.

5.22. Write a program named test_sort that works like test_search except it should test sort functions instead of search functions. To make accurate comparisons, you should create a random list of strings, make a copy of the list, and then sort one copy with isort and the other copy with msort. If a is a RandomList object this statement will define b to be a copy of a:

```
>>> b = a.clone()
```

5.23. Write your own version of the merge function, and test it using random lists of numbers and strings:

```
>>> a = RandomList(5, 'elements', sorted=True)
>>> a
['In', 'Md', 'Th', 'Tl', 'Zr']
>>> b = RandomList(5, 'elements', sorted=True)
>>> b
['Be', 'Cf', 'Cl', 'Np', 'Zn']
>>> merge(a,b)
['Be', 'Cf', 'Cl', 'In', 'Md', 'Np', 'Th', 'Tl', 'Zn', 'Zr']
```

5.24. ♦ Write a function named rsum that uses recursion to compute the sum of values in a list a. If a has just one item then the sum is just the value of a[0]. Otherwise the sum is a[0] plus the sum of the rest of the values, which is computed by the recursive call.

5.25. ◆ Write a function that uses a recursive algorithm to compute the greatest common divisor (GCD) of two integers a and b. First check to see if b is 0, and if it is, return a as the result. If $a > 0$ the GCD is defined recursively to be the GCD of b and a mod b. To compute a mod b in Python use the % operator, that is, evaluate a % b. To test your function repeat the tests used for the iterative version of gcd (Problem 4.21 on page 116).

5.26. ◆ Write a recursive function named rev that will return a copy of a string s but with all the characters in reverse order. Let n be the length of s. In the base case $n = 1$; because s has only one character just return s. Otherwise when $n > 1$ return a string consisting of the last letter in s followed by the reverse of the first $n - 1$ letters.

5.27. ◆ Write a function that returns a list of Gray codes of a specified length. A Gray code is a special type of binary code organized so that any two successive bit patterns differ at only one location. For example, a 2-bit Gray code is ['00', '01', '11', '10']. Notice how in the transition from the second code to the third code only the first bit in the code changes.

Here is a recursive algorithm for generating an n-bit Gray code:

- If $n = 1$ return the list with two 1-bit codes ['0','1']
- Otherwise let a be the list of Gray codes with $n - 1$ bits, and let r be a second list that has the same codes as a but in reverse order; the output should be a list of codes with a 0 in front of every code in a followed by codes that have a 1 in front of every code in r.

Here is an example that shows what your function should produce for $n = 3$:

```
>>> graycode(3)
['000', '001', '011', '010', '110', '111', '101', '100']
```

Notice how the first four codes have 0's in front of the four 2-bit codes and the last four codes have a 1 in front of the 2-bit codes in the opposite order. Some hints:

- The + operator can be used to concatenate two strings and to append two lists.
- A built-in function named reversed will invert the order of items in a list.

Chapter 6

Spam, Spam, Spam, Mail, and Spam

A machine learning approach to filtering junk mail

Anybody who uses e-mail knows the problem: we are constantly bombarded by unsolicited junk mail, or "spam." Organizations that monitor network traffic estimate that between 80% and 90% of all e-mail transmitted over the Internet is spam, most of it generated by "spambots" that blindly send out a massive number of copies of the same message.

Internet service providers and other groups that manage mail servers fight a constant battle with spammers by setting up filters that try to detect spam messages before they are sent on to users. A straightforward approach to filtering out junk mail is to create a set of rules, for example, "if a message was sent from location x redirect it to a Junk folder" or "delete any message that contains a link to a known virus." Unfortunately, it seems that every time someone develops an effective set of rules spammers come up with new tricks to get around the rules.

An alternative approach uses techniques from a branch of computer science known as **machine learning** to teach a system how to recognize spam. When the filtering application is first installed it knows nothing about the type of mail it will be receiving. As messages arrive they are used to train the program. If a user thinks a message is spam it is flagged so the system learns what spam looks like. Eventually, after a sufficient number of both spam and non-spam messages have been processed, the program will be able to take over and automatically identify spam by itself.

The machine learning approach is based on the frequencies individual words appear in messages. The spam filter examines each word in every incoming message. During the learning phase the program keeps track of the number of times words appear in each type

Why is Junk Mail called Spam?

The use of the word "spam" to refer to unsolicited e-mail has its origins in a famous Monty Python skit. In an episode broadcast on the BBC in 1970 a waitress at a diner read from a menu where breakfast choices all featured a serving of spam (including a dish called "spam, spam, spam, egg, and spam"). While she was reading the menu an annoying chorus in the background kept chanting the word "spam."

Many years later, mischief makers who wanted to disrupt Internet chat rooms started typing the word "spam" over and over again. Eventually someone figured out how to write a program that typed the word for them and the first spambot was born. By the mid 1990s the term was being used to describe e-mail, especially messages sent by unprotected computers that had been taken over and converted to robots that did nothing but repeatedly send out junk mail.

of message. After seeing enough messages it will have the data it needs to estimate probabilities for words. For example, if "diet" appeared in 63 out of 500 messages flagged as spam the spam probability is $63/500 = 0.126$. If the same word was seen only 10 times in 300 good messages the non-spam probability would be 0.033. Although both of these probabilities are small numbers, the ratio, $0.126/0.033 = 3.82$ predicts that messages containing "diet" are almost four times more likely to be spam.

One of the reasons this approach is so effective is that users trains the spam filtering application using their own mail. A college professor who gets a lot of spam promising low interest rates on mortgage refinancing would have relatively high probabilities in her spam database for words like "mortgage" and "loan." But a mortgage broker will be sent valid mail containing those same words, so his word frequency database would have lower probabilities.

The project for this chapter will explore some of the algorithms used in the machine learning approach to spam filtering. We will write a function named pspam (the name stands for "probability of spam") that takes the name of a file as an argument. The value returned by the function will be a probability value between 0 and 1, where a number close to 1 means the file probably contains a spam e-mail message.

The spam filtering project provides an opportunity to introduce some important constructs in Python. We will see how to read lines of text from a file and learn more about string objects and methods that carry out some very useful operations. We will also have a chance to work with a new type of object called a **dictionary**, a container that is used in a wide variety of problems besides string processing algorithms.

6.1 A Closer Look at Strings

Strings were introduced in Chapter 2 as a way of representing text data in a program. The easiest way to create a string in Python is to simply enclose a sequence of characters between quotes:

```
>>> s1 = 'hello'
>>> s2 = 'bonjour'
```

A very important attribute of strings was left out of the discussion in earlier chapters: in Python, strings are **immutable**. Once we create a string we cannot change the contents in any way.

We can easily verify that strings are immutable by trying to substitute one of the characters. With lists we can update the contents by putting an index expression on the left side of an assignment statement; for example, this statement will set the first item in a list a to 0:

```
>>> a[0] = 0
```

But if we try the same thing with a string, for example, if we try to change the first letter in s1 to 'j', we get an error message:

```
>>> s1[0] = 'j'
TypeError: 'str' object does not support item assignment
```

If you explore the documentation of str (the name of the class for string objects) you will find several methods that would appear to modify a string. For example, there is a method named capitalize that converts the first letter in a string to uppercase:

```
>>> s1.capitalize()
'Hello'
```

It seems like this expression is telling Python to change the first character in s1 to a capital letter. But if we take a closer look we can see that Python does not actually change s1. Instead, it makes a copy of s1, and the first letter in the copy is the uppercase version of the first character in the original. After typing the statement above, we can ask Python to show us the value of s1 and we will see it is unchanged:

```
>>> s1
'hello'
```

Here is another example, using a method named upper that converts lowercase letters into uppercase. If we store the result returned by the method we can see that after the call there are two separate objects:

```
>>> s3 = s2.upper()
>>> s2
'bonjour'
>>> s3
'BONJOUR'
```

Another construct that appears to modify a string is the increment operator. We've seen that Python allows use to write s1 + s2 when we want to create a string with all the characters in s2 following the characters in s1. It turns out the += form can also be used with strings:

```
>>> s1 += '?'
```

If we look at the value of s1 it will seem like Python did in fact modify the string:

```
>>> s1
'hello?'
```

But remember that += is simply a shorthand notation for a longer expression, and the statement above is equivalent to

```
>>> s1 = s1 + '?'
```

So while it looks like += modified s1, what really happened is that Python made a new string object with all the letters in s1 and with the question mark added at the end. It then assigned this new string as the value of s1 (and discarded the old value of s1).

A method we will use extensively for the projects in this chapter is named split. It is used to break a string into smaller pieces; for example, we can call it to split a sentence into individual words:

```
>>> s4 = 'strings are sequences of characters'
>>> s4.split()
['strings', 'are', 'sequences', 'of', 'characters']
```

The result of a call to split is a list of strings, one for each word in the sentence. But in spite of its name, this method is like the other string operations described in this section: it makes a list of new string objects, one for each word in the string it operates on, but it does not modify the original string.

The split method does not have a very sophisticated definition for a "word." It simply looks for spaces between characters, as shown in this example:

```
>>> s5 = '100% all natural!'
>>> s5.split()
['100%', 'all', 'natural!']
```

Note how the strings in the result contain everything but the spaces from the original string. All of the letters and punctuation marks should be present in the list of words.

So far, all the examples of strings have been short pieces of text that contain letters, digits, and punctuation. However, it is not uncommon to have other types of characters in a string, for example, characters that specify how the text should be displayed on the terminal. Suppose we want to create a string object that represents a short phrase that spans two lines:

> I think,
>   therefore I am.

If we want to put this text in a string object, we need to include a character that marks the end of the first line. When Python prints the text it will appear as two lines in our shell window.

Python and other programming languages use a **newline character** to indicate the end of a line of text. A newline is a single character, like a letter or a digit, but it serves a special purpose: when a program is displaying the text and it sees this character it starts printing the characters that follow on a separate line. The technique for including a newline character in a string is to type a backslash followed by the letter n:

```
>>> quote = 'I think,\n  therefore I am.'
```

The backslash is known as the **escape character**: it is a signal to Python that the next character has a special meaning. When Python sees a slash followed by an n it puts a single newline character in the string (see the sidebar below on "Escape Sequences" for other things besides n that can follow a backslash).

The expression above created a string object and saved a reference to it in the variable quote. If we ask Python to show us the value of the variable we will see it is a single string:

```
>>> quote
'I think,\n  therefore I am.'
```

But notice what happens when we print the string:

```
>>> print(quote)
I think,
   therefore I am.
```

Just what we were looking for: the print function saw the newline character and started printing the characters after the comma on a separate line.

Notice what happens if we use the split method with our new string:

```
>>> quote.split()
['I', 'think,', 'therefore', 'I', 'am.']
```

All the words from both lines are collected into a single list. That's because newline characters are treated just like spaces. Python simply gathers characters to form a word, and when it finds a newline or space it skips ahead to the next non-space character and begins a new word.

Escape Sequences

We often need to include special characters in strings, but the rules for typing statements in an interactive shell session make it difficult. A good example is the character that marks the end of a line of text -- we can't simply hit the return key because Python would think we are done with the command:

```
>>> quote = 'I think,
SyntaxError: EOL while scanning string literal
```

To include a newline in a string type a backslash followed by the letter n. Some other common escape sequences are shown in the table below.

A backslash followed by a u is the escape sequence for entering Unicode characters or symbols (Unicode is discussed in more detail in Chapter 8).

```
>>> card = 'A\u2665'
>>> print(card)
A♥
```

\n	newline
\r	return
\t	tab
\uNNNN	Unicode character NNNN

In programming jargon, characters like space and newline are called **whitespace** characters. Other whitespace characters include the tab character and, in documents created on a Microsoft Windows® system, a character called "return" that is included as a second character at the end of a line.

The bottom line: when Python evaluates a call to s.split() for some string s it creates a list of individual words, where a word is defined as any sequence of non-whitespace characters.

Tutorial Project

T1. Make an object to use in tests with strings:
```
>>> s = 'As seen on TV!'
```

T2. Use the len function to count the characters in the string:
```
>>> len(s)
14
```
Notice how the count includes spaces and punctuation.

T3. Type these expressions to access characters in the string using the index operator:
```
>>> s[0]
'A'
>>> s[-1]
'!'
```

T4. Try changing one of the letters:
```
>>> s[-1] = '?'
TypeError: 'str' object does not support item assignment
```
As expected, this statement generated an error.

T5. Call a method named upper, which converts lowercase letters into uppercase letters:
```
>>> s.upper()
'AS SEEN ON TV!'
```
Notice how all the other characters besides lowercase letters were not changed.

T6. Ask Python to show you the value of s:
```
>>> s
'As seen on TV!'
```
So the call to upper did not change s.

T7. To verify the claim that upper returns a new string object, save the result of the call in a new variable:
```
>>> t = s.upper()
```

T8. If you look at the values of the two variables you will see you have two different string objects:
```
>>> s
'As seen on TV!'
>>> t
'AS SEEN ON TV!'
```

T9. Call the split method:
```
>>> s.split()
['As', 'seen', 'on', 'TV!']
```
Note the result contains four strings, one for each word in s.

T10. This for statement will print the length of each word:
```
>>> for w in s.split():
>>> ...    print(len(w))
...
2
4
2
3
```

Notice how the last word has three characters. The split method simply gathers all the characters between spaces and considers them as words.

The goal for the next set of exercises is to create a Python string object to represent a short poem where the title (*Lines on the Antiquity of Microbes*) is longer than the poem itself:

> *Adam*
> *Had 'em.*

T11. If you hit the return key on your keyboard after typing the "m" in "Adam" you will get an error message:
```
>>> poem = 'Adam
SyntaxError: EOL while scanning string literal
```
The "EOL" in this message stands for "end of line." Python did not know you wanted to continue the string on the next line.

T12. Type this statement to create a string that has a newline character in the middle:
```
>>> poem = "Adam\nhad 'em"
```

T13. Call the len function to count the characters in the string:
```
>>> len(poem)
12
```
Count the characters in poem yourself. Do you see why there are twelve characters? The count consists of nine letters, one space, one punctuation mark, and one newline character.

T14. Ask Python to print the poem. You should see two separate lines:
```
>>> print(poem)
Adam
had 'em
```

T15. Use split to break the poem into individual words:
```
>>> poem.split()
['Adam', 'had', "'em"]
```
Notice how the newline character is considered "white space" by the split method.

The final set of experiments in this section illustrates how Unicode characters specified by multi-character escape sequences are in fact just a single letter inside the string object.

- The sidebar on escape sequences (page 151) showed how to define a string that contains a Unicode character. Type this statement to make a string with a heart symbol:
  ```
  >>> s = 'I \u2665 cats'
  ```

- Ask Python to show you the contents of the string:
  ```
  >>> s
  'I ♥ cats'
  ```
  If you do not see a heart symbol in your terminal window check to see if your IDE or terminal emulator allows you to define an "encoding." If so, set the encoding to UTF-8.

◆ How many characters do you think are in this string?

◆ Use `len` to verify your answer to the previous problem:

```
>>> len(s)
8
```

Does that seem right? Do you see how the escape sequence \u2665 turned into a single character?

◆ Do a web search to find a table with Unicode symbols and create some some other strings with the \u escape sequence.

6.2 Reading Text from a File

All our projects up to this point have used objects that reside in a computer's main memory. When Python executes an assignment statement that defines a new object, it carves out some space for the object and associates the variable name with the object. The steps of an algorithm might introduce new objects, or change the contents of existing objects, but all the information is present in memory.

Another way to create an object is to read a piece of data stored in a file. When we select the Open command from an application's File menu, we are telling the program to read some data from a file and load it into memory. After updating the information we can write it back out to a file using a Save command (Figure 6.1).

The advantage of this approach, of course, is that data stored in files can be reused at some point in the future. All the objects created in an interactive session with Python are lost when we exit from our development environment, but data saved in a file can be loaded back into memory some time in the future.

A modern operating system must deal with dozens of different file formats, and it is not uncommon for each application to define its own unique format. Word processors, spreadsheets, graphic design programs, music libraries, and other applications create and manage their own types of files. Python has library modules that read and write these types of files, but for the project in this chapter we will work with a very simple type of file called a **plain text** file.

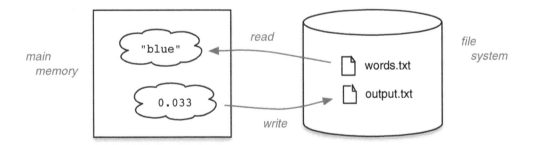

Figure 6.1: *Programs often read data from a file and use the information to create objects in memory. The program can also write data out to a file so it can be used again some time in the future.*

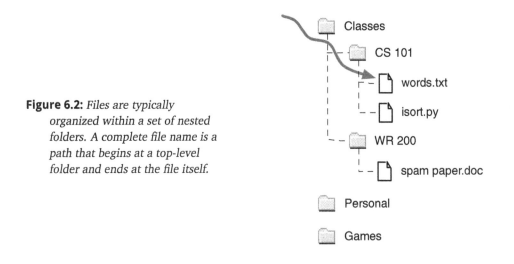

Figure 6.2: *Files are typically organized within a set of nested folders. A complete file name is a path that begins at a top-level folder and ends at the file itself.*

A program that reads data from a file uses the same steps an office worker might use to get information from a file cabinet. The first step is to **open** the file to make the data accessible. Then the data is read, and after the task is complete the file is **closed**.

In order to open a file so we can read its contents we need to tell Python where the file is located. Because operating systems allow users to organize their files inside folders, a file name typically specifies one or more folder names as well as the name of the file itself.

For example, suppose we have a file named words.txt in a folder named CS 101. Suppose also that this folder is itself inside another folder named Classes (Figure 6.2). In Python, the full name of this file is a string object that contains the folder names and the file name:

```
>>> filename = "Classes/CS 101/words.txt"
```

A string that contains a sequence of folder names preceding a file name is called a **path**. The character between folder names is known as a **separator**. Operating systems descended from Unix, including Linux and Mac OS X®, use a slash (/) separator; Microsoft Windows uses a backslash (\). Within Python programs, however, we can use a forward slash no matter which operating system we use, and Python will translate forward slashes into back-slashes or other symbols if it needs to (if you use Microsoft Windows, make sure you read the "Path Strings on Microsoft Windows" sidebar on page 156).

After we have a string that specifies the path to a file we want to access, we can call a function named open in order to access the contents of the file:

```
>>> f = open(filename)
```

Because open is a built-in function, like print, it does not need to be imported from any module.

The value returned by a call to open is an object from a class named io. In order to get information from the file we just have to call methods defined for the object. One common technique is to read one line at a time, calling the method named readline. After we have opened a file, this statement will read the first line:

```
>>> s = f.readline()
```

Path Strings on Microsoft Windows

If you have been using the Command Prompt application on a Microsoft Windows system you know that Windows uses a backslash character to specify folder names. For example, if you have a folder named CS for your computer science projects, and inside that you made another folder named Sieve for the programs you wrote for Chapter 2, you might have used this command to navigate to your project folder:

```
> cd My Documents\CS\Sieve
```

Python programs running on Windows systems can use forward slashes, so a string representing this path can be written:

```
>>> p = "My Documents/CS/Sieve"
```

There are two reasons to prefer forward slashes. For one thing, programs are more portable. If you are working on a group project, it won't matter if some group members use Windows and others have different operating systems. Second, if you want to use backslashes, you will need to use a double slash because the backslash is Python's escape character. To write the path shown above with backslashes you would have to write the assignment statement like this:

```
>>> p = "My Documents\\CS\\Sieve"
```

If we look at the value of s we should see the first line from the file, including the newline character that marked the end of the line. The next call to readline will get the second line from the file, and so on.

When we are done using the file all we need to do is call the close method:

```
>>> f.close()
```

In Python all three of these operations—open a file, read lines, close the file—can be performed using a simple for loop. We just put the call to open in a for statement:

```
for line in open(filename):
```

On each iteration of the loop, Python will read a line of text from the file, store it in a string named line, and execute the statements in the body of the loop. After the last line has been read from the file the loop terminates and the file is automatically closed.

A trivial example of a function that uses a for loop to read data is filesize (Figure 6.3). This function simply counts the number of characters in a file. The argument passed to the function should be a string containing the path to the file. The function initializes a counter that will hold the number of characters, and on each iteration the number of characters in the current line is added to the count. When the for loop terminates, the total number of characters is returned as the value of the function call.

There are many other ways to get information from text files. It is possible to read individual characters one at a time, or to load the entire file into a single string variable. For the projects in this book, however, we will typically read lines of text one at a time using a for loop like the one in Figure 6.3.

```
1   def filesize(fn):
2       "Count characters in file fn."
3       nchars = 0
4       for line in open(fn):
5           nchars += len(line)
6       return nchars
```

Figure 6.3: *This function counts the number of characters in a file. When the function is called the argument should be a string containing a path to the file.*

Download: spam/`filesize.py`

To set up and run experiments with files the PythonLabs software includes a few simple text files that can used for test data. These files are stored in the same folders as the Python modules, so if you want to open one if these files you need to find out where it was saved when you installed PythonLabs. The easiest way to do this is to use a function from PythonLabs named `path_to_data`. If you pass this function the name of the file you want it will return a string containing the full path to the file.

This example shows how to test the `filesize` function on a file named quote1.txt, which has the four lines of text shown in Figure 6.4:

```
>>> filesize(path_to_data('quote1.txt'))
188
```

As you work through the tutorial for this section you will have a chance to examine each line of the file in detail and verify the result is correct.

Tutorial Project

T16. Start a new shell session in your IDE and import the `path_to_data` function:
```
>>> from PythonLabs.Tools import path_to_data
```

T17. Find out where the data file named quote1.txt is stored on your computer:
```
>>> path_to_data('quote1.txt')
'PythonLabs/data/text/quote1.txt'
```
The string you get may differ, depending on your operating system and where Python installed the lab modules on your system.

T18. Open the file and save the object returned by open:
```
>>> f = open(path_to_data('quote1.txt'))
```

T19. If you call this object's `readline` method you should get the first line in the file:
```
>>> f.readline()
'If you have no confidence in self,\n'
```

T20. Call `readline` again, but this time save the result in a new variable named s:
```
>>> s = f.readline()
```

T21. Ask Python to show you the value of s, which should be the second line in the file:
```
>>> s
'  you are twice defeated in the race of life.\n'
```
Notice how the string starts with two space characters and ends with a newline character.

```
  If you have no confidence in self,
    you are twice defeated in the race of life.
  With confidence, you have won even before you have started.
    -- Marcus Tullius Cicero (106 BC -- 43 BC)
```

Figure 6.4: *The file named* quote1.txt *has four lines of text. Notice that the second and fourth lines are indented two spaces. These spaces will be included in the count of the number of characters in the file.*

T22. Call len to count the number of characters in s:
```
>>> len(s)
46
```

T23. Repeat the calls to readline and len two more times to count the number of characters in the last two lines of the file.

T24. If you call readline again you should get back an empty string:
```
>>> f.readline()
''
```
Python returns a string with 0 characters as a signal that we have reached the end of the file.

T25. Call the close method to close the file:
```
>>> f.close()
```

T26. If you try to read a line from a closed file you should get an error message:
```
>>> f.readline()
ValueError: I/O operation on closed file.
```

T27. Use your IDE to create a new file named filesize.py. Enter a preliminary version of the definition of filesize that has these three lines:
```
    def filesize(fn):
        for line in open(fn):
            print(line, end = "")
```
This first version should simply print each line in the file (see the sidebar on "extra lines" on page 159 to see why there is an additional argument being passed to print).

T28. Save your file, load it into your shell session, and call filesize:
```
>>> filesize(path_to_data('quote1.txt'))
```
Did it print all four lines in the file?

If your version of filesize is not printing all four lines exactly as they appear in Figure 6.4, go back to your IDE and try to locate the error. Are the three lines in your definition indented properly? Did you spell the variable names (fn and line) the same way each time they are used?

T29. Edit your definition of filesize so it looks like the one in Figure 6.3: add the line that initializes nchars to 0, replace the print statement with the assignment that adds the length of the line to nchars, and add the return statement.

T30. Save the file, reload it into your shell session, and repeat the statement that calls the function. This time you should see the number of characters in the file:
```
>>> filesize(path_to_data('quote1.txt'))
188
```

Why Are There Extra Lines in the Output?

When we call `print(s)` to ask Python to print a string s the system normally adds a newline to the end of s. If s is a string that was read from a file it already has a newline, so when we print the line it looks like there is an extra space.

To tell Python not to print a newline at the end of an output line pass an optional parameter in the call to print:

```
>>> print(s, end = "")
hello!
>>>
```

```
>>> s = f.readline()
>>> s
'hello!\n'
>>> print(s)
hello!
                ←——— extra line
                     added by print
>>> len(s)
7
```

6.3 Counting Words in a Text File

Continuing our introduction to processing data in a text file, we are going to extend the `filesize` function so it returns three pieces of information about a file: the number of lines, the number of words, and the number of characters. The techniques introduced for this part of the project will set the stage for the spam filtering function, which also relies on identifying individual words in a file.

Keeping track of the number of lines is trivial. We simply need to initialize a variable named `nlines` to 0 and then add 1 to it on each iteration of the `for` loop.

Counting the number of words is also straightforward. We will define another counter, named `nwords`. On each iteration we can use the `split` method to break the line into individual words and then simply use the length of the resulting list as the number of words.

As we saw in Section 6.1, `split` has a very simple definition for a word. It includes punctuation marks and other symbols as parts of words, and it also considers numbers and other sequences of characters to be words:

```
>>> s = "Has Python's prompt been >>> since 1991?"

>>> s.split()
['Has', "Python's", 'prompt', 'been', '>>>', 'since', '1991?']
```

For the word counting project, however, this behavior will actually work quite well. It will count contractions (things like "isn't" or "won't"), hyphenated words ("big-time"), and other unusual cases as single words, which is probably what we want.

The definition of the new function that counts lines, words, and characters is shown in Figure 6.5. The first line in the body of the function uses a type of assignment statement called a **multiple target** assignment. We could have written three separate assignment statements, one for each counter, but the form shown here is an idiom often used by Python programmers. A multiple target assignment fits on one line, which is a benefit (shorter programs are often easier to understand), and it also emphasizes that all three values are set to the same initial value.

```
1   def wc(fn):
2       """
3       Count the number of lines, words, and
4       characters in file fn.
5       """
6       nlines = nwords = nchars = 0
7       for line in open(fn):
8           nlines += 1
9           nwords += len(line.split())
10          nchars += len(line)
11      return nlines, nwords, nchars
```

Figure 6.5: *This function works like the Unix utility (also named wc) that counts lines, words, and characters in a file.*

Download: *spam/wc.py*

There is one remaining detail to cover: how does a function return multiple values? Our specification calls for a function that computes the three values shown in the figure and then returns all three as the value of the function call. As the final line of Figure 6.5 shows, we simply have to write a return statement that contains all three values separated by commas:

```
    return nlines, nwords, nchars
```

This is what the result looks like if we test the new function using the short quotation from the previous section:

```
>>> wc(path_to_data('quote1.txt'))
(4, 35, 188)
```

The three values returned show the file has 4 lines, 35 words, and 188 characters.

If we want to save these values in variables so we can use them later, we just put three variables on the left side of the assignment statement:

```
>>> lines, words, chars = wc(path_to_data('quote1.txt'))
```

This type of assignment statement, where there are multiple names on the left side of the assignment operator and also multiple values on the right side, is known officially as a **tuple assignment**, but often called a **parallel assignment**.

Tutorial Project

The tutorial project for this section is to implement the function named wc. The recommended way to start this project is to duplicate your file from the previous section (filesize.py) and name the copy wc.py.

T31. Type this statement to verify multiple target assignments work as described in this section:
```
>>> a = b = 3
```

T32. Confirm the fact you have two new variables by using them in some expressions:
```
>>> a + b
6
>>> a * b
9
```

◆ **Tuples**

You may have noticed that Python printed parentheses around the three values returned by a call to wc. When a return statement lists two or more values Python creates an object called a *tuple* to hold the values.

A tuple is another type of container object. It is basically a list that cannot be modified in any way — we can't add items or change the value of any item.

```
>>> x = wc(path_to_data("quote1.txt"))
>>> type(x)
<class 'tuple'>
>>> x[0]
4
>>> x[0] = 1
TypeError: 'tuple' object does not
support item assignment
```

Other than through functions that return multiple values we will not be using tuple objects in any of our projects in this book. They are a very useful construct in Python, however, and readers who plan to study Python further are encouraged to learn more about tuples.

T33. Open your new wc.py file with your IDE and make the changes needed to transform filesize into wc:
   - Change the function name.
   - Edit the line that initializes nchars so it also initializes nlines and nwords.
   - Add the assignments that update nlines and nwords.
   - Modify the return statement so all three values are returned.

T34. Load your wc function into an interactive session and test it on quote1.txt. This is what you should see if the function is working correctly:
```
>>> wc(path_to_data('quote1.txt'))
(4, 35, 188)
```

T35. Repeat the call, but this time save the results in three variables:
```
>>> lines, words, chars = wc(path_to_data('quote1.txt'))
```

T36. Verify the fact that Python assigned values to all three variables by asking for their values:
```
>>> lines
4
>>> words
35
>>> chars
188
```

6.4 Dictionaries

Before we move on to the next step in the development of our spam filtering program, we are going to take a short detour to learn about a type of object that is quite useful in a variety of situations. These types of objects are very widely used and are part of all modern programming languages, but the terminology varies from one language to the next. The first languages to use them called them "tables"; some other more recent names are "maps" or "associative arrays".

In Python the objects are called **dictionaries**. These objects have much in common with lists, because both are collections that contain references to other objects. The difference between lists and dictionaries is in how we insert an item or look up the value of an item using an index expression. With a list we access an item by its location, but with a dictionary items are accessed by their *name*, not their location.

As an example of how to use a dictionary, suppose we want to keep track of how many feet are in various distances in Imperial units (Figure 6.6). If the dictionary is named d, we will use an expression like d[x] to look up the number of feet in the unit named x. For example, to find out how many feet are in a yard, the expression is

```
>>> d['yard']
3
```

Note the difference between lists and dictionaries: if d is a list, an expression like d[x] means "return the value at location x in d," and Python expects x to be an integer between 0 and one less than the length of d. But when d is a dictionary, x is the name of some value in the dictionary, and we're asking Python to look up the value associated with this name.

One way to create a new dictionary object is to use an assignment statement with an empty dictionary on the right side:

```
>>> d = { }
```

This is the same as the statement that creates an empty list, except Python knows we want a dictionary object because we're using braces ("curly brackets"). As always, we can find out what kind of object we have by calling the type function:

```
>>> type(d)
<class 'dict'>
```

We can add values to a dictionary by using an assignment statement. The left side specifies the dictionary and the name of the item we want to add, and the right side has the value we want to associate with this name. To record the fact that there are 3 feet in a yard, we would write

```
>>> d['yard'] = 3
```

Here 'yard' is the name of the item being added to the dictionary, and 3 is its value.

unit	distance
yard	3
fathom	6
furlong	660
mile	5280

```
d = {
    'yard': 3,
    'fathom': 6,
    'furlong': 660,
    'mile': 5280
}
```

Figure 6.6: *At left is a table showing the distance (in feet) in various units of distance. On the right is an assignment that will create a dictionary object that represents the information in the table.*

To continue this example, these statements show how to add three more values to the dictionary. There are 6 feet in a fathom, 660 feet in a furlong, and 5,280 feet in a mile:

```
>>> d['fathom'] = 6
>>> d['furlong'] = 660
>>> d['mile'] = 5280
```

Note that in each of these examples the index, the object between the square brackets, is a string, not an integer.

Our new dictionary object is essentially a table. To find the number of feet in a certain unit we just look up the value in the table. To look up the number of feet in a furlong,

```
>>> d['furlong']
660
```

The Kentucky Derby is 10 furlongs, so to find out how far that is in feet, we can ask Python to evaluate this expression:

```
>>> 10 * d['furlong']
6600
```

For a slightly more complicated example, to convert a distance from feet into miles, we need to divide the distance by the number of feet in a mile. Here is how we can use Python to convert the distance of a 10-furlong race from feet into miles:

```
>>> (10 * d['furlong']) / d['mile']
1.25
```

If we ask Python to show us the value of the dictionary object this is what we'll see:

```
>>> d
{'mile': 5280, 'yard': 3, 'fathom': 6, 'furlong': 660}
```

We added four names to the dictionary, so the output shows four items, where each item is a name and its associated value. The names of the items in a dictionary are known as **keys**, so we often say a dictionary is a collection of key-value pairs.

Note that the items are not always printed in the same order in which they were inserted into the dictionary. If you type the previous statements on your computer and ask Python to show you the result, you may very well get a different order of key-value pairs. What is important about a dictionary is that we can retrieve a value by name; we leave it up to the system to figure out the most efficient way to store the items or what order to display them.

For the projects in this chapter the dictionary keys will be strings, but in fact we can use almost any type of object. The next example shows how to make a dictionary that associates integers with their Roman numeral equivalent. In this case, the keys will be integers and the values will be strings. This statement also shows how to create a dictionary object by giving a set of key-value pairs enclosed in braces using the same format Python uses when it prints a dictionary on the console:

```
>>> roman = { 1: 'I', 5: 'V', 10: 'X', 50: 'L', 100: 'C' }
```

The statement above creates a dictionary object where the number 1 is associated with the string 'I', the number 5 corresponds to the string 'V', and so on.

We can find the roman numeral for an integer (as long as that integer is a key in the dictionary) by using the integer as the key is an expression that looks up a value in the table:

```
>>> roman[5]
'V'
```

It is important to note the expression roman[5] does not mean "what is in location 5 of the list named roman?" Because roman is a dictionary, and not a list, the expression means "what object is associated with the key 5?"

Dictionaries have much in common with lists, and many of the operators and functions we saw earlier for lists and strings can be applied to dictionaries. To see if a name has been added to a dictionary use the in operator:

```
>>> 'mile' in d
True
>>> 1000 in roman
False
```

The len function returns the number of key-value pairs in a dictionary:

```
>>> len(roman)
5
```

Finally, we can iterate over a dictionary to perform some operation on every item by using a for loop. The for statement sets the control variable to each key in the dictionary:

```
for n in roman:
...    # do something with roman[n]
```

There are many other useful operations, but as always we will postpone discussing them until they are needed for a project.

Tutorial Project

T37. Make a new empty dictionary named e2f (the name stands for "English to French"):
```
>>> e2f = { }
```

T38. Use the built-in function named type to get some information about the new object:
```
>>> type(e2f)
<class 'dict'>
```
So the previous statement created an instance of the class named dict (Python's name for the dictionary class).

T39. The new dictionary should be empty, which means it should have 0 items:
```
>>> len(e2f)
0
```

T40. Add a few items to the dictionary. In these statements, the ket is an English word and the value on the right side is its French equivalent:
```
>>> e2f['cat'] = 'chat'
>>> e2f['dog'] = 'chien'
>>> e2f['duck'] = 'canard'
>>> e2f['cow'] = 'vache'
```

T41. Check the length again. We should have four items now:
```
>>> len(e2f)
4
```

T42. If you ask Python to print the dictionary, it will display all the keys and their corresponding values:
```
>>> e2f
{'duck': 'canard', 'cow': 'vache', 'cat': 'chat', 'dog': 'chien'}
```
Python does not print the items in the order they were created or in any other specified order. The dictionary is simply a set of key-value pairs.

T43. Look up a name in the dictionary:
```
>>> e2f['duck']
'canard'
```

T44. Try an experiment where the name is not in the dictionary:
```
>>> e2f['wolf']
KeyError: 'wolf'
```
Python generates a runtime error if we try to look up a value with a key that is not in the dictionary.

T45. An easy way to see if an item is in a dictionary is to use the in operator:
```
>>> 'dog' in e2f
True
>>> 'wolf' in e2f
False
```

T46. Use a for statement to print each item in the dictionary (make sure you indent the second line by typing one or more space characters so Python knows the print statement is in the body of the loop):
```
>>> for w in e2f:
...     print('The French word for ' + w + ' is ' + e2f[w])
...
The French word for duck is canard
The French word for cow is vache
The French word for cat is chat
The French word for dog is chien
```

The next exercises show that instead of creating an empty dictionary and adding items one by one, we can specify the names and their values when the dictionary is created.

T47. Type this expression to create a dictionary that associates letter grades with the values used to compute a grade point average:
```
>>> points = { 'A': 4, 'B': 3, 'C': 2, 'D': 1, 'F': 0}
```

T48. Look up the point value for some grades:
```
>>> points['A']
4
>>> points['C']
2
```

One of the programming projects at the end of this chapter is to modify the GPA program from Chapter 2 so it uses a dictionary like the one above instead of an if-elsif-else statement.

6.5 Word Frequency

The project in Section 6.3 used a simple strategy that relied on Python's split method to count the number of words in a file. In this section we will develop a similar function, but in this case we are interested in computing the *frequency* of each word, that is, we want to know how often each word appears in a file. This is an important milestone in the spam filtering project, because our approach to deciding whether or not a message is valid is based on how often words are used in different types of messages.

The plan for the word frequency function is to use a dictionary that associates strings with integers, where the strings are words found in the input file and the integers are the number of times a word has been seen. We'll call this dictionary count, and initialize it as an empty dictionary:

```
count = { }
```

Then, inside a loop that iterates over all the strings created by a call to split, we just need to add one to the count for each word. If the variable x is a reference to a word, the statement that updates the number of times we have seen x is:

```
count[x] += 1
```

If we are going to use this approach we have to make sure Python does not generate an error message the first time it sees a word in a file. As we saw in the previous section, if x is not currently in the dictionary, a reference to count[x] will cause a runtime error. We need to structure the loop body so the first time a word is seen its count is initialized, and after that we can increment the count using the += operator.

Figure 6.7 shows two ways to check if a word is already in the dictionary before incrementing its count. The first method, shown on the left, simply uses the in operator, with a slight variation: the word not means the Boolean expression will be true if the word has not been added to the dictionary yet. In that case we simply initialize the count to 1. But if the word was previously added we just have to update the count.

The right side of Figure 6.7 shows an alternative. Because the situation of testing to see if an item is in a dictionary and setting an initial value is so common, the dict class has a method that combines both operations into a single method call. A call of the form setdefault(x,y) means "check to see if the item x is in the dictionary, and if not, insert it and give it the value y." If the item already is in the dictionary the method doesn't do anything. An important detail is that with this approach the initial count for a word is 0.

```
1    if word not in count:
2        count[word] = 1
3    else:
4        count[word] += 1
```

```
1    count.setdefault(word, 0)
2    count[word] += 1
```

Figure 6.7: *(left) Using an if-else statement to check if a word is already in the dictionary. (right) A method named setdefault combines the testing and initialization steps.*

That's because the next statement will also be executed, and that statement will increment the count to 1, which is what we want.

To count the frequency with which words appear in a file, we will again use split to break each line into words. The program will have a loop that iterates over each word:

```
for word in line.split():
    ...
```

The two statements from the right side of Figure 6.7 will be in the body of the loop so the words are added to the dictionary and we can count how often each work appears in the input file.

When we try this approach with the example text in quote1.txt, however, we will quickly discover a new problem. The dictionary will have two different entries for the word "confidence":

```
{ ... 'confidence,': 1, ... 'confidence': 1, ... }
```

If you look carefully at the ends of these strings you'll notice one of them has a comma. On the third line of the quote (*"With confidence, you have won..."*) there is a comma on the right side of the word "confidence." Because split keeps all the punctuation marks from the input text the comma was included as part of the string created for this word. The other place "confidence" appears in the text (*"If you have no confidence in self"*) the word is surrounded by spaces. Because these are two different strings the word is added to the dictionary twice.

A second problem is that some words in the text are capitalized but others are not:

```
{ ... 'With': 1, ... 'in': 2, ...}
```

In a longer piece of text we will probably find cases where words are included twice, once starting with a capital letter and once with a lower case letter. It would be better to ignore capitalization and just save the lowercase form of each word.

These are fairly common situations in text processing applications and, not surprisingly, Python has several features that will help us address them. The first is a method named strip that removes characters from the end of a string. When we call strip we can pass it a character, and all occurrences of this character on either end of the string are discarded. Here is an example from an interactive session that shows how to remove periods from the ends of strings:

```
>>> s1 = '...hello...'
>>> s1.strip('.')
'hello'
```

Remember, however, that Python isn't really stripping characters away from the ends of s1; rather, it is returning a new string that is a copy of s1 but without the periods at the ends.

The argument passed to strip can be a longer string, in which case any of the characters in the argument are removed. If we want to remove digits from the ends of a word we just have to pass the string '0123456789' in a call to strip:

```
>>> s2 = '25 or 6 to 4'
>>> s2.strip('0123456789')
' or 6 to '
```

```
1   tmp = w.strip(punctuation)
2   word = tmp.lower()
```

```
1   w.strip(punctuation).lower()
```

Figure 6.8: *Before inserting a word into a dictionary that counts the number of times a word occurs we need to strip punctuation marks from the ends and convert it to lowercase. (**left**) These operations can be done in separate assignments, using a temporary variable to save the result of the call to* strip. *(**right**) Both steps can be combined into a single statement.*

In order to remove all the punctuation marks that appear in English text we could define a string that has a period, a comma, and anything else we might expect to find in a text file. Python's string library already has such a string, so all we have to do is import it from the library and pass it in a call to strip:

```
>>> from string import punctuation
>>> s1.strip(punctuation)
'hello'
```

Finally, one way to handle the situation where some words will be capitalized is to convert all characters to lowercase before we put a word in the dictionary. That is exactly what the string method named lower does:

```
>>> s3 = 'PythonLabs'
>>> s3.lower()
'pythonlabs'
```

So our new strategy is that for every string w in the list created by a call to split, first call strip to remove the punctuation marks and then call lower to convert the letters to lowercase. The code on the right side of Figure 6.8 shows that both of these operations can be combined into a single assignment statement. Although this construction may appear to be more complicated, there is a benefit to keeping programs short and concise (in this case the benefit is avoiding coming up with a name for a variable that is only used once).

After putting all these pieces together we have the final version of a function that computes the frequency of words shown in Figure 6.9. A helper function named tokenize takes a single line from the input file and returns a list of individual words, where each word has been stripped of any punctuation and all the letters converted to lowercase. The loop in the main function wf is basically the same as the one we used before in the word counting function of Section 6.3: it simply opens a file and uses a for loop to iterate over all lines in the file. In this new program, the body of the main loop contains an inner loop that iterates over the list of words created by tokenize to update the count for how often each word was seen.

```
1    from string import punctuation
2
3    def wf(fn):
4        "Make a dictionary of word frequencies"
5        count = { }
6        for line in open(fn):
7            for w in tokenize(line):
8                count.setdefault(w, 0)
9                count[w] += 1
10       return count
11
12   def tokenize(s):
13       "Return the list of words in string s"
14       a = [ ]
15       for x in s.split():
16           a.append( x.strip(punctuation).lower() )
17       return a
```

Figure 6.9: *The word frequency function wf counts how often words are used in a text file. tokenize is a helper function that makes a list of words on one line. Each word is stripped of punctuation and converted to lowercase.*

Download: *spam/wf.py*

Tutorial Project

T49. Create a new empty dictionary:
```
>>> count = { }
```

T50. As we saw earlier, trying to access an item that is not in the dictionary will cause an error:
```
>>> count['spam'] += 1
KeyError: 'spam'
```

T51. Use the setdefault method to initialize a count:
```
>>> count.setdefault('spam',0)
0
```

T52. Ask Python to print the contents of the dictionary to verify the new item has been added:
```
>>> count
{'spam': 0}
```

T53. Repeat the statement that updates the count and print the dictionary again:
```
>>> count['spam'] += 1
>>> count
{'spam': 1}
```

T54. Repeat the call to setdefault but this time with a different value:
```
>>> count.setdefault('spam',7)
1
```

T55. The return value from the previous call is a hint the new value was ignored. Print the dictionary again to make sure:
```
>>> count
{'spam': 1}
```

Make sure you understand how setdefault works. Do you see that it adds a word to the dictionary if it is not already there, but if the word is in the dictionary the call has no effect?

T56. Make a string to use in tests of split, strip, and lower:
```
>>> s = '--Spam-and-Eggs--'
```

T57. To remove the dashes on each side of the string, pass the character '-' to strip:
```
>>> s.strip('-')
'Spam-and-Eggs'
```
Notice how the dashes in the middle of the string are still there: strip just removes characters from the ends of a string.

T58. If you ask Python to print s you will see that strip, like other string methods, does not modify s (strings are immutable):
```
>>> s
'--Spam-and-Eggs--'
```

T59. Calling the lower method will return a copy with all characters in lowercase:
```
>>> s.lower()
'--spam-and-eggs--'
```

T60. Combine the calls to strip and lower into a single statement:
```
>>> s.strip('-').lower()
'spam-and-eggs'
```
Do you see why this single statement works? The call to s.strip() returns a string object, and Python then calls lower for this object. The result is a string that has been processed by both methods.

T61. Import the string named punctuation from Python's string library:
```
>>> from string import punctuation
```

T62. Tell Python to print the string:
```
>>> print(punctuation)
!"#$%&'()*+,-./<=>?@[\]^_'{|}~
```

T63. Because the dash character is in punctuation we can also pass this string to strip to remove dashes:
```
>>> s.strip(punctuation).lower()
'spam-and-eggs'
```

The next phase of this project uses the incremental development strategy to test parts of the word frequency function before combining them into a single word counting function.

T64. Make a copy of wc.py, the file with your word counting program, and name the copy wf.py.

T65. Add the definition of the helper function named tokenize shown in Figure 6.9.

T66. Make a string to test the tokenizer:
```
>>> s = 'Tyger Tyger, burning bright, in the forests of the night:'
```

T67. Notice how split includes the comma with the second appearance of the word "Tyger":
```
>>> s.split()
['Tyger', 'Tyger,', 'burning', 'bright,', ... 'night:']
```

T68. Load your definition of tokenize into the interactive session and pass the function your test string:
```
>>> tokenize(s)
['tyger', 'tyger', 'burning', 'bright', ... 'night']
```
Compare the words returned by tokenize with the strings produced by split. You should see that all punctuation marks have been removed, and all letters have been converted to lower case.

Internationalization

Although Python strings can contain Unicode characters, and Unicode has encodings for every modern language (and several ancient languages), many subtle issues arise when writing programs that can be used around the world. The punctuation symbols defined in Python's string library are a good example:

```
>>> string.punctuation
'!"#$%&\'()*+,-./:;<=>?@[\\]^_`{|}~'
```

This string isn't really a comprehensive list of punctuation marks. It's basically just the set of characters found on an American keyboard. It's missing marks used in other languages (symbols like « and ¿) as well as symbols from other English-speaking countries (€ and £).

There are other Python libraries that will query the host operating system to find out what language the user prefers and local conventions for printing currency, dates, and so on. An actual spam filtering program would take advantage of these libraries instead of simply assuming every message is in American English.

T69. Edit the main function in wf.py so it looks like the one in Figure 6.9:
- Set the function name to wf.
- Instead of initializing three counters to 0, create a new empty dictionary named count.
- Create a new inner loop that iterates over all words returned by tokenize and increments the count.

T70. Load your new function into an interactive session and test it on the quote by Cicero:
```
>>> wf(path_to_data('quote1.txt'))
{'tullius': 1, 'self': 1, '43': 1, ... 'confidence': 2, ... }
```
Does the word "*confidence*" appear as many times as you expected?

T71. William Blake's poem named *The Tyger* is in a file named tyger.txt. Test the wf function on the complete poem, saving the result in a variable named count:
```
>>> count = wf(path_to_data('tyger.txt'))
```

T72. Find out how often "Tyger" appears in this file:
```
>>> count['tyger']
4
```
Note that when we look up the count for a word, we have to write the word in lower case because that's how the counts are stored.

Open tyger.txt with your IDE (or find a copy of the poem on the Internet). Print the counts for some other words. Is the function returning the counts you expect?

♦ Do you think it's important that we call strip before calling lower when creating tokens? In other words, instead of writing
```
>>> s.strip(punctuation).lower()
```
could the statement be
```
>>> s.lower().strip(punctuation)
```
Explain why we could (or could not) write the method calls in either order, then test your hypothesis on several different test strings.

6.6 Spamicity

The word frequency algorithm from the previous section is a key component of the Bayesian approach to identifying and filtering spam. During the training phrase, when users tell the system which messages are good messages and which are bad, the system uses the word frequency function to keep track of how often words appear in each type of message.

In order to put this data to work to predict whether new messages are spam the word counts are converted into probabilities. A common approach for converting word counts into probabilities is to divide the number of messages that contain a word by the total number of messages of each type. For example, suppose we are training the system and we have shown it 1000 spam messages. If the word "*secret*" was found in 252 of them the spam probability of "*secret*" is $252/1000 = 0.252$. If we have shown the system 600 non-spam messages, and the same word was seen in 31 of those messages, the non-spam probability is $31/600 = 0.052$.

These two probabilities are a special type of probability known as a **conditional probability**. In mathematical notation, the probability a word w appears in a spam message is written $P(w \,|\, \text{spam})$, which stands for "the probability of seeing word w given a spam message." The item to the left of the vertical bar is an event we want to predict and the item to the right of the bar is something we already know. In this example the probability tells us how likely it is that we will find the word w in a message we know is spam. The other probability, of finding a word in a non-spam message, is written $P(w \,|\, \text{good})$.

After we have looked at a sufficient number of messages we will have reliable estimates for the two probabilities. Our goal is to use the two conditional probabilities to predict whether or not any new incoming message is spam. In order to do this, we need to switch the order of the items in the probability equations. Instead of $P(w \,|\, \text{spam})$ we want to know $P(\text{spam} \,|\, w)$, that is, the probability a message is spam given the fact we see a given word w somewhere in the message.

This inverted probability $P(\text{spam} \,|\, w)$, which is sometimes referred to as the **spamicity** of the word w, can be derived using Bayes' Theorem (see the sidebar on page 173 for details). Spamicity is simply a ratio defined in terms of the two probabilities from the training set:

$$\text{spamicity}(w) = P(\text{spam} \,|\, w) = \frac{P(w \,|\, \text{spam})}{P(w \,|\, \text{spam}) + P(w \,|\, \text{good})}$$

Note that since spamicity is itself a probability it is also a number between 0 and 1, where the higher the value the more likely it is that a message containing this word is spam.

For our project we are going to use training data from a website that makes spam data available for people who are interested in researching and testing spam filtering algorithms. This data, along with several functions we will use in our interactive experiments, is part of a module named SpamLab, so at the start of each session we need to import the module:.

```
>>> from PythonLabs.SpamLab import *
```

Details about the public data set and how it was used to develop word probabilities can be found in the SpamLab section of the Lab Manual.

A function named `load_probabilities` will read the data from one of the training sets and return it in a dictionary object that associates a word with its probability. The two files

◆ **Using Bayesian Inference to Filter Spam**

If you have studied probability theory you are familiar with the notation $P(X)$ to mean "the probability of event X." For example, $P(\text{ace})$ might refer to the probability of drawing an ace from a deck of cards. If we have a standard deck with 52 cards, and 4 of them are aces, it's easy to calculate $P(\text{ace}) = 4/52 = 1/13$.

Bayesian inference is named for Thomas Bayes, an 18th century English mathematician and clergyman, whose work was based on the idea of **conditional probability**. Suppose we have two decks of cards, one a standard deck and the other a deck used to play pinochle. A pinochle deck has 48 cards, and 8 of them are aces. So now the probability of drawing an ace depends on which deck we select from. A conditional probability is written $P(X \mid Y)$, which means "the probability of X given Y." In the card example, we would write $P(\text{ace} \mid \text{standard})$ to mean "the probability of drawing an ace given we have a standard deck" and $P(\text{ace} \mid \text{pinochle})$ to mean "the probability of drawing an ace given a pinochle deck."

For spam filtering we are concerned with the probability of words. When we train a program we give it messages that we have decided are spam. By counting the number of times a word w appears in these messages it is possible to compute a value for $P(w \mid \text{spam})$, that is, the probability of seeing word w given a spam message. Similarly, using a set of good (non-spam) messages, we can compute $P(w \mid \text{good})$, the probability of seeing word w given a good message.

Now as a new message arrives, we want to "turn things around" and compute $P(\text{spam} \mid w)$, that is, the probability the message is spam given that we see some word w in the message. We can do this using an equation known as Bayes' rule:

$$P(\text{spam} \mid w) = \frac{P(w \mid \text{spam}) \cdot P(\text{spam})}{P(w)}$$

The three terms on the right side of this equation are

$P(w \mid \text{spam})$ The probability of seeing word w in a spam message, which we have already computed using the training set.

$P(\text{spam})$ The "prior probability" a message is spam. This is our estimate of how likely it is that the message is spam before we look at it. If we suspect that 80% of our incoming mail is spam we can use 0.8 as the value of $P(\text{spam})$.

$P(w)$ The overall probability of seeing word w, no matter which type of message it occurs in. We can compute this using the number of times we see a word in either training set.

◆ **Spamicity**

Many applications that use Bayesian filtering do not try to anticipate what proportion of incoming messages might be spam. Instead they assume each type of message is equally probable, which leads to a simple formula known as the "spamicity" of a word:

$$\text{spamicity}(w) = \frac{P(w \mid \text{spam})}{P(w \mid \text{spam}) + P(w \mid \text{good})}$$

Spamicity is also a probability. It is closer to 1 when a word appears in more spam messages than good messages and is closer to 0 when a word is found in more good messages than spam.

are named good.txt and bad.txt. To create the two dictionaries we just find the paths to the files and pass them to load_probabilities:

```
>>> pgood = load_probabilities(path_to_data('good.txt'))
>>> pbad = load_probabilities(path_to_data('bad.txt'))
```

Now if we want to know $P(w \mid \text{spam})$ for a word w we just have to look up the word in the dictionary named pbad. This example shows how to determine the spam probability of "*money*":

```
>>> pbad['money']
0.127
```

In terms of percentages, a probability of 0.127 is equivalent to 12.7%, which means this word appeared in about one out of every eight spam messages used to train the system.

Similarly, we can look up the same word in the dictionary named pgood to find $P(w \mid \text{good})$, the probability of the word appearing in a non-spam message:

```
>>> pgood['money']
0.0164
```

So "*money*" was seen in only 1.64% of the non-spam training messages.

An important detail we haven't considered yet is what to do if a word in an incoming message was never seen during training. This is clearly something we have to anticipate. New words are continually being added to the English language, and spammers like to try to disguise words, using tricks like writing "*mOney*" instead of "*money.*"

To address this problem, the spamicity function (Figure 6.10) first needs to make sure the word is defined in both dictionaries. As we saw earlier, in Section 6.4, the in operator will tell us if a word has been added to a dictionary. The if statement on line 6 tests to see if the word is in both dictionaries, and if so the spamicity equation is used to compute the value returned by the function. If the word is missing from one (or both) of the dictionaries the special object None is returned as a signal that spamicity is not defined for this word.

After we load the function definition into an interactive session we can test it:

```
>>> spamicity('money', pbad, pgood)
0.8856345885634589
```

This value is pretty close to 1.0, which is not surprising. We saw earlier that the spam probability for "*money*" is much higher than the non-spam probability. One way to interpret this result is that it predicts almost nine out of every ten incoming messages that have the word "*money*" will be spam. But note that this prediction is only for the person or group of people who provided their e-mail to train the system; a different set of training messages will no doubt lead to a different prediction.

An example of a word that has a low spamicity is "*object*":

```
>>> pbad['object']
0.0049
>>> pgood['object']
0.0472
>>> spamicity('object', pbad, pgood)
0.09404990403071017
```

```
1   def spamicity(w, pbad, pgood):
2       """
3       Compute the probability a messaage is spam when it contains a word w. The
4       dictionaries pbad and pgood hold p(w|spam) and p(w|good), respectively.
5       """
6       if w in pbad and w in pgood:
7           return pbad[w] / ( pbad[w] + pgood[w] )
8       else:
9           return None
```

Download: spam/spamicity.py

Figure 6.10: *The "spamicity" of a word w is defined in terms of $P(w \mid \text{spam})$, the probability the word appears in a spam message, and $P(w \mid \text{good})$, the probability of the word appearing in a good (non-spam) message.*

This training set was based on messages sent to researchers who were involved with several open source software projects, so it's not surprising that terminology from computer programming will have a low spamicity.

Here is a third example, this time with a word that was never seen in any of the good messages, so it is not in the pgood dictionary:

```
>>> spamicity('paycheck', pbad, pgood)
```

It may look like Python didn't do anything for this call to spamicity, but as we saw in Chapter 4, that is how Python normally handles a function that returns None.

Tutorial Project

The projects in this section use functions defined in the SpamLab module. Make sure you import the module at the start of an interactive session:

```
>>> from PythonLabs.SpamLab import *
```

T73. Call load_probabilities to create the two dictionaries:
```
>>> pbad = load_probabilities(path_to_data('bad.txt'))
>>> pgood = load_probabilities(path_to_data('good.txt'))
```

T74. Try looking up a few words you might expect to find in spam messages:
```
>>> pbad['diet']
0.0024
>>> pgood['diet']
0.0005
>>> pbad['refinance']
0.0095
>>> pgood['refinance']
0.0002
```
Even though the individual probabilities are not that high, notice how the spam probability is higher in each case.

T75. Repeat the experiment, but this time with technical words related to computer programming:

```
>>> pbad['variable']
0.0011
>>> pgood['variable']
0.0563
>>> pgood['iterate']
0.0024
>>> pbad['iterate']
```

Notice how the non-spam probabilities are higher.

Use your IDE to create a new file named spamicity.py and enter the definition from Figure 6.10. Load your definition into the interactive session so you can test it.

T76. Call your spamicity function using the same words you looked up earlier:

```
>>> spamicity('diet', pbad, pgood)
0.8275862068965517
>>> spamicity('refinance', pbad, pgood)
0.979381443298969
>>> spamicity('variable', pbad, pgood)
0.019163763066202093
>>> spamicity('iterate', pbad, pgood)
0.04000000000000001
```

Use a calculator to verify these numbers are correct given the word probabilities you found earlier.

T77. Are the spamicity values generally what you expected to see? Are the values higher for words you expect to find in spam?

T78. Call the spamicity function with a few strings that might not be in one or both of the dictionaries.

```
>>> spamicity('hobbit', pbad, pgood)
>>> spamicity('klingon', pbad, pgood)
```

If a word is not in a dictionary the spamicity function returns None.

T79. Try some experiments with words from good and bad messages in your own mailbox. Keep in mind, however, that this training set was put together in 2008, and words common today might not have been around or were very rare.

6.7 An Algorithm for Identifying Junk Mail

We're now ready to tackle the final step in our spam filtering project. Previous sections laid the foundations, showing how to read a file and split it into individual words, and we also saw how to compute the "spamicity" of a word, which predicts how likely a message is spam when that word occurs somewhere in the text. In this section we will put the pieces together to develop a function that will read a file and compute the probability it contains a spam e-mail message.

The spamicity function tells us something about a single word, but classifying an entire message as spam just because it has one highly spammy word is not going to be very reliable. A typical message will have dozens of words, some that will predict the message is spam

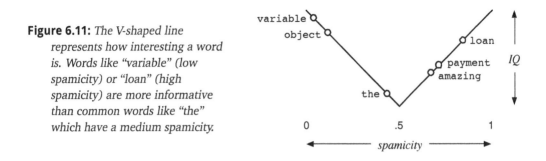

Figure 6.11: *The V-shaped line represents how interesting a word is. Words like "variable" (low spamicity) or "loan" (high spamicity) are more informative than common words like "the" which have a medium spamicity.*

and others pointing in the opposite direction. What we need is a method for combining probabilities of individual words into a single overall probability for the entire message.

One approach, introduced by computer scientist Paul Graham,[1] is to consider only the most "interesting" words. The idea is to use words that have either very high or very low spamicity, as those are the ones that give us the most information about the message. A word with a spamicity of 0.5 is not very informative because it appears equally often in spam and good messages. Words farther away from 0.5, either closer to 0 or closer to 1, will tell us more about the message.

If we know the spamicity of a word, we can define the "interestingness quotient" (IQ) of the word using this formula:

$$IQ = |\,0.5 - s\,|$$

The vertical bars signify the absolute value of the expression. Because the spamicity s has a value between 0 and 1, IQ will range from 0 to 0.5. Boring words, with a spamicity around 0.5, will have an IQ close to 0. But words with a very high spamicity near 1.0, or a very low spamicity close to 0, will lead to higher IQ values (Figure 6.11).

In order to find the interesting words in a message we can store the words in a special type of list known as a **priority queue**. Structurally, a priority queue is just like a list: it's a linear container with references to other objects. What distinguishes a priority queue from a regular list is that the objects in the priority queue are always sorted. Each time we add a new item to a priority queue, the item is automatically saved in a location that preserves the order. In our spam filtering program we will store words according to how interesting they are, so that words with a high IQ value will always be toward the front of the list.

The SpamLab module has a class named WordQueue that implements this behavior. The size of the queue is specified when the WordQueue object is created. For example, if we want to keep track of the ten most interesting words in a message we should create a WordQueue with room for ten words:

```
>>> pq = WordQueue(10)
```

Like many other objects defined in PythonLabs we can display the queue in a graphics window and it will automatically be updated as words are added. To display the new WordQueue object, pass it to a function named view_queue:

```
>>> view_queue(pq)
```

---

[1]http://www.paulgrahm.com/spam.html

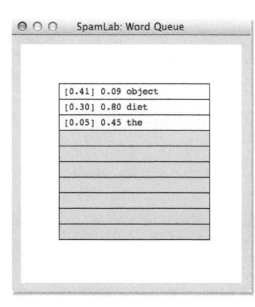

Figure 6.12: *A priority queue is a list that is always sorted. In a WordQueue object, the words are sorted so the most interesting words are at the front of the list. In this example, the word "object" at the front of the queue (the line at the top of the display) has a spamicity of 0.094, which gives it a priority of* $|0.5 - 0.094| = 0.406$.

After a queue has been created, words are added by a method named `insert`. When we call `insert` we need to pass it the word to add to the queue and the spamicity of the word. The method will figure out how interesting the word is and store the word in a location that keeps all the words sorted according to their IQ.

As an example of how the priority queue works, let's first add the word *"the"*, which is very common and occurs almost equally often in each dictionary:

```
>>> spamicity('the', pbad, pgood)
0.4521124184600378
>>> pq.insert('the', 0.45)
```

Because we told the method the spamicity of this word is 0.45, it should be saved with a priority of $w = |0.5 - 0.45| = 0.05$.

Next let's add a second word to the queue. According to Exercise T76 at the end of the previous section the spamicity of *"diet"* is 0.828:

```
>>> pq.insert('diet', 0.8)
```

Because *"diet"* is more interesting than *"the"* it should be placed at the front of the list. If the queue is being displayed on the canvas, we should see *"diet"* in the top slot and see *"the"* moved down one place. The new word is at the front of the list because its IQ is $|0.5 - 0.828| = 0.328$, which is higher than 0.05.

Now let's add a second interesting word:

```
>>> pq.insert('object', 0.094)
```

The IQ for this word is $|0.5 - 0.094| = 0.406$, which means it is more interesting than "diet", so it should be in the top slot. At this point the canvas should look like the screen snapshot in Figure 6.12.

```
1    from PythonLabs.SpamLab import *
2    from wf import tokenize
3    from spamicity import spamicity
4
5    def pspam(fn):
6        "Compute the probability the message in file fn is spam"
7        queue = WordQueue(15)
8        pgood = load_probabilities(path_to_data('good.txt'))
9        pbad = load_probabilities(path_to_data('bad.txt'))
10       for line in open(fn):
11           for word in tokenize(line):
12               p = spamicity(word, pbad, pgood)
13               if p != None:
14                   queue.insert(word, p)
15       return combined_probability(queue)
```

Download: spam/pspam.py

Figure 6.13: *A function for identifying spam email messages. The word probabilities and the function that loads them are imported from SpamLab. The two helper functions (*tokenize* and *spamicity*) were defined in Figures 6.9 and 6.10, respectively.*

After we have processed all the words in a file, we will have a list of the ten most interesting words. The final step is to compute a single probability based on the spamicity of these ten words. SpamLab has a function that will do the calculation for us. We simply have to pass the priority queue to this function, and it will return the probability the message is spam according to the words in the queue. An optional set of exercises at the end of the tutorial will give you a chance to see how this overall value is calculated.

The final Python code for our function, named pspam, is shown in Figure 6.13. By now the overall structure of the function should be familiar: we iterate over every line in a file and break each line into words. The statements in the body of the inner loop get the spamicity of a word and, if spamicity is defined, save the word in a priority queue. After all the lines have been processed, the function calls combined_probability to compute the overall spam probability using all the words in the queue.

Lines 2 and 3 in Figure 6.13 have a type of import statement we haven't seen before. Every time we've used import in previous projects it was to tell Python to include a function defined in one of its libraries. These libraries are either standard libraries like math and string that are always installed along with Python, or, as shown on the first line, are modules that we have installed ourselves.

These new import statements show that it is possible to include functions from our own programs. Line 2 says pspam is going to use the tokenize function we wrote for the wf (word frequency) project, and line 3 tells Python to import the spamicity function we wrote in the previous section. Be sure to read the sidebar on Importing Functions on page 180 to make sure you understand important details about how Python finds these other programs so it can import your functions.

To experiment with the algorithm we can use e-mail files distributed as part of the Python-Labs data directory. This example shows how the function analyzes the message in the file named `msg1.txt` (the complete text is shown in Figure 6.14):

```
>>> pspam(path_to_data('msg1.txt'), 0.01)
0.9293048326577117
```

The message is pretty suspicious, and the `pspam` function says there is a 92% chance it is spam.

Here is another example of a call to `pspam`:

```
>>> pspam(path_to_data('msg4.txt'))
3.758445064217253e-15
```

At first it might look like the result is a mistake, because probabilities are supposed to be numbers between 0 and 1, and this number appears to be around 3.75. But notice how the value printed by Python ends with $e-15$. This is an example of how Python prints numbers in scientific notation. The output is Python's way of writing 3.758×10^{-15}, which is a very small number. In other words, there is an extremely low probability that this file contains a spam message.

The first few exercises in this section will illustrate how priority queues work and how probabilities from the queue are combined to form the final result. We will use a version of `pspam` implemented in SpamLab; this version draws the queue on the PythonLabs canvas and updates the queue as it processes a message. The function will initialize the canvas with the drawing of an empty priority queue. Then it will display the words from the text of the message and show how they are handled by the algorithm. At any point in the algorithm the most interesting words will be displayed in order, with the words having the highest weight at the top of the list.

Importing Functions from Other Files

When Python executes an import statement, for example,

```
from wf import tokenize
```

it automatically attaches `.py` to the end of the module name and looks for a file with that name. In this example, Python will look in `wf.py` to find a definition of a function named `tokenize`.

Importing definitions from other files is easy as long as all the files — the file doing the importing and the files that contain the definitions — are in the same directory.

It is possible to import from modules located in other folders but doing so introduces complications. For the projects in this book we recommend keeping all the source files for a project in a single directory.

```
Hurting for funds right now?

It doesn't have to be that way. Here is 1,500 to ease
your pain:
http://bulk.hideorganic.com/171026390236329103248372180

Transfer immediately to the account of your choice:
http://bulk.hideorganic.com/171026390236438808248372180

Take your time to pay off this amazing loan. Small
payment due in late September
or early October (and not all in one payment!).
```

Figure 6.14: *These two messages are in files that are part of the PythonLabs data directory. The first message* (msg1.txt) *includes several words that suggest it might be a financial scam. The lower box is the first part of a message* (msg4.txt) *that should not be identified as spam.*

```
Hi John,

I meant to ask you if you tried the revised cat
command.  Were you able to do what you needed?

Regarding your lab meetings... sure, I could come and
give a brief description and answer any questions your
group members might have.  My assistant, Erik, has just
put up more information from Chris' slides onto the
wiki that might be helpful.
```

Tutorial Project

If you start a new shell session, remember to import SpamLab and load the word probabilities:

```
>>> from PythonLabs.SpamLab import *
>>> pbad = load_probabilities(path_to_data('bad.txt'))
>>> pgood = load_probabilities(path_to_data('good.txt'))
```

You will also need the tokenize and spamicity functions. You can either load your own versions or use the definitions included in SpamLab.

T80. Create a new WordQueue object:

```
>>> pq = WordQueue(10)
```

T81. If you tell Python to print the queue it will look like an empty list:

```
>>> pq
[]
```

T82. But if you ask to see what type of object this is you will see it's a WordQueue:

```
>>> type(pq)
<class 'PythonLabs.SpamLab.WordQueue'>
```

T83. Initialize the canvas. You should see an empty queue with room for ten words:

```
>>> view_queue(pq)
```

T84. Get the spamicity values for a few words:

```
>>> spamicity('you', pbad, pgood)
0.4862119013062408
>>> spamicity('doctor', pbad, pgood)
0.8904761904761905
>>> spamicity('spam', pbad, pgood)
0.1652892561983471
```

T85. Add these words to the queue and watch how the display is updated when each word is placed in the queue:

```
>>> pq.insert('you', 0.48)
>>> pq.insert('doctor', 0.89)
>>> pq.insert('spam', 0.16)
```

When you look at a queue entry on the screen, you will see a word next to its spamicity value. The number in brackets at the front of the line is the "interestingness quotient." Do these IQ values look correct? Do you see how the words are sorted so the highest IQs are at the top?

T86. Pass the word queue to the `combined_probability` function:

```
>>> combined_probability(pq)
0.5872164948453609
```

This result is as expected: the queue has three words, one with high spamicity, one with low spamicity, and one in the middle, so the overall probability should be close to 50%.

T87. Reinitialize the queue and the canvas:

```
>>> pq = WordQueue(10); view_queue(pq)
```

T88. Define a variable s to be a short sentence:

```
>>> s = 'I am a virus. Please delete all your data.'
```

T89. This for loop will insert all the words in the sentence into the queue:

```
>>> for w in tokenize(s):
...     pq.insert(w, spamicity(w, pbad, pgood))
...
```

Don't forget to type one or more spaces at the front of the second line.

T90. Get the combined probability of the words:

```
>>> combined_probability(pq)
0.000682879941598876
```

Does this value seem right to you? Are there more words with low spamicity than high spamicity?

T91. Call the pspam function to have it evaluate the first message shown in Figure 6.14:

```
>>> pspam(path_to_data('msg1.txt'))
0.9293048326577117
```

Look at the final queue on the canvas. Do most of the words have high spamicity values?

T92. Repeat the call, but this time evaluate the message in the file named `msg4.txt`.[2] If you want to speed up the display pass a second argument to pspam (the number is the amount of time, in seconds, to pause between words):

```
>>> pspam(path_to_data('msg4.txt'), 0.01)
3.758445064217253e-15
```

Look at the queue again. Are these mostly low spamicity words?

If you want to try some more experiments, `msg5.txt` is an example of a spam message that will slip past the system, and `msg6.txt` is an example of a valid message that is erroneously identified as spam. Both of these examples are short messages. You may have noticed on your own system the spam messages that get past your spam filter tend to be very short, and as you can see from these examples short messages are often harder to characterize.

---

[2]If you want to see the complete text, you can use `path_to_data` to see where the file is stored on your computer so you can open the file in your IDE.

- If you want to experiment with your own version of pspam either create a new file named pspam.py with your IDE and type in the program shown in Figure 6.13 or download the file from the book website.

- Load your file into a new shell session. You should be able to call your function using the same commands shown above, and your results should match the output from the SpamLab version (but your function will not display the queue on the canvas).

6.8 Summary

One way to fight spam is to sit down and define a set of rules for what characterizes junk e-mail. If you find yourself getting a lot of mail telling you to "act now to take advantage of this amazing offer" you could tell your e-mail client to redirect any incoming mail containing that phrase to a special mailbox set up for trash. Similarly, if you see a pattern in the names of the mail servers sending you spam, you could redirect mail from these servers to the same folder.

The Bayesian approach we took for the project in this chapter is based on a different idea. Instead of explicitly defining rules, we gradually train a mail filtering program by showing it examples of good message and spam messages. After seeing a sufficient number of each type of message, this sort of "machine learning" program will be able to reliably distinguish good mail from bad. The key idea is to take advantage of the information contained in the data, in this case in all the incoming mail we receive.

Spam filtering using Bayesian inference is just one example of a growing number of applications that rely on large amounts of data. The phrase "big data" is often used to describe the use of huge amounts of information in the sciences. Biology labs in universities use gene sequencing technology capable of generating over 10^9 bytes of DNA sequence data in a single week. Oceanographers, geophysicists, and other earth scientists are taking detailed measurements that also generate hundreds of terabytes of data, and the high-energy colliders built by physicists generate massive data sets that are painstakingly analyzed for traces of elusive particles.

Data-driven analysis is playing in increased role in other fields, including retail and finance. Algorithms very much like the ones illustrated in this chapter were used to predict results of the 2012 presidential elections in the United States, using large amounts of polling date from previous elections, and were very accurate in predicting the eventual outcome in each state.

The spam filtering project also served to introduce two very important features in Python. We saw how to access data stored in a file, and we learned about a new data structure for organizing information in an application.

File processing in Python follows the same basic procedure we use when working with file folders in an office: after we locate the file, we open it, read the contents, and close it. Locating a file in Python is a matter of specifying a path to the file, where a path is a string that contains a directory (folder) names and the name of the file itself. A simple "idiom" for reading data from a file is to use a for statement to iterate over the file. The for statement automatically opens the file and fetches lines one by one, executing the statements in the body of the loop once for each line, and then automatically closes the file after the last line.

Concepts and Terminology Introduced in This Chapter

spam	Unsolicited junk e-mail, typically mass produced by a "spambot"
conditional probability	A probability that depends on a previous event or prior information
Bayesian inference	A mathematical process of decision making involving conditional probabilities based on previously collected data
tokenizer	A function that splits a string into separate words, ignoring spaces and punctuation
text file	A file containing plain text without font size, color, or other formatting information
path	A string representing a file name, possibly including names of folders the file is enclosed in
open	A Python built-in function used to access information in a file
dictionary	A collection of key-value pairs where items are accessed by name instead of location
priority queue	A data structure, similar to a list, except objects are always in order

The new type of object introduced in this chapter is a called a "dictionary" in Python. A dictionary is similar to a list, in that it is a collection of references to other objects. The difference is that items in a dictionary are accessed by their name, not their location.

Dictionaries were used in the spam filtering project to hold probabilities. The training data included with the PythonLabs library was loaded into dictionary objects named pbad and pgood. To find $P(w \,|\, \text{spam})$ (the probability that word w would be found in a spam message) we just have to look it up in the dictionary by asking Python to find pspam[w].

Python's dictionaries will prove very useful throughout the rest of this book. The programming projects at the end of the chapter provide a chance to use dictionaries in several other small projects.

Exercises

Figure 6.15 shows a few mathematical and financial symbols used in the exercises in this section.

6.1. Suppose a variable s has been defined with this assignment statement:
```
>>> s = "To be, or not to be, that is the Question:"
```
What will Python print for each of the following statements?
```
>>> print(s)
>>> print(len(s))
>>> print(s.split())
>>> print(tokenize(s))
>>> print(s.strip() + '...')
```

6.2. Repeat the previous question using these definitions of s:
```
>>> s = "April showers\n  bring May flowers."
>>> s = "\u2200x: human(x) \u2192 mortal(x)."
```

6.3. Write assignment statements that will create Python strings for the following sentences:
- The current exchange rate is £1 = ¥167.
- We paid €12 each!
- x ← y ÷ z.

6.4. Suppose a dictionary object is defined with the following statement:
```
>>> d = {'M':1000, 'D':500, 'C':100, 'L':50, 'X':10, 'V':5, 'I':1}
```
What will Python print as the value of the following expressions?
```
>>> len(d)
>>> d['X']
>>> d['Z']
>>> 'Q' in d
>>> 5 in d
>>> list(d.keys())
```

6.5. Write assignment statements that create dictionaries for the following sets of data:
- a) The months of the year, using numbers from 1 to 12 as keys and month names as values.
- b) The colors of the rainbow, using the letters in the acronym ROY G. BIV as keys and the corresponding colors as values.
- c) The first 7 elements in the periodic table, where keys are chemical symbols ("H", "He", "Li", etc.) and values are the names of the elements.

Figure 6.15: *The Unicode IDs in this table are 4-digit hexadecimal (base 16) numbers. To include a symbol in a string use the \u escape sequence followed by the 4-digit code number.*

Symbol	Code	Symbol	Code
£	00A3	←	2190
¥	00A5	→	2192
€	20AC	∀	2200
÷	00F7	∃	2203
∞	221E	Σ	2211

6.6. The first `for` loops we saw used a `range` expression to specify values for an iterator variable:

```
for i in range(10)
```

Later we learned `range` can also be used to create a list of values:

```
list(range(10))
```

This chapter showed how `open` can be used in a `for` statement to iterate over the lines in a file:

```
for line in open(fn)
```

Do you suppose we can use `open` as an argument to `list` to make a list of lines from a file? What do you think Python will do if you ask it to evaluate an expression like this?

```
list(open(fn))
```

Test your hypothesis in an interactive session.

6.7. Assume the dictionaries based on training data have been loaded into an interactive session:

```
>>> pbad = load_probabilities(path_to_data('bad.txt'))
>>> pgood = load_probabilities(path_to_data('good.txt'))
```

Write Python statements that carry out the following operations:

a) Look up $P(\text{doctor} \mid \text{spam})$ (the probability of seeing "doctor" in a spam message)

b) Look up $P(\text{happy} \mid \text{good})$ (the probability of "happy" in a non-spam message)

c) Compute the spamicity of "depressed"

6.8. Here is the output of a `for` loop run in an interactive session after loading the two probability files:

```
>>> for x in ['get', 'your', 'free', 'money', 'here']:
...     print(pbad[x], pgood[x])
...
0.1705 0.2447
0.5655 0.2983
0.052 0.0691
0.127 0.0164
0.2107 0.2023
```

For each word in the list

- Compute the spamicity of the word.
- Compute the IQ ("interestingness quotient") of the word.

6.9. Given the probabilities printed in the previous problem, do you think a message containing the sentence "get your free money here" will be flagged as spam? Explain why or why not.

6.10. The method for converting word counts into probabilities is explained at the beginning of Section 6.6. Given this method, does it matter how many times a word appears in a message? For example, if a spam message has the phrase "money, money, money!" does it increase the estimate of $P(\text{money} \mid \text{spam})$? Explain why or why not.

6.11. ♦ [Essay question] The algorithm that decides whether a message is spam by combining the probabilities of the fifteen most interesting words is called a "naive Bayes" classifier because it assumes the probabilities are all independent of one another. How would the spam filter be more effective if it could use combinations of words? Should it take into account where a word is seen, for example, if the word is in the subject line or in the body of a message?

6.12. ♦ The equation for $P(\text{spam} \mid w)$, the probability that a message is spam given it contains a word w, is shown in the sidebar on page 173. Here is an expanded form of $P(w)$, the term in the denominator of that equation:

$$P(w) = P(w \mid \text{spam}) \cdot P(\text{spam}) + P(w \mid \text{good}) \cdot P(\text{good})$$

Prove that when $P(\text{spam}) = 0.5$ the formula for $P(\text{spam} \mid w)$ is equivalent to the spamicity equation shown at the bottom of the sidebar.

Programming Projects

6.13. Write a function named `listing` that will read a text file and print it with a line number at the front of each line. For example, here is a listing of the file that contains the filesize function:

```
>>> listing('filesize.py')
1 : def filesize(fn):
2 :     "Count characters in file fn."
3 :     nchars = 0
4 :     for line in open(fn):
5 :         nchars += len(line)
6 :     return nchars
```

6.14. Write a function that creates an acronym from the first letter of each long word in a list, where a long word is any word with more than three letters. Some examples:

```
>>> acronym('operating system')
'OS'
>>> acronym('association for computing machinery')
'ACM'
```

6.15. Modify your `acronym` function so it takes an optional argument that tells it to include the first letter of all words, but short words should not be capitalized:

```
>>> acronym('department of motor vehicles')
'DMV'
>>> acronym('department of justice', True)
'DoJ'
```

6.16. Add a dictionary of special cases to the `plural` function you wrote in Chapter 2. For example,

```
odd_words = { 'mouse' : 'mice', 'goose' : 'geese' }
```

When your function is called it should check to see if the word is in this dictionary, and if so return the associated plural form, otherwise apply one of the rules you wrote earlier.

6.17. Modify the gpa function you wrote in Chapter 2 so it looks up grade values in a dictionary instead of using an `if` statement. Define the dictionary named `points` shown in Problem T47 (page 165), and then for each item x in a list of grades, look up `points[x]` to find the point value for that grade.

6.18. Add code to your gpa function that adds additional grades to the dictionary so it has keys for A+, A-, B+, and so on. Can you solve this problem using a loop that iterates over the grades instead of simply writing a bunch of assignment statements?

6.19. Write a function named `blackjack_points` that will create a dictionary that associates a card symbol with the number of points that card is worth in the game of blackjack. Use one-letter strings for card names: A for ace, K for king, Q for queen, J for jack, T for 10, and then digits 2 through 9 for the remaining cards. Aces are worth 11 points, face cards (10 through king) are worth 10, and the other cards are worth their natural value.

```
>>> pts = blackjack_points()
>>> pts['J']
10
>>> pts['3']
3
```

Instead of writing a list of thirteen key-value pairs see if you can use a loop to "automate" the process as much as possible. Hint: If s is a one-character string containing a digit, `int(s)` will turn the digit into an integer.

6.20. ◆ Write a function named `blackjack_total` that will return the total number of points in a hand, where a hand is represented by a string of one-letter card symbols defined in the previous exercise. Here are some examples:

```
>>> blackjack_total('AK')
21
>>> blackjack_total('K4Q')
24
>>> blackjack_total('AA7')
19
```

Here is an outline of the algorithm to use: iterate over the hand and compute the sum of points in all cards, using the `blackjack_points` dictionary to look up the value of each card. In the same loop count the number of aces. Then, because aces can be either 11 points or 1 point, use a second loop to convert some aces to 1: while the sum is greater than 21 and the ace counter is greater than 0 subtract 10 from the total and 1 from the ace counter. When the second loop terminates the sum will be the point total for the hand.

6.21. Write a function named `encrypt` that will be passed two arguments, a string and a dictionary that represents a cipher that specifies how letter substitutions should be performed. The return value should be a new string where every letter from the input string has been translated according to the cipher. Here is an example with a small cipher for only three letters:

```
>>> cipher = {'a':'x', 'b':'y', 'c':'z'}
```

The cipher replaces "a" with "x", "b" with "y", and "c" with "z". This call to `encrypt` shows an example of how the cipher is applied to an input string:

```
>>> encrypt("abba", cipher)
'xyyx'
```

Define the cipher using the statement shown above and test your function on several other strings that contain only the letters "a", "b", and "c".

6.22. Test your `encrypt` function using a dictionary named `caesar_cipher` defined in SpamLab. In the Caesar cipher, each letter is replaced by the letter three places later in the alphabet, that is, "a" becomes "d", "b" turns into "e", and so on. At the end of the alphabet, the cipher "wraps around" so "x" is replaced by "a".

```
>>> from PythonLabs.SpamLab import caesar_cipher
>>> encrypt('brute', caesar_cipher)
'euxwh'
```

Do some more tests with strings that contain only lowercase letters.

6.23. Modify your `encrypt` function so that (1) if the input string contains a character that is not in the cipher, the character is just copied to the output string and (2) case is preserved, that is, if the input has an uppercase letter the corresponding letter in the output is also upper case.

```
>>> encrypt('Et tu, Brute?', caesar_cipher)
'Hw wx, Euxwh?'
```

6.24. Write a function named `crypto_quote` that will read the text in a file, encrypt it with your `encrypt` function, and print the result on the terminal window. The two arguments to your function should be the name of a text file and the cipher to use. Here is an example that encrypts the quote by Cicero (Figure 6.4) using the Caesar cipher:

```
>>> crypto_quote(path_to_data("quote1.txt"), caesar_cipher)
 Li brx kdyh qr frqilghqfh lq vhoi,
   brx duh wzlfh ghihdwhg lq wkh udfh ri olih.
 Zlwk frqilghqfh, brx kdyh zrq hyhq ehiruh brx kdyh vwduwhg.
     -- Pdufxv Wxoolxv Flfhur (106 EF -- 43 EF)
```

6.25. ♦ This statement shows how to use two of Python's built-in functions to define the three-letter cipher in Problem 6.21:

```
>>> cipher = dict(zip("abc","xyz"))
>>> cipher
{'c': 'z', 'b': 'y', 'a': 'x'}
```

Can you write a similar statement that defines the Caesar cipher? Note: ascii_lowercase, a string you can import from Python's string library, contains all 26 lowercase letters of the English alphabet.

6.26. ♦ Write a function named random_cipher that will create a dictionary where each letter of the alphabet is associated with a random letter; for example,

```
>>> random_cipher()
{'j': 'y', 'k': 'l', 'h': 'f', ... 't': 'b', 'u': 'z'}
```

Python's random library has a function named shuffle that will make a random permutation of a list. One way to solve this problem would be to make a list of letters, shuffle it, and use the result in the expression that creates the dictionary. Note that this approach would allow a letter to be associated with itself, but most cryptogram puzzles use ciphers that make sure every letter changes into a different letter. Can you write your function in a way that ensures every letter will be changed when the cipher is used to encrypt a file? Test your program on the quote by Cicero:

```
>>> crypto_quote(path_to_data("quote1.txt"), random_cipher())
Rd wfi tjcy pf xfpdroypxy rp qyud,
...
```

6.27. ♦ Write a function named roman_to_int that will translate a string of roman numerals into an integer value:

```
>>> roman_to_int('MMXIV')
2014
>>> roman_to_int('CMXCIX')
999
```

A straightforward algorithm is as follows. Create a dictionary that associates the roman digits with their numeric value:

```
digits = { 'M' : 1000, 'D' : 500, ... 'I' : 1 }
```

Use a while loop to iterate over the input string using an index variable i initially set to 0. Let s be the name of the input string. On each iteration, if the digit at s[i] has a lower value than the digit at s[i+1], subtract the value of s[i+1] from the value of s[i], add the result to the output value, and increment i by 2 (e.g., if "I" is followed by "X" add 9 to the output value). Otherwise just add the value of s[i] to the output value and increment i by 1.

6.28. Write a function named common that will take two parameters, a dictionary of word probabilities and a cutoff value, and return a list of all words in the dictionary that have a probability above the cutoff. This example shows how to find all words from pbad (the word probabilities recorded from spam messages) that have a probability higher than 0.5:

```
>>> pbad = load_probabilities(path_to_data('bad.txt'))
>>> common(pbad, 0.5)
['and', 'you', 'for', 'of', 'your', 'the', 'to', 'is', 'a']
```

6.29. ◆ If you call the common function from the previous problem, passing it pgood (word probabilities gathered from non-spam messages) and a cutoff of 0.5, you will probably get a list similar to the one shown in the previous problem. That's not too surprising, because we should expect very common words like "the" to appear frequently in each type of message.

Write a function named outliers that will look for words that are common in one list but not the other. The arguments to the function should be two dictionaries and a cutoff value. It should use common as a helper function to build a list of words that are in the first dictionary with a cutoff above the specified value but not in the second dictionary using the same cutoff value. Here is an example, looking for words that are in pbad but not in pgood using a cutoff of $p = 0.1$:

```
>>> pgood = load_probabilities(path_to_data('good.txt'))

>>> outliers(pbad, pgood, 0.1)
['buy', 'right', 'were', 'broker', ... 'money', 'before', 'our', 'profit']
```

Experiment with different cutoff values. Does this function help identify "spammy" words from the training set?

Chapter 7

Now for Something Completely Different

An algorithm for generating random numbers

Most popular games involve some element of chance. Players roll dice, shuffle a deck of cards, spin a wheel, or use some other method for making a random selection in the game. Computer-based versions of these games are played the same way, but instead of using real dice or using a real deck of cards a computer program manages the game. Somewhere inside the application is an algorithm that generates the next roll of the dice or shuffles the deck of cards.

As you might imagine, a computer-based game would not be very popular if the algorithm that generates moves is not realistic. If you're playing a board game like backgammon, you expect rolls of the dice to be similar to rolls of real dice, and you would start to become suspicious if your opponent rolls doubles far more often than you do. Some people like to practice playing poker against a computer to prepare for tournaments. If the program they train with doesn't deal the same kinds of hands that will be seen in the tournament, the software is not going to be very useful.

The word that best describes our expectations for the computer-based games is **random**. When we play a board game, we want the computer to generate a pair of numbers that is just as random as rolling a pair of real dice, and when we play cards, we want the computer to generate an ordering for a deck of cards that is as random as what we would get if we carefully shuffle a real deck.

The natural question, of course, is how to define what we mean by "random." Colloquially, random often means "unusual" or "unexpected." But in games, and in other situations where random values are required, something is random if it is **unpredictable**. To be more precise, what we are looking for is an algorithm that generates a sequence of values in which there is no apparent pattern or relationship between one value and the next. If the algorithm is used to simulate rolls of a six-sided die, each new number should be independent of the previous number, and if it is used to deal cards, each new hand should be unrelated to the previous hand. In other words, to use the phrase in the title of this chapter, each value produced by the algorithm should be "completely different" from the previous value.[1]

The goal for this chapter is to explore techniques for using a computer to generate random values. We will take a closer look at the idea that a sequence of values is random if successive values are independent of one another, and explore various ways of trying to determine whether the values are, in fact, random, or whether there are some unexpected connections between them.

This chapter also provides an opportunity to introduce some important ideas in Python programming. Projects in this chapter will use objects from a class named Card, where each object is a single card from a standard deck of 52 playing cards. In the first sections we will use the Card class defined in PythonLabs and see how to make a deck of cards, shuffle the deck, and so on. Because these objects are so simple, they are a good candidate for introducing basic concepts of object-oriented programming, and in the final two sections we will learn how to use Python's `class` statement to define our own Card class.

7.1 Pseudorandom Numbers

Before you read this section, here is a simple experiment to try. Create a new text file with your IDE. Type 50 numbers between 1 and 6, putting one number on each line. Try to write the numbers as if you were rolling a die; in other words, each number should be unrelated to the one on the previous line.

As you were thinking of numbers to write, were you able to concentrate on each new number, forgetting about what you had written before? Or did you find yourself thinking something like, "I haven't written a 3 in a while, I'd better write one now" or "Hmmm, that's three 6s in a row, I'd better write something else." If you gave in to the temptation to think about previous values you were starting to add a bias to your results. It is very hard for people to generate a truly random sequence of values.

Engineers, statisticians, and other professionals have used random sequences for many years. Before the algorithms described in this chapter were available, people who needed a random sequence would look in a book of random numbers. A well-known reference book, published in 1955 by the RAND Corporation, was *A Million Random Digits*, a 400-page book with 2,500 random digits between 0 and 9 on each page. The authors used what they called an "electronic roulette wheel" to generate random electronic pulses. The electronic circuit was connected to a computer, and a device that measured the pulses converted them to digital form. Another way to generate random signals is to use a white noise generator, an audio device that produces something that sounds like static.

To play a game with a computer we do not need to connect to a roulette wheel, white noise generator, or other physical device that behaves randomly. Instead, applications use an algorithm that produces a different value each time a function is called. For example, the

---

[1]The chapter title is inspired by the Monty Python movie *And Now for Something Completely Different*.

`randint` function from Python's `random` module will return a random integer between two specified values *a* and *b*:

```
>>> from random import randint
>>> randint(1,10)
4
>>> randint(1,10)
9
```

There is an interesting paradox here. According to the definition given in Section 1.3, an algorithm has a well-defined set of inputs, and a precise and unambiguous sequence of steps that leads to the output. If that's the case, how can an algorithm produce a different result each time it is run? Or, in terms of Python programs, how is it possible for a function to return a different value when we call it several times with the same arguments?

The answer to these questions is that the function *appears* to be random, but in fact it follows a predefined set of rules, just like any function. The general idea is to have the function produce a long sequence of numbers, and if we look at a small set of numbers in the middle of the sequence, they will appear random. Even though they are not truly random, many applications, such as computer games, can use these numbers in place of actual random values. Because the numbers are created by an algorithm, and not an external source of random data like the RAND corporation's roulette wheel, we refer to them as **pseudorandom**, and the algorithm that produces the sequence is a **pseudorandom number generator**, or PRNG.

As an introduction to how a PRNG might work, imagine a situation where an event needs to be scheduled at regular intervals throughout a day. Perhaps a nurse in a hospital needs to periodically administer medications to a patient, or a researcher needs to collect data from an experiment, and it is our job to schedule these events. If the events occur every eight hours the schedule is simple and easy to remember. One plan would be to schedule the first three events at midnight, 8 A.M., and 4 P.M. The next event would be at midnight again, and the schedule would repeat. Because the schedule is the same every day it is easy to remember, and we can simply tell people what the schedule is (Figure 7.1).

However, if the events need to occur every seven hours, the schedule is more complicated. If the first event is at midnight, the next three events would be 7 A.M., 2 P.M., and 9 P.M. The first event on the next day will be at 4 A.M., not midnight, and the second day would have a different schedule. Because the days are so different from one another, we would have to write out a complete chart for each day of the week to make sure people can follow the schedule.

It's very straightforward for Python to figure out the schedule using an operator that computes the remainder of a division operation. The operator is called the **mod operator**, and its symbol is a percent sign. To compute the remainder of x divided by y, we write x % y. Here are some examples:

```
>>> 13 % 5
3
>>> 27 % 9
0
```

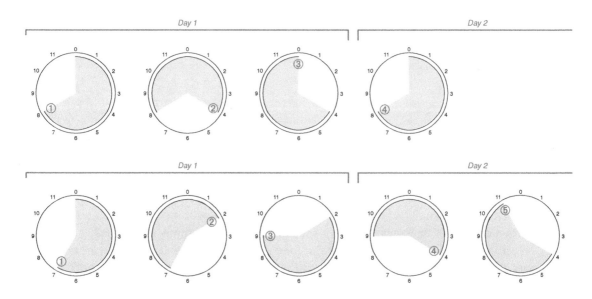

Figure 7.1: *If events occur every eight hours (top row), a schedule is very predictable, as events happen at the same time every day. However, if events are seven hours apart (bottom row), successive days are different, and it is more difficult to keep track of when events are scheduled.*

The name "mod" comes from number theory, which calls this function *modulo*. We use this function every day, even though we might not call it by that name. It is sometimes introduced to school children as "clock arithmetic" because operations are similar to figuring out the time on an analog clock. If it is currently 5 o'clock and we want to know what time it will be eight hours from now, we can get the answer with the mod operator:

```
>>> (5 + 8) % 12
1
```

(assuming, of course, we are using a 12-hour clock).

To have Python create a schedule of events that occur every seven hours, start by making a list named t and initializing it with the time for the first event:

```
>>> t = [0]
```

Let's assume we can tell by context whether "0" is midnight or noon, and whether a "4" means 4 A.M. or 4 P.M. (one of the exercises at the end of the chapter will be to modify the schedule to use a 24-hour clock).

To add the time for the next event, the list needs to be extended with a time that is seven hours later than the one currently at the end of the list. Recall that a negative number used as a list index refers to items at the end of a list, so if we want to know the last number in the list t, we can check the value of t[-1]. The statement that adds the next appointment time to the schedule is

```
>>> t.append( (t[-1] + 7) % 12 )
```

Python will execute this statement by getting the value currently at the end of t, adding 7, and keeping the remainder after dividing by 12. The remainder is passed to the append method, and the new event time is added to the end of the list.

If the above operation is repeated several more times, this is what the list would look like:

```
[0, 7, 2, 9, 4, 11, 6, 1, 8, 3, 10, 5, 0, 7, 2]
```

That's a pretty difficult schedule to remember, but it matches the specification, and we could either put it on a poster or ask people to store it in the calendar on their cell phone.

To get back to the subject of generating random numbers, at first it looks like the list above might be random. If we give the list to people who do not know the numbers were times from a 12-hour clock they will probably have a hard time guessing the list was generated by a simple rule, or what that rule is.

On closer inspection, however, we can see some regularity in the list. For one thing, every hour appears exactly once in the first 12 places in the list (assuming we interpret 0 as 12 o'clock). When we are rolling a six-sided die it would be very rare to see a sequence of rolls in which every number between 1 and 6 showed up before any number appeared again. In fact, it's not at all uncommon to have the same number appear twice in a row. The fact that we used a rule that forces every number to occur once before the first number appears again means the list is not truly random.

If we add a few more items to the list, using the same rule, we will see conclusive evidence that the list is not random. As soon as any number appears a second time in the sequence, the pattern will start to repeat itself. In other words, as soon as the rule attaches a 0 to the end of the list, all the values that followed 0 the first time will occur again, and in exactly the same order. If we want to know what follows 2 at the end of the list shown above, we can either apply the rule again and evaluate $(2 + 7)$ mod 12, or we can find the 2 earlier in the list and look at the number that came after it. Either way, the next number in the sequence will be a 9.

The rule of adding 7 and taking the remainder mod 12 does not create a very useful list of pseudorandom numbers, but it is a good starting point. A more general formula for adding a new value x_{i+1} to the end of a list x is

$$x_{i+1} = (a \times x_i + c) \bmod m$$

where a, c, and m are all constants and x_i is the item at location i in the list. The three parameters that define the sequence are known as the multiplier, increment, and modulus. The "add seven to the current time" rule follows this general pattern: it has a multiplier $a = 1$, an increment of $c = 7$, and modulus $m = 12$.

If the multiplier and increment are defined properly, this rule will place every value from 0 to $m - 1$ in the list before it repeats (Figure 7.2). As we saw previously, with $a = 1$ and $c = 7$, the rule makes a list of all twelve values between 0 and 11. But if we use $c = 8$ (which is what we did originally, when the schedule called for events every eight hours), the list starts repeating much sooner:

```
[0, 8, 4, 0, 8...]
```

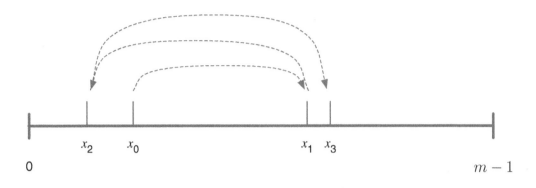

Figure 7.2: *This number line shows the first four items in a pseudorandom sequence defined by the rule $x_{i+1} = (a \times x_i + c)$ mod m. If values of a, c, and m are chosen carefully, every value between 0 and $m - 1$ will appear in the first m numbers in the sequence. If we draw a bar for every value in the sequence, the number line will be completely filled in.*

Mathematicians who study pseudorandom number generators refer to the number of items in the repetitive part of the list as the **period** of the list. The period when $c = 8$ is 3, because the same three numbers keep repeating. The period when $c = 7$ is 12 because all twelve numbers are in the list before it repeats.

If we use a large value of m and carefully selected values of a and c we can make a very long list that contains all values from 0 to $m - 1$. Furthermore, if we look at small portions of the list, it will be very difficult to figure out what rule is used to generate the numbers. In practical terms, we will have a list of random numbers. Even though they were produced by a pseudorandom number generator, and will not be truly random, for many applications they might be "random enough."

The formula shown above was used in some of the earliest pseudorandom number generators, and is still widely used because it is very easy to implement and does a reasonable job for games and other casual applications. In current implementations, m is typically 2^{32}, or roughly 4×10^9, so this technique will generate a list of over four billion numbers before it repeats. The PRNG built into Python is based on a newer and more sophisticated algorithm that has a period of 2^{19937}.

The following tutorial project will use a pseudorandom number generator defined in RandomLab, the PythonLabs module for this chapter. A function named prng_sequence uses the general form of the equation to create a list of pseudorandom numbers. The three parameters passed to the function are the values of a, c, and m to plug into the equation. The list we get back will have m numbers, and with the right combination of a and c all values from 0 to $m - 1$ will be in the list.

For our experiments with longer pseudorandom sequences we will draw a number line on the PythonLabs Canvas and plot the numbers in the sequence (Figure 7.3). If a sequence has a period of m, every number between 0 and $m - 1$ will be in the sequence, and the number line display will be completely filled in. A sequence with a shorter period, however, will have large gaps, because many of the numbers will never be generated.

Figure 7.3: *The first 100 values produced by a PRNG with a = 81, c = 337, and m = 1,000.*

Tutorial Project

T1. Import the functions defined in RandomLab:
```
>>> from PythonLabs.RandomLab import *
```

T2. To make a schedule of events, initialize a list with the time for the first event:
```
>>> t = [0]
```

T3. Apply the rule that adds a new event that will occur eight hours after the previous event:
```
>>> t.append((t[-1] + 8) % 12)
```

T4. Look at the schedule:
```
>>> t
[0, 8]
```
Do you see why Python added a new event at 8 o'clock?

T5. Apply the rule three more times (use your IDE's command line editing feature to repeat the statement three times). This is what the schedule should look like:
```
>>> t
[0, 8, 4, 0, 8]
```
Do you see how the schedule is simply going to keep repeating, and how the period for this sequence is only 3?

T6. Start a new schedule by typing the first expression again:
```
>>> t = [0]
```

T7. Repeat the expression that adds new events, but change the 8 to a 7 to schedule the next event seven hours later:
```
>>> t.append((t[-1] + 7) % 12)
>>> t
[0, 7]
```

T8. Enter a for loop to repeat the statement fifteen more times:
```
>>> for i in range(15):
...     t.append((t[-1] + 7) % 12)
...
```

T9. Look at this new schedule:
```
>>> t
[0, 7, 2, 9, 4, 11, 6, 1, 8, 3, 10, 5, 0, 7, 2, 9, 4]
```
Can you see how the period for this list is 12, that is, how all twelve numbers between 0 and 11 appear in this list before it starts to repeat?

Next we'll do some experiments with longer lists, using the prng_sequence function from RandomLab. This function uses the for loop shown in Problem T8 to make a list of m numbers. A good combination of parameters will lead to a random number generator that makes lists with all m numbers between 0 and $m - 1$.

T10. Call `prng_sequence` to make a list of twelve numbers using $a = 1$, $c = 8$, and $m = 12$:

```
>>> prng_sequence(1,8,12)
[0, 8, 4, 0, 8, 4, 0, 8, 4, 0, 8, 4]
```

It's pretty easy to see the list starts repeating with a period of 3.

T11. Call the function again, with $c = 7$ and $m = 12$:

```
>>> prng_sequence(1,7,12)
[0, 7, 2, 9, 4, 11, 6, 1, 8, 3, 10, 5]
```

Now the period is equal to m, and the list has every value between 0 and $m - 1$.

T12. Initialize the canvas to show a number line display for the numbers from 0 to 1,000:

```
>>> view_numberline(1000)
```

T13. A function named `tick_mark` will display a mark at a specified point on the line. Type this statement to display a tick mark in the middle of the line:

```
>>> tick_mark(500)
```

T14. Use a for loop to draw 100 marks, at locations 0 through 99:

```
>>> for i in range(100):
...       tick_mark(i)
```

You should see a solid blue wall of marks covering the first 100 spaces on the number line.

T15. Use `prng_sequence` to make a list of 1,000 numbers, using the parameters $a = 3$, $c = 337$, and $m = 1000$:

```
>>> seq1 = prng_sequence(3, 337, 1000)
```

T16. Tell Python to print the sequence in the shell window:

```
>>> seq1
[0, 337, 348, 381, 480, 777, ... 97, 628, 221]
```

Does this list look "random" to you?

T17. Reinitialize the display by calling `view_numberline` again, and then use a for loop to display all the numbers in the sequence:

```
>>> view_numberline(1000)
>>> for n in seq1:
...       tick_mark(n)
...
```

There are lots of gaps in the number line. It turns out this sequence only has a period of 100.

T18. Print the first 5 numbers in the sequence:

```
>>> seq1[0:5]
[0, 337, 348, 381, 480]
```

T19. Now print a slice that goes from `seq1[99]` to `seq1[105]`:

```
>>> seq1[99:105]
[221, 0, 337, 348, 381, 480]
```

Sure enough, after generating 100 numbers, the sequence started over again with 0 in location seq[100]. This combination of a, c, and m yields a list that has the same 100 numbers repeated ten times—not very random at all.

T20. Here is a better combination. Make a second list, but use 81 for a instead of 3:

```
>>> seq2 = prng_sequence(81, 337, 1000)
```

T21. Draw a new number line and plot the values in this new sequence:

```
>>> view_numberline(1000)
>>> for n in seq2:
...      tick_mark(n)
...
```

Now your number line should be completely filled in. There are 1,000 numbers in the list, and each value from 0 to 999 occurs exactly once.

The expression in Exercise T21 filled in each point in the number line. But we could have done that using a for loop to draw 1,000 tick marks (see Exercise T14). When you were watching the display, were the numbers created by prng_sequence added in a random order?

If you're curious to see how we can verify seq1 has only 100 numbers but seq2 has all 1,000 values the next set of exercises will use a Python data type known as a **set** to count the number of unique items in each list.

♦ Type this statement to make a short list of numbers that contains duplicates:

```
>>> a = [2,3,7,2,6,3,1,2]
```

♦ If we pass this list to the set constructor it will build a set object containing all the unique numbers from the list:

```
>>> set(a)
{1, 2, 3, 6, 7}
```

Notice how the set is printed with curly braces instead of square brackets.

♦ This statement will make a set from the numbers in first pseudorandom sequence:

```
>>> set(seq1)
{0, 520, 17, 21, 537, 28, ... 501, 508}
```

♦ You can tell at a glance the set has far fewer values than the list. If you want to know exactly how much shorter, call len to count the number of items in the set and the list:

```
>>> len(seq1)
1000
>>> len(set(seq1))
100
```

♦ The same expressions confirm the second sequence consists of all 1,000 numbers between 0 and 999:

```
>>> len(seq2)
1000
>>> len(set(seq2))
1000
```

7.2 Numbers on Demand

As a practical matter, applications do not generate a list of all numbers in a pseudorandom sequence. A game may need only a few hundred rolls of the dice, and it would be a big waste of time and space to generate the full list of billions of pseudorandom numbers defined by the best random number generators. Instead, games and other applications use a programming technique that creates random numbers "on demand."

As an analogy for how this might work, imagine a scenario where a statistician has a lab assistant who is in charge of random numbers. The situation where the full sequence is generated ahead of time would correspond to the lab assistant carrying around the RAND book of random digits with a bookmark to show the location of the next random number to give his boss. Each time the statistician needs a random value, the assistant would look up the current number in the book and then move his bookmark to the next number.

When the numbers are generated by an algorithm, the assistant just needs to keep track of one number on a small piece of paper. Let's call this value x. When the statistician needs a random number, the assistant evaluates $((a \times x + c) \bmod m)$, reports the result to the statistician, and then erases the old value of x and replaces it with the value of the equation.

If we want to implement this idea in Python, we can define a function named rand that will return a single number at a time instead of an entire pseudorandom sequence. Because we have been referring to the pseudorandom sequence as $x_0, x_1, ... x_{m-1}$ we will use a variable named x to hold the current value of the sequence. An assignment statement that updates x to the next value in the sequence is easy enough:

```
x = (a * x + c) % m
```

The expression on the right side of the assignment is how we write the equation $(a \times x + c) \bmod m$ in Python. This statement simply uses the current value of x to calculate the next value.

When we go to put this assignment statement in our function, however, we discover a problem. How are we going to initialize x? We can't put the assignment that creates x inside the function because that would initialize x to 0 each time function is called:

```
def rand(a, c, m):
    x = 0                        # wrong -- don't set x to 0 each time
    x = (a * x + c) % m
```

If you enter that definition into your shell session and try calling rand a few times you'll see you always get back the same result.

The solution is to make x a **global variable**. A global variable is defined outside any function, unlike local variables (described on page 32 in Chapter 2), which are created when the statements in the body of a function are executed.

One important difference between global and local variables is that local variables are erased from memory when the function returns, but global variables live on, meaning they are still in memory the next time the function is called. A second difference is that global variables are intended to be shared by two or more functions instead of just being used in the context where they are defined.

The Python code for a working version of rand is shown in Figure 7.4. The global statement on line 4 tells Python the function wants to use the global variable, which is defined on line 1. When this program is loaded into an interactive session, Python will execute the assignment statement that initializes x to 0, and then it will process the lines that define the rand function. Now each time we call rand it will use the current value of x to compute the next value:

```
>>> rand(1, 8, 12)
8
>>> rand(1, 8, 12)
4
```

```
1   x = 0          # current value of sequence
2
3   def rand(a, c, m):
4       "Generate a value between 0 and m-1"
5       global x
6       x = (a * x + c) % m
7       return x
```

Figure 7.4: *The rand function uses a global variable named x to hold the current value in the sequence. Each call to rand updates x to the next value in the sequence and returns the new value.*

Download: random/prng-v1.py

Tutorial Project

The project for this section uses the definition of rand shown in Figure 7.4. You can either create a new file named prng.py and type in the definition, or download prng-v1.py from the book website and rename it prng.py (we'll see in the next section why the file is named prng.py and not rand.py).

T22. Load the contents of prng.py, including the assignment statement that defines x, into a shell session.

T23. Call rand a couple of times to see that it works as expected:

```
>>> rand(1, 8, 12)
8
>>> rand(1, 8, 12)
4
```

Are these the first two values after the initial 0 returned by prng_sequence for the same values of *a*, *c*, and *m*?.

T24. Initialize a number line plot (view_numberline is defined in RandomLab, so import that module if you started a new shell session):

```
>>> view_numberline(1000)
```

T25. Use this for loop to plot the first 50 values from the sequence that uses $a = 81$, $c = 337$, and $m = 1,000$:

```
>>> for i in range(50):
...     tick_mark(rand(81, 337, 1000))
...
```

Do you see how we were able to make one random number at a time, instead of all 1,000? And that the for loop was able to plot each number as soon as it was returned by rand?

♦ In the exercises above we didn't reinitialize x to 0 before calling rand with $a = 81$, $c = 337$, and $m = 1,000$. Will rand return values from this new sequence, or did we need to reinitialize x first? Predict the values you think were generated on the first three iterations of the for loop, then check your answer by repeating all the exercises, except this time have the for loop print the value returned by rand instead of plotting it on the canvas.

♦ What will happen if we switch back to calling rand(1,8,12) after making 50 calls with $m = 1,000$? Will the function crash with an error, or will it return valid results from the sequence with $a = 1$, $c = 8$, and $m = 12$?

7.3 Modules and Encapsulation

Our new `rand` function is ready to be used in the projects described later in the chapter. One of these projects will show how to convert the numbers generated by `rand` into integers between 1 and 6 to represent rolls of six-sided dice, and the other project will show how to use random numbers to shuffle a deck of cards.

A simple way to use `rand` in these programs is to copy the text from `prng.py` and add it to the source code for the other programs. There are several problems with this approach, however. If we ever find a bug in `rand`, or want to make improvements, we'll have to find all the programs that have a copy and change each copy. There is also a danger of a "name clash." What if the other programs also have a global variable named x?

Another problem is that the way we wrote `rand` forces these other programs to know about the parameters a, c, and m. Each call to `rand` must supply values for the three parameters, and as we saw, the results can be pretty dismal if the program does not use a good combination. When we write the programs that use `rand` we may have forgotten what parameter combinations work best, or if we give the function to someone else to use we will have to explain what the arguments are.

A better design would keep information about good combinations of a, c, and m together with the definition of the function. The person who designs the `rand` function should be an expert in random numbers who is familiar with how the function works and the best combination of parameters. People who use random numbers should just be able to call a function that returns a new random value without having to know what values of a, c, and m work best. This is the approach used by Python's `randint` function. If we want a number between 1 and 10 we just call `randint` with those arguments:

```
>>> randint(1,10)
4
```

We don't have to understand the algorithm Python uses, or what special parameters that algorithm might need.

If we want to follow this same basic design, we should create our own module so that programs that need random numbers can import the function from the module. By putting all the code related to `rand` in one module, we only have to edit one file if we ever decide to make changes. Global variables will be safely "hidden" inside the module so there is no danger of a name clash. And as we will see, it will be possible to specify default values for the parameters that define the sequence. Other programs will simply call a function that gets the next random value without having to worry about any extraneous details.

The strategy of gathering all the knowledge related to a set of operations into a single package is known as **encapsulation**. It is one of the most important principles in computer programming, and learning how to break a large task into manageable chunks so they can be implemented in separate modules is a key step in becoming a proficient programmer.

Recall from earlier projects that when we want to use functions defined in a module, we need to use an `import` statement to get them from the library. This statement shows one way of telling Python we want to use items defined in the math module:

```
>>> import math
```

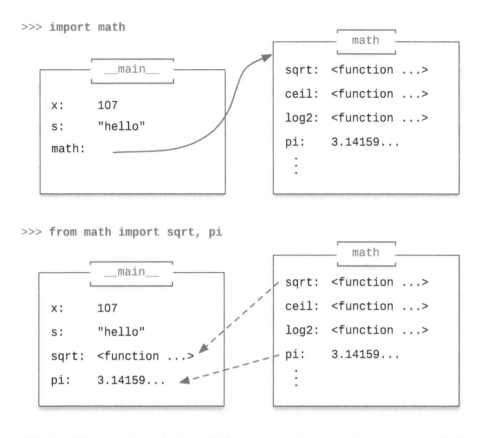

Figure 7.5: *The object store (or "whiteboard") for an interactive session is a namespace called __main__. (Top) If we import a module, its name becomes a reference to the namespace for that module, and we access items in the module using a qualified name. (Bottom) Importing individual items with a from statement adds those items to the object store.*

After executing this statement the name math is defined in the current session. If we want to access a variable or call a function defined in the module we use the qualified name syntax, which specifies the module name along with the name of the item we want to use:

```
>>> math.pi / 2
1.5707963267948966
>>> math.sqrt(100)
10.0
>>> math.ceil(math.log2(100))
7
```

A module like math that is a collection of function and variable definitions is an example of a **namespace**. A namespace is simply a collection of names, which can be names of functions, objects, other modules, or anything else in Python that has a name. The "whiteboard" that contains all the objects that have been defined for an interactive session is a namespace, and the math module and other library modules are also namespaces.

An import command simply adds the name of a module to the whiteboard and makes the name a reference to the module's namespace, as shown in the top half of Figure 7.5. It is basically just an assignment statement that defines a module name to be a reference to a namespace.

When the math module was introduced back in Chapter 2, we used a from statement to import only part of the module. For example, this statement imports two names from the math module:

```
>>> from math import sqrt, pi
```

In this case, instead of making the name math part of the current namespace, the system imports only the two names specified in the from statement. After executing this statement, two names, one a function and the other a floating point number, are added to the whiteboard (see the bottom half of Figure 7.5). Because the names are now defined for the current session they can be used directly, without the math qualifier:

```
>>> sqrt(2)
1.4142135623730951
>>> pi
3.141592653589793
```

Note the difference in the two types of imports. If we import the entire module, the name of the module is defined, and we access items in the module using a qualified name such as math.sqrt(2). We don't have to tell Python about sqrt or ceil or any other name in the module. Only when Python evaluates an expression of the form math.x does it search the namespace called math to see if x is defined there. On the other hand, when we import individual items with a from statement, Python searches the module right then, and it copies the definitions over to the current namespace. After that, the names can be used directly, without referring to the original module.

Let's return now to the pseudorandom number generator and see how we can use Python's rules for importing from modules to encapsulate our own rand function inside a module. With Python we don't have to do anything special to create a new module because every file is a module. Because the rand function is in the file named prng.py this is the statement we use to import the module:

```
import prng
```

When Python executes an import statement, it adds .py to the end of the module name, and then reads and executes all the statements in that file.[2]

The complete module for our pseudorandom number generator is shown in Figure 7.6. The first thing to notice about the module is that the three parameters that determine the behavior of the pseudorandom sequence are now global variables. They, along with the current value of the sequence, are created and initialized when the module is loaded.

Notice also that the new version of rand, the function that computes the next value in the sequence, has an empty parameter list. It does not need to be passed the *a*, *c*, and *m*

---

[2]For the projects in this book we will assume modules we write are in the same folder where the main program resides, and that our modules do not have names that conflict with any of Python's own library modules. Refer to your Lab Manual for information about how to put modules in other folders.

```
1    x = 0           # current value of sequence
2    a = 81          # multiplier
3    c = 337         # increment
4    m = 1000        # modulus
5
6    def rand():
7        "Generate a random value"
8        global x
9        x = (a * x + c) % m
10       return x
11
12   def reset(mult, inc, mod):
13       "Define new a, c, and m values"
14       global x, a, c, m
15       x = 0
16       a = mult
17       c = inc
18       m = mod
```

Figure 7.6: *A module containing functions and variables that comprise the pseudorandom number generator.*

Download: random/prng.py

parameters that were specified in the earlier version. Instead, those values are defined as global variables inside the module.

If we want to use the new version of the function to generate random numbers we can import it from the module with a `from` statement:

```
>>> from prng import rand
>>> rand()
337
>>> rand()
634
```

The values of a, c, and m are kept inside the module, and we don't have to know their values in order to obtain a new random number—we just have to call `rand()`.

Line 8 in Figure 7.6 (the first line in the body of `rand`) has the statement that tells Python x is a global variable. But notice that a, c, and m are not declared as global. In Python, only variables that will be modified from inside a function need to be specified as global variables. If all we are doing is accessing the value of a variable Python will automatically find the variable.

The module defines a second function, named `reset`. Programmers who know about random number generators may want to specify their own values for the three parameters of the pseudorandom sequence. They can call `reset`, passing it the values to use for subsequent calls to `rand`. Because the function is assigning to a, c, and m those three variables are all listed in the `global` statement on line 14.

Compared to other programming languages, Python has a slightly different meaning for the word "global." In most languages, a global variable is one that is accessible by all modules in a program. In Python, "global" just means "defined outside a function and accessible to all functions in this module." So even though a, c, and m are global variables defined in the prng module they are not automatically accessible to the rest of the world. Importing rand into an interactive session only allows us to call that function; it does not import the global variables defined in the module. In this case that's exactly the behavior we're looking for, because we want to exploit Python's module structure to encapsulate all the parameters that control the pseudorandom sequence inside the prng module.

As a final note to close out this introduction to namespaces, there is a built-in function in Python named dir, which is short for "directory." This function will return a list of names in a namespace. If we call it with no arguments, it will return a list of names defined in the interactive shell session. You should see something like this if you start a new session and import rand and reset from prng:

```
>>> from prng import rand, reset
>>> dir()
['__builtins__', '__doc__', ... 'rand', 'reset']
```

We can ignore the names that start and end with underscores—they're special variables that are always defined in each session. The two names created when we imported functions from prng are at the end of the list.

We can also pass a module name as an argument to dir to find out what names are defined in that module's namespace. This example imports math and then gets a list of names in the module:

```
>>> import math
>>> dir(math)
[... 'ceil', ... 'pi', 'pow', ... 'sqrt', ...]
```

Tutorial Project

T26. Start a new shell session in your IDE. Call dir to verify the only names defined are the special system variables with names that start with underscores.

```
>>> dir()
['__builtins__', '__doc__', '__loader__', '__name__', '__package__']
```

The list above was printed by Python 3.3 running on Mac OS X. Other systems or IDEs might have a different set of names.

T27. Define a variable and call dir again:

```
>>> s = 'hello'
>>> dir()
['__builtins__', '__doc__', ... 's']
```

Note that s has been added to the list.

T28. Try listing the names in the math module:

```
>>> dir(math)
NameError: name 'math' is not defined
```

That failed because the name math is not in the namespace for the shell session.

T29. Importing math will add the name math to the set of names on our "whiteboard":

```
>>> import math
>>> dir()
['__builtins__', '__doc__', ... 'math', 's']
```

T30. Now we should be able to get a list of names in that module:
```
>>> dir(math)
['__doc__', ... 'ceil', ... 'trunc']
```
The math module has its own special names like __doc__ shown above, but you should also see some familiar names.

T31. If math is a name on the whiteboard, do you suppose you can just type the name to ask Python to show you the value associated with the name? Make a prediction about what will happen if you type math, then test your hypothesis:
```
>>> math
???
```

T32. The previous exercise shows that math is the name of a module. Recall that the type function tells us what type of object a name refers to:
```
>>> type(math)
<class 'module'>
```

The previous experiments should convince you that import is basically just an assignment statement. Importing the math module defined a new variable named math and made it a reference to a module.

T33. Restart your shell session so no variables are defined.

T34. Use a from statement to import selected names from the math module:
```
>>> from math import sqrt, pi
```

T35. Use dir to get a list of names. Do you see how sqrt and pi are there, but math is not?

T36. Ask Python to show you the values of the imported items:
```
>>> sqrt
<built-in function sqrt>
>>> pi
3.141592653589793
```

Note that typing sqrt without parentheses is not the same as typing sqrt() with parentheses. The expression above is simply asking Python to tell you the value of sqrt, and Python replied that it is the name of a function.

Before you start the interactive session for the next exercises make sure you are in the same directory where you saved your prng.py file.

T37. Either download prng.py (Figure 7.6 from the book website or modify your prng.py file so it looks like the version in the figure:
- Add the assignment statements that define a, c, and m.
- Edit the def statement for the rand function so it has no parameters.
- Type in the definition of the reset function.

T38. Start a new shell session and import rand:
```
>>> from prng import rand
```

T39. Call rand a few times. Is it producing the sequence of values you expect?
```
>>> rand()
337
>>> rand()
634
```

T40. Restart the session, and this time import the entire prng module:
```
>>> import prng
```

T41. Print the contents of the module's namespace:

```
>>> dir(prng)
['__builtins__', ... 'a', 'c', 'm', 'rand', 'reset', 'x']
```

Do you see how all six names—two functions and four global variables—are all defined in that namespace?

T42. Use the qualified name syntax to call rand:

```
>>> prng.rand()
337
```

T43. Do you think the global variables are also accessible? Try asking for a value:

```
>>> prng.m
1000
```

Yes, all names in the module are accessible.

T44. Call reset to tell the module to start making values from a different sequence:

```
>>> prng.reset(1,8,12)
>>> prng.rand()
8
>>> prng.rand()
4
```

Do you see how the new parameters affected the calls to rand?

7.4 Games with Random Numbers

If the three parameters a, c, and m that define the relationship between successive numbers in a pseudorandom sequence are chosen carefully the sequence will have values from 0 up to $m - 1$. However, if we are writing a program to play a game like backgammon, which is based on rolling a pair of dice, we want a series of numbers between 1 and 6.

A simple approach to simulating random rolls of a six-sided die is to get a value from the pseudorandom sequence, find its remainder after dividing by 6 (which will yield a number between 0 and 5), and then add 1. Here is a Python expression that transforms a number returned by our prng module:

```
>>> prng.rand() % 6 + 1
2
```

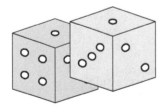

Figure 7.7: *If pseudorandom numbers are used by a program that plays a game with dice, the values in the pseudorandom sequence are mapped to numbers between 1 and 6.*

```
1    def roll():
2        "return a random value between 1 and 6"
3        return rand() % 6 + 1
```

Figure 7.8: *A function that uses the pseudorandom sequence to get a roll of a die.*

The mod operator % has the same precedence as the divide operator, so the expression first gets a number from rand, computes its remainder after dividing by 6, then adds 1. The result should always be a value between 1 and 6:

```
>>> prng.rand() % 6 + 1
5
```

If we evaluate this expression several more times and save the results in a list, we will see something interesting:

```
>>> rolls = []
>>> for i in range(10):
...     rolls.append(prng.rand() % 6 + 1)
...
>>> rolls
[2, 3, 4, 3, 4, 5, 2, 1, 4, 3]
```

As expected, all the numbers in the list are between 1 and 6. But notice the numbers that follow the 2s in this list: the first 2 is followed by 5, the second by 3, and the third by 1.

The fact that a number can be followed by different values adds to the illusion that this sequence is truly random, not just pseudorandom. Because $m = 1,000$ and we are using a combination of a, c, and m that we have tested we know the pseudorandom sequence generated by calls to rand will have every number between 0 and 999. When we divide these numbers by 6, the remainders will be between 0 and 5. Each remainder will show up hundreds of times, but any two successive remainders can be any of the numbers between 0 and 5.

To make it easier to write programs that simulate rolls of dice we can add a new function to our module. The definition of a function named roll that maps values from the pseudorandom sequence into numbers between 1 and 6 is shown in Figure 7.8. An important detail is that roll is in the same namespace as the rand function because they are both defined in the same file. That means the call to rand on line 2 does not need to use a qualified name.

There is one more issue that needs to be addressed if we want to use our prng module to play games. How should we initialize the pseudorandom sequence? Each time the module is loaded at the start of a new shell session, the variable x, which represents the current state of the sequence, it initialized to 0.

The same will be true if we write a game program that imports the module. Every time the game starts, the prng module will be imported, and the variable x in the module will be set to 0. Because values of a, c, and m are also defined inside the module, and they are also given the same value each time the module is loaded, the same sequence of numbers will be generated at the start of each game. Our backgammon program will always begin with the same rolls of the dice: 2, 5, 2, 3, 4, 3, 4, 5, 2, etc. Players will soon recognize that every game starts the same way, and they will lose interest.

One way to address this problem is to assign a different value to x each time the module is loaded. Because x is the variable that holds the current number in the sequence, setting it to a different value each time will lead to a unique sequence for each game. In the terminology of pseudorandom number generators this initial value is called the **seed**.

But now the question becomes, which value do we use for a seed? If we're using $m = 1,000$ we can choose any number from 0 to 999. But if we just choose our favorite number, and write that number into the module, we will have the same problem. The game will start out with a different set of rolls than if the sequence started with 0, but because the sequence always starts in the same state players are again going to see the same set of rolls at the start of each game.

A common solution is to use the system clock. Python has a module named time that has a collection of functions for dealing with dates and times. Inside that module is a function, also called time, that returns the number of seconds since January 1, 1970 (the date of the "Big Bang" when time started in the Unix world). Here is an example of how to import and call this function:

```
>>> from time import time
>>> time()
1376150835.068832
```

The first line might look strange, but all it means is that somewhere in Python's library there is a module named time, and inside this module there is a def statement that defines a function named time.

The example shown above was taken from an operating system that keeps time accurate to a millionth of a second, so the number returned by time is a floating point value. Other systems might return an integer, or report time values in terms of thousandths of a second. If we want to use the system time as the seed for our pseudorandom sequence we want to convert it to an integer and then map that integer into a number between 0 and $m - 1$:

```
x = int(time()) % m
```

If we modify the assignment statements at the start of prng.py to look like Figure 7.9 the module will use the system clock to initialize x. Note that the statement that defines x has been moved down to line 7, and that it now follows the statement that creates m, because the expression on the right side of line 7 refers to m. This will work because Python executes the statements in the module file as the file is read. By the time it executes line 7, the variables a, c, and m will all have values, and the time function will have been imported so we can call it to get the system time.

This technique of seeding the pseudorandom number generator with the system clock will give us the behavior we are looking for. The values returned by time are increasing in a predictable way, and so is the initial value assigned to x. But the values that follow from this initial assignment will be very different, and the games will start out with a unique sequence of rolls each time.

```
1    from time import time
2
3    a = 81          # multiplier
4    c = 337         # increment
5    m = 1000        # modulus
6
7    x = int(time()) % m
```

Figure 7.9: *Modifying the assignment statements at the beginning of the prng module so x is initialized according to the system clock will lead to a different sequence each time the module is loaded.*

Tutorial Project

T45. Start a new shell session. Import the `time` function and call it a few times:
```
>>> from time import time
>>> time()
1376153746.442665
>>> time()
1376153747.198874
```
If you use your IDE's command line editing feature (just hit the up arrow and return) you can probably get times that differ very little.

T46. Use `int` in a few expressions to make sure you know what it does:
```
>>> int(4.7)
4
>>> int(time())
1376153935
```

T47. Define a value for `m` and test the expression that uses the system clock to create a seed value:
```
>>> m = 1000
>>> int(time()) % m
658
```

T48. Edit the assignment statements at the front of your `prng.py` file so they look like the ones in Figure 7.9.

T49. Save your new version of the module and load it into a new interactive session:
```
>>> from prng import *
```

Python has a construct called **list comprehension** that makes it very easy to create a list where each value is the result of a call to a function (see the sidebar on List Comprehension on page 212).

T50. This expression uses list comprehension to create a list of ten values from the pseudorandom sequence:
```
>>> [rand() for i in range(10)]
[652, 149, 406, 223, 400, 737, 34, 91, 708, 685]
```

T51. Start a new shell session, import the module, and repeat the previous exercise. Did you get a different sequence in this new session?

Do you see how a slight change in the seed leads to a big change in the pseudorandom sequence? If the second shell session starts just a few seconds later, the time used to set the seed will lead to a completely different sequence of random numbers at the start of each game.

List Comprehension

A convenient alternative to writing a for statement to create a list in Python is to use a construct known as **list comprehension**. Write a pair of brackets, and inside the brackets put an expression that describes each list item. For example, to make a list of numbers from 1 to 9 write

```
>>> [i for i in range(1,10)]
[1, 2, 3, 4, 5, 6, 7, 8, 9]
```

This expression makes a list of the squares of the numbers in the same range:

```
>>> [i**2 for i in range(1,10)]
[1, 4, 9, 16, 25, 36, 49, 64, 81]
```

In general, a list defined this way has an expression to the left of the word `for` and a specification of a set of values to the right of the `for`. The expression will be evaluated using each value in the set, and the result of the expression will be saved in the new list.

T52. Edit prng.py so it includes the function named roll shown in Figure 7.8.

T53. Reload the prng module and make a list of 10 rolls:

```
>>> [roll() for i in range(10)]
[5, 2, 1, 4, 3, 2, 3, 6, 5, 2]
```

If you start a new session and repeat this last exercise you should see a different sequence of values between 1 and 6.

7.5 Random Shuffles

If we want to write a program to play bridge, poker, or some other card game we are going to need a function that makes random collections of cards. Simulating the roll of six-sided die just involved computing a number between 1 and 6, but to play a game of cards it is not enough to generate a list of random cards.

It would be easy enough to label each card in the deck with a number between 0 and 51, and then use a pseudorandom number generator to compute a number between 0 and 51 as a way of choosing a card at random. But if we want to deal a hand there is a chance that we could end up with two copies of the same card. In bridge, each hand has thirteen cards, and if we make a hand by just calling randint(0,51) 13 times, odds are we will have a collection where the same number appears twice.

One way to solve this problem is to use a function that shuffles a deck of cards. We will start with a collection of all 52 cards, and then each time we play a game we can rearrange the items in a random order. The goal for this section will be to develop a function that rearranges objects in a list using a pseudorandom number generator to produce a random ordering.

To make the project more realistic, the RandomLab module defines a class named Card. A Card object will have a rank (ace, king, queen, etc.) and a suit. We will represent a deck of cards as a list of 52 objects that each represent a different card from a standard deck.

As with other kinds of objects, the name of the class is also a constructor, a function that creates objects of that class. If we call Card without passing it any arguments we will get back a random card:

```
>>> Card()
J♣
>>> Card()
9♥
```

Two methods defined for Card objects return the attributes of a card. If we call the method named suit we will get back a number between 0 and 3, for clubs, diamonds, hearts, and spades, in that order (Table 7.1). A call to rank will return a number between 0 and 12. Aces have the highest rank, and 2's have the lowest rank. Here is an example:

```
>>> c = Card()
>>> c
3♥
>>> c.rank()
1
>>> c.suit()
2
```

The rank is 1 because a 3 is the second lowest card, and according to Table 7.1 hearts is suit number 2.

Table 7.1: *Card objects have two attributes, their rank and their suit. Ranks are integers between 0 and 12, and suits are integers between 0 and 3.*

Ranks		Suits	
2	0	♣ / C	0
3	1	♦ / D	1
4	2	♥ / H	2
5	3	♠ / S	3
6	4		
7	5		
8	6		
9	7		
10	8		
J	9		
Q	10		
K	11		
A	12		

When we make a deck of cards, we don't want 52 random cards, but instead we want one of each possible card. We can pass an integer between 0 and 51 to the constructor to tell it which card to make:

```
>>> Card(0)
2♣
>>> Card(1)
3♣
>>> Card(50)
K♠
>>> Card(51)
A♠
```

Note that Card objects have a defined order: the first 13 cards are clubs, the next 13 are diamonds, and so on. Within a suit, cards are ordered from 2 up to A.

To create a complete deck of cards we can use Python's list comprehension to specify a list with all 52 card objects:

```
>>> [Card(i) for i in range(0,52)]
[2♣, 3♣, 4♣, ... Q♠, K♠, A♠]
```

Mathematicians refer to an ordering of items as a **permutation**. The goal for the project in this section is to define a function named permute that will make a new random ordering of the items in a list. For example, after making a full deck of cards as shown above, we can call permute to shuffle the deck. Each time we call permute it will make a new random permutation:

```
>>> deck = [Card(i) for i in range(0,52)]
>>> permute(deck)
>>> deck
[7♥, 9♥, 3♣, ... 2♥, 3♥, 5♣]
```

To make it easier to see the effects of a call to permute, we can write two Python statements on one line, separated by a semicolon:

```
>>> permute(deck); deck
[A♣, 2♠, J♦, ... 2♣, 4♥, A♦]
>>> permute(deck); deck
[2♦, A♥, 8♥, ... 6♦, J♠, K♥]
```

A program that plays poker would probably "deal" the cards the same way we would in a real game. After calling permute to shuffle the deck, it would give deck[0] to the first player, deck[1] to the next player, and so on. For the experiments in Section 7.6, however, we'll just need one hand after each shuffle, and we can make this hand using the first five cards in the deck.

To see what that hand would be, we can modify the statement above to print a slice containing the first five cards:

```
>>> permute(deck); deck[0:5]
[A♠, 6♣, 10♦, 6♠, 8♦]
```

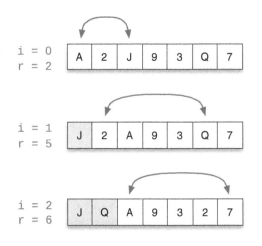

Figure 7.10: *On each iteration of the algorithm that permutes a list the item at location i changes places with an item at a random location to the right.*

A straightforward algorithm for permuting the items in a list is based on an iteration that exchanges two items at each step. Begin by picking up the first item and exchanging it with a random item somewhere to the right. Then exchange the second item with a random item somewhere to its right, then exchange the third item with a random item somewhere to its right, and so on until you reach the end of the list. This algorithm is reminiscent of the insertion sort algorithm: on each iteration the list has two regions, where items in the left part of the list have been exchanged and items to the right are waiting to be moved.

An example showing the progress of this algorithm as it scrambles a list of seven items is shown in Figure 7.10. A variable named i indicates the position of the item to move on the current iteration. On each iteration, a second index variable r was set to a random value between i and 6, the last location in the list.

On the first iteration, i was 0, the random value r was 2, and the program exchanged a[0] with a[2]. Note that at the start of the second iteration the item that used to be in a[2] was moved to the front of the list. In the second round, i was 1, and r could have been any location between 1 and 6. On the next round, i was 2, and the exchange could be with any location between 2 and 6, and so on.

Note that r, the location to move to, can be any value between i and len(a) $-$ 1, which means it is possible to have i $=$ r. In that case the item at a[i] is not moved, it just stays where it is. It is also possible for an item to be moved several times. The A that originally started out in location 0 in Figure 7.10 is going be moved again on the third iteration.

It's easy to exchange the values of two items in Python by using a parallel assignment. As we saw in the previous chapter, a parallel assignment has two or more variable names on the left, as in

```
>>> x, y, = 3, 4
```

The variables on the left are assigned the corresponding values from the right, so in this example x is set to 3 and y is set to 4. It's called a "parallel" assignment because, from the programmer's point of view, the two assignments happen at the same time.

```
1    from random import randint
2
3    def permute(a):
4        "Rearrange list a in a random order"
5        for i in range(0,len(a)-1):
6            r = randint(i, len(a)-1)
7            a[i], a[r] = a[r], a[i]
```

Figure 7.11: *A function that permutes a list, placing the items in a new random order.*

Download: random/permute.py

To use a parallel assignment to exchange the value of two variables, write the names of the variables in different orders on the left and right sides of the assignment operator:

```
>>> x, y = y, x
```

Python executes this statement by fetching the current values of y and x and then storing them back in the opposite place. It might help to think of a juggler picking up two balls and tossing them to different hands in the same motion: while the value of y is being tossed to x the value of x is being tossed to y.

Parallel assignment also works for locations in a list. For example, to exchange the values in the first two locations of a list a we can just write

```
>>> a[0], a[1] = a[1], a[0]
```

Now that we know how to exchange two items in a list, writing the permute function is straightforward (Figure 7.11). The algorithm is an iteration in which a variable named i takes on values from 0 to $n - 2$, where n is the length of the list. At each step, we just need to set a variable r to a random value between i and $n - 1$, which is the last location in the list, and then exchange the items at locations i and r. The random location is chosen by the randint function from Python's standard library; a call to randint(i,j) returns a number between i and j. After we have a value for a random location to the right of i we exchange the items at locations i and r using a parallel assignment.

Tutorial Project

Either use your IDE to create the file named permute.py shown in Figure 7.11 or download the file from the book website.

T54. The exercises in this section use functions and classes defined in RandomLab:
```
>>> from PythonLabs.RandomLab import *
```

T55. Make a small list of strings and ask Python to display the list on the terminal window:
```
>>> a = RandomList(5, 'colors')
>>> a
['khaki', 'yellow', 'salmon', 'forest green', 'sienna']
```

T56. Type a parallel assignment expression that exchanges the values in the first and third locations, and then print the list again:

```
>>> a[0], a[2] = a[2], a[0]
>>> a
['salmon', 'yellow', 'khaki', 'forest green', 'sienna']
```

Can you see how Python exchanged the strings at a[0] and a[2]?

T57. Make a list of numbers to use to test the permute function:

```
>>> a = list(range(0,10))
```

T58. Load the definition of permute into your interactive session. This statement will pass a to permute and then show the new value of a:

```
>>> permute(a); a
[5, 9, 3, 6, 8, 4, 7, 0, 1, 2]
```

T59. Repeat the statement a few times; you should see a new random permutation after each call to permute.

```
>>> permute(a); a
[0, 8, 3, 2, 6, 1, 4, 7, 9, 5]
>>> permute(a); a
[2, 8, 3, 0, 7, 6, 5, 4, 9, 1]
```

T60. If you want to monitor the progress of permute, add the following print statement to the code, right after the assignment to r selects the location to swap:

```
print(r, a[:i], a[i:])
```

The print statement will display the location to exchange, the part of the list already finished, and the portion still to be rearranged.

T61. Here is an example of what is printed when permute scrambles the list of numbers:

```
>>> permute(a); a
3 [] [8, 4, 7, 1, 5, 6, 0, 9, 2, 3]
8 [1] [4, 7, 8, 5, 6, 0, 9, 2, 3]
5 [1, 2] [7, 8, 5, 6, 0, 9, 4, 3]
...
```

The number at the front of a line is the value of r, which is the location where the front of the "to do list" will be moved. On the first iteration, the number 8 will be exchanged with a[3], and you can see the result of this swap on the second line. On each round the list of items already moved grows longer while the list of items still to be moved grows shorter.

The actual result you get will be different, because the location used for the exchange is chosen at random. But you should be able to follow the steps of the algorithm by looking at the front of each line to see which location was chosen, and then noticing on the following line how the item at the front of the second list was swapped with the item at the chosen location.

T62. Call Card with no argument a couple of times to make some random cards:

```
>>> Card()
3♥
>>> Card()
J♣
```

If you do not see a card symbol in your terminal window, check to see if your IDE or terminal emulator allows you to define an "encoding" and set it to UTF-8. If you want to print letters instead of card symbols type this command:

```
>>> Card.print_ascii()
```

T63. Define a variable named c to be a reference to a specific card, then call methods of the Card class to get attributes of the card:

```
>>> c = Card(24)
>>> c
K♦
>>> c.suit()
1
>>> c.rank()
11
```

Do these values agree with Table 7.1?

T64. Type this command to make a list named a with 13 random cards and then print the list:

```
>>> a = [Card() for i in range(13)]; a
[8♣, A♦, 10♣, ... 3♠, 3♥, 2♠]
```

T65. It's easier to see if a list contains duplicates if the list is sorted:

```
>>> a = [Card() for i in range(13)]; print(sorted(a))
[5♣, 10♣, 10♣, ... Q♥, 3♠, 9♠]
```

T66. There is a chance that by simply calling Card() 13 times to deal a hand we will get a duplicate card. Does your list have any duplicates? Repeat the expression above a few times. How often do you get duplicate cards?

T67. Make a full deck of cards containing each of the 52 cards, this time passing i as an argument in the call to Card:

```
>>> d = [Card(i) for i in range(0,52)]
```

T68. Repeat the two statements used for earlier tests with permute, but this time permute the list of cards:

```
>>> permute(d); d
[4♥, Q♣, J♣, ... 10♣, 7♣, 9♠]
>>> permute(d); d
[7♥, A♦, 6♥, ... 4♣, 10♦, 10♥]
```

Does it look like a random shuffling after each call?

T69. The following statement is like the one above, except it deals a bridge hand (13 cards) and it sorts the hand before it is displayed:

```
>>> permute(d); print(sorted(d[0:13]))
[5♣, 6♣, 7♣, 8♣, J♣, A♣, 4♦, 8♦, 10♦, 4♥, Q♥, K♠, A♠]
```

Repeat the expression a few times. Because the hand is made after a random shuffle of a full deck you should never see any duplicates.

7.6 Tests of Randomness

So far we've been relying on our intuition that the output from our pseudorandom number generators looked random. In this section we'll perform some tests on the sequence of numbers produced by our rand function to investigate the question of whether or not the sequences are random.

The two techniques we will use are both informal tests based on graphical displays. Visualization is a very powerful method for seeing whether there are any biases or hidden patterns in data. These informal tests will not give us a definitive answer, of the form "yes, this sequence is random." However, if a sequence of numbers has some hidden problems,

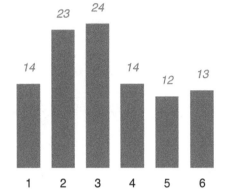

Figure 7.12: *A histogram (bar chart) showing the frequency of random rolls of a die based on the first 100 numbers from the pseudorandom sequence with $a = 81$, $c = 337$, and $m = 1,000$.*

the nonrandomness often shows up in the form of patterns in the display, so we can say with confidence "no, this sequence is not random."

The first type of display is a **histogram** (often called a "bar chart"). To test a sequence of simulated rolls of a die, we can just count the number of times each number from 1 to 6 occurs in the sequence. The histogram will have one bar for 1s, another bar for 2s, and so on, where the height of a bar indicates how often that number showed up in the random sequence (Figure 7.12).

The algorithm we have been using for making pseudorandom sequences should create what is known as a **uniform distribution**. In an experiment with 1,000 rolls of the die, we expect each number will occur roughly $1000 \div 6 = 167$ times. But suppose there is a problem with the parameter combination, and it gives us 300 6s. If all the other rolls occur equally often they will show up $(1000 - 300) \div 5 = 140$ times. The bar for 6s will be over twice as high as the others, and we can tell at a glance that the sequence is biased. But if all six bars are roughly the same height, we could then do some further statistical tests if we wanted to know for certain whether or not the sequence is random.

The RAND corporation used a similar approach to test the numbers in their book of random digits. One of their tests was called the "poker test." If we want to do something similar with our random number generator, we can use the rand function to deal 1,000 poker hands. According to the laws of probability, we should see 420 hands that have two cards with the same rank (a pair), 21 hands with three cards of the same rank (three of a kind), and so on. If we plot the results with a bar chart, and see a lot more straights and full houses than pairs, then something is clearly wrong.

To help count the number of rolls of each type or the number of times each poker hand is seen RandomLab defines a class named Histogram. The argument passed to the constructor for this class is a list of labels. If we want to count rolls of a six-sided die the labels are a list of numbers from 1 to 6:

```
>>> hist = Histogram(range(1,7))
```

A Histogram object is essentially just a standard Python dictionary where each value is initialized to 0. If we ask Python to show us the object we just created this is what we would see:

```
>>> hist
{1: 0, 2: 0, 3: 0, 4: 0, 5: 0, 6: 0}
```

The items in a histogram are called "bins." In this example, we have six bins. The dictionary keys are the bin labels, and the values are the corresponding counts.

There are two important differences between a regular dictionary and a Histogram object. First, the Histogram class does not allow us to arbitrarily set a value for one of the bins. This is what happens if we try to set the count for the number of 6s:

```
>>> hist[6] = 10
RandomError: can't assign to histogram; use count to update a bin
```

Instead, we need to call a method named count, passing it the name of a bin we want to update:

```
>>> hist.count(6)
```

If we look at the histogram again we will see the bin labeled 6 has been incremented:

```
>>> hist
{1: 0, 2: 0, 3: 0, 4: 0, 5: 0, 6: 1}
```

If we want to know the count for this one bin just access the value the way we normally do:

```
>>> hist[6]
1
```

The second difference is that if a Histogram object is displayed on the PythonLabs canvas the drawing is automatically updated when we call count. Use the view_histogram function to display the histogram:

```
>>> view_histogram(hist)
```

Now if we call the count method to update a bin we will see the size of the bar for that bin grow by a small amount:

```
>>> hist.count(3)
```

Another technique for testing whether the sequence of values in a pseudorandom sequence is random is to look for patterns or correlations between successive numbers in the sequence. Here is the expression we typed to make a list of 1,000 numbers:

```
>>> seq = prng_sequence(81, 337, 1000)
```

and these are the first 10 numbers in the sequence:

```
>>> seq[0:10]
[0, 337, 634, 691, 308, 285, 422, 519, 376, 793]
```

From this short sample our sequence looks random, but in fact there is a pattern. The pattern is easier to spot if the numbers are converted into a value between 1 and 6, which is what we would do if we're using the numbers in a game based on rolling dice. Here are the first 20 numbers in the sequence, converted to values between 1 and 6. Do you see the pattern?

```
>>> [ (x % 6) + 1 for x in seq[0:19] ]
[1, 2, 5, 2, 3, 4, 3, 4, 5, 2, 1, 4, 3, 2, 3, 4, 5, 2, 3]
```

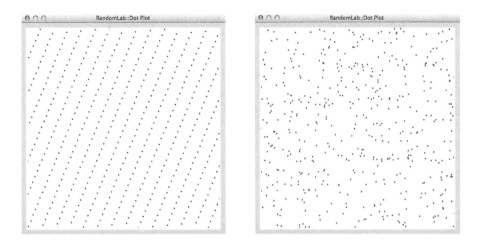

Figure 7.13: *In these snapshots of the PythonLabs Canvas, the x and y coordinates of a dot are determined by two successive numbers from a pseudorandom number generator. The dots in the window on the left are from our rand function with a = 81, c = 337, and m = 1,000. The window on the right is based on the randint function from Python's random module (part of the standard library).*

Based on this short list it looks like the numbers alternate between even and odd. If the computer used this pseudorandom sequence to play a game, every time it rolled an even number it would follow with an odd number, and vice versa. It wouldn't take long for a person playing a game based on this sequence to suspect something was wrong. For one thing, in a game with two dice, the computer would never roll "doubles."

A visual display that makes this type of pattern easy to see is known as a **dot plot**. As the name implies, the drawing will be a set of dots on a canvas. To make a dot plot on the PythonLabs canvas, call a method named `view_dotplot`:

```
>>> view_dotplot(1000)
```

The argument is the maximum value that will be used for *x* or *y* coordinates, where (0,0) is in the upper-left corner.

To display a dot on the canvas, call `plot_point`, passing it the *x* and *y* coordinates of the dot. For example, this statement will draw a dot in the middle of the canvas:

```
>>> plot_point(500, 500)
```

One way to look for patterns in values from a pseudorandom sequence is to plot a set of points in which the *x* and *y* coordinates of a dot are determined by getting two successive numbers from the sequence. Here is a `for` loop that uses numbers from even locations (`seq[0]`, `seq[2]`, etc.) as *x* coordinates and numbers from odd locations (`seq[1]`, `seq[3]`, etc.) as *y* coordinates to place 500 points on the dot plot:

```
>>> for i in range(0, 1000, 2):
...     plot_point(seq[i], seq[i+1])
...
```

If there is no correlation between successive numbers, the points will be scattered at random all over the canvas. But if there are any hidden patterns in the data, even if they are very subtle, the dots will line up or form some other distinctive shapes.

It's obvious from the dot plot in Figure 7.13 that our sequence of numbers generated using $a = 81$, $c = 337$, and $m = 1,000$ is not at all random. Even if we had not noticed that the sequence alternated between even and odd numbers, this plot would have told us there was some sort of pattern, and we would not use this set of parameters in an application that requires random numbers.

As was the case with histograms, a dot plot visualization is convenient for getting an initial impression. When there are hidden patterns in the data they often jump out when the data is displayed with a dot plot. But to truly check to see whether the data is random it would be necessary to perform a detailed statistical analysis of the x and y coordinates. For this chapter, however, we will simply do informal tests using visualizations.

Tutorial Project

The first set of projects will plot a histogram for rolls of a die produced by the pseudorandom number generator we used in previous projects.

T70. If you started a new shell session import the definitions in RandomLab:
```
>>> from PythonLabs.RandomLab import *
```

T71. Create a new Histogram object to count the rolls of a die:
```
>>> hist = Histogram(range(1,7))
```

T72. Ask Python to print the object in the shell window:
```
>>> hist
{1: 0, 2: 0, 3: 0, 4: 0, 5: 0, 6: 0}
```

T73. This looks like a regular Python dictionary object, but if you ask to see its type you will see it belongs to a class defined in PythonLabs:
```
>>> type(hist)
<class 'PythonLabs.RandomLab.Histogram'>
```

T74. Call the count method to update a bin and print the object again:
```
>>> hist.count(5)
>>> hist
{1: 0, 2: 0, 3: 0, 4: 0, 5: 1, 6: 0}
```
Do you see how the value of bin 5 increased by 1?

T75. Call view_histogram to draw the object on the canvas:
```
>>> view_histogram(hist)
```
You should see six lines in the middle of the screen. Five of the bins will have zero items, and the bar for bin 5 will be slightly taller.

T76. Type this command to increment bin number 2:
```
>>> hist.count(2)
```
Do you see how the rectangle for this bin got slightly taller?

T77. Repeat the previous statement a few times, using different bin numbers, until you're confident you know how this expression updates the histogram.

T78. Reinitialize the histogram by making a new object and displaying it:
```
>>> hist = Histogram(range(1,7))
>>> view_histogram(hist)
```

T79. The next experiment will use the `roll` function from our `prng` module:
```
>>> from prng import roll
```

T80. This `for` statement will get the first 100 rolls of the die from the our PRNG and update the histogram for each roll:
```
>>> for i in range(100):
...     hist.count(roll())
...
```

T81. Look at the counts:
```
>>> hist
{1: 20, 2: 13, 3: 14, 4: 15, 5: 16, 6: 22}
```
Do the sizes of the bars on the canvas agree with the counts you see?

If any of the bins reaches the top of the canvas, the drawing methods will automatically rescale the histogram to make it smaller. So if you see the histogram suddenly shrink, don't worry. It just means the drawing methods are "zooming out" and making room for more data.

T82. In an earlier exercise we saw how the functions in the `prng` module generate every number between 0 and 999. That means there should be the same number of 1s, 2s, 3s, and so on when those numbers are converted to values between 1 and 6. To verify this claim, repeat the previous loop 900 more times, so a total of 1,000 numbers are in the histogram:
```
>>> for i in range(900):
...     hist.count(roll())
...
```
What did you see? Are the results what you expected?

T83. Display the Histogram object in the shell window:
```
>>> hist
{1: 167, 2: 167, 3: 167, 4: 167, 5: 166, 6: 166}
```
Do the counts add up to 1000?

The next set of exercises makes another histogram, this time using poker hands.

T84. The names of the different kinds of poker hands are in a list named poker_hands that was defined when you imported RandomLab:
```
>>> poker_hands
['high card', 'pair', 'two pair', 'three of a kind', 'straight',
  'flush', 'full house', 'four of a kind', 'straight flush']
```
Note the hands are ordered, with the most common hands on the left.

T85. Pass the list of names to the constructor that creates a new Histogram object and display the histogram on the canvas:
```
>>> hist = Histogram(poker_hands)
>>> view_histogram(hist)
```
Because there are nine names in the list, you should see nine empty bins.

T86. To verify the count method works correctly when bin labels are strings try incrementing the count for some of the bins:
```
>>> hist.count('high card')
>>> hist.count('flush')
```

T87. Make a new deck of cards:

```
>>> d = [Card(i) for i in range(52)]
```

T88. A function defined in RandomLab will figure out what type of poker hand is in a list of five cards and return one of the strings in poker_hands. Here is a quick test to see how it works:

```
>>> d[0:5]
[2♣, 3♣, 4♣, 5♣, 6♣]
>>> poker_rank(d[0:5])
'straight flush'
```

Because a new deck has the five lowest clubs as the first five cards the result is what we expect.

T89. This set of commands will shuffle the deck and print the type of poker hand in the first five cards:

```
>>> permute(d); d[0:5]; poker_rank(d[0:5])
[J♥, 5♥, 9♠, 7♠, 6♣]
'high card'
>>> permute(d); d[0:5]; poker_rank(d[0:5])
[7♣, 7♠, 9♥, 3♥, 3♦]
'two pair'
```

T90. This for loop will shuffle the deck 1,000 times, classifying the poker hand at the front of the deck and updating the histogram after each shuffle:

```
>>> for i in range(1000):
...     permute(d)
...     hist.count(poker_rank(d[0:5]))
...
```

Note that the result of a call to poker_rank is one of the histogram labels so we can pass it to hist.count to keep track of how many times that type of hand was dealt.

T91. Here is a loop that will print the histogram in order of decreasing probability:

```
>>> for x in poker_hands:
...     print(x, hist[x])
...
high card 519
pair 417
...
four of a kind 0
straight flush 0
```

♦ Find a table of poker probabilities on the Internet and compare your results with the expected frequency for each type of hand. Do you think there is any bias in these hands, or does it seem like permute did a good job of shuffling the deck?

7.7 Defining New Objects: class Statements

The cards used in the exercises in the previous sections are very simple objects. Each card has two attributes, a suit and a rank, which can be accessed by calling methods defined in the Card class. If we dig a little deeper, it's apparent there must be some additional code associated with these objects. Somewhere there is a specification of how to compare two cards, because we were able to sort a list of Card objects. There must also be some code that decides how to print the name of a card, for example printing the string 2♣ when we ask to see the object created by a call to Card(0).

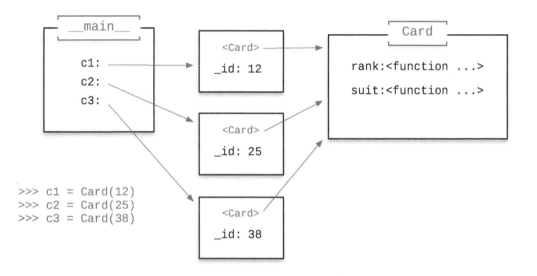

Figure 7.14: *Each instance of a class is a namespace that contains variables unique to each object and a reference to the object's class. A class is also a namespace; it contains definitions of methods and other items shared by all objects in the class.*

We conclude this chapter with a brief introduction to creating new types of objects in Python. We will learn how to collect all of the behavior associated with Card objects into a single module by defining our own class. Classes are important examples of how encapsulation is used to manage the complexity of a large program. Programmers who design an application to play a card game will think about everything their program needs to do with cards, and then organize all those operations into a single module that can be imported by other parts of the program.

To set the stage for how objects are defined in Python, consider the state of the object store in an interactive session after three Card objects have been created. Figure 7.14 shows that three names have been added to the whiteboard, and that each is a reference to a Card object. The new detail added by this drawing is that an object is a small namespace. In the terminology of object-oriented programming, each object is an **instance** of the Card class, and any variables that appear in an instance are called **instance variables**.

A programming convention is to give instance variables names that begin with a single underscore, so the variable that represents the ID of each card is named `_id`. The fact that each instance has its own `_id` variable is what distinguishes each object from the others.

Let's consider now what happens when we type an expression that calls a method, for example

```
>>> c3.suit()
2
```

Python first looks in the main namespace to find c3. Because c3 is a reference to an object, Python follows the link and discovers the object is an instance of the Card class. The link from the object instance to its class leads to a namespace that contains a function named suit, so Python calls that function.

The question at this point is, how does the suit function know we want the suit of the object pointed to by c3 and not some other object? To answer this question let's go back to an example from Chapter 2, where methods were first introduced. If s is a string object, and we want to know how many spaces are in the string, we would typically call the count method using this syntax:

```
s.count(' ')
```

What we saw in Chapter 2 was that this syntax is a convenient shorthand for a more formal notation that uses a qualified name with both the class and the name of the function:

```
str.count(s, ' ')
```

Python makes the same transformation when we use our own objects. If a program contains the statement

```
c3.suit()
```

Python treats it as if we had written

```
Card.suit(c3)
```

This second form is exactly what we want: a call to the function named suit defined in the class named Card, telling it to do something with the object named c3.

Normally we don't care about any behind-the-scenes magic that Python might be doing, but in this case it's important because it helps explain why function definitions inside a class appear to have an extra argument. Another programming convention is to give this argument the name self, and in the body of the method any reference to self is a reference to the object being worked on by the method.

Structurally, a class definition is very similar to a function definition: it begins with a class statement that specifies the name of the class, and is followed by a set of indented lines that define the methods that belong to the class. The Card class in Figure 7.15 shows three function definitions. Each def statement is indented with respect to the class statement on the first line, so each function is included in the body of the class, i.e., the function becomes a method that will work on objects of the class.

The first method defined in the class, __init__, has a name that begins and ends with a pair of underscore characters. We saw special names like this earlier in the chapter when we used the dir function to list the contents of a module. Python classes have several functions with names of this type that are called in a variety of special circumstances. All of them are optional, but as we will see they help make our Card objects easier to use.

The function named __init__ is called whenever we create a new instance of the class. When we type Card(12) Python does two things: it creates a new namespace to represent the new instance, and it makes a call to Card.__init__. For the Card class, the __init__ function is passed two arguments: a reference to the new namespace and the argument passed to the constructor, in this case the number 12.

The namespace that will become our new object instance is initially empty. The role of the __init__ method is to initialize the namespace by defining any new instance variables that should become part of the object. In the case of our Card class, we just need to save the card number. The assignment statement in the body of __init__ is how each instance in Figure 7.14 got its own unique value for its own copy of _id.

```
1   class Card:
2       "An instance of the Card class is a single playing card."
3
4       def __init__(self, n):
5           "Assign the ID for a new card"
6           self._id = n
7
8       def rank(self):
9           "A card's rank is an integer between 0 (2) and 12 (ace)."
10          return self._id % 13
11
12      def suit(self):
13          "Return a suit number: 0 = clubs, 1 = diamonds, 2 = hearts, 3 = spades."
14          return self._id // 13
15
16  def new_deck():
17      "Create a list of 52 cards objects."
18      return [Card(i) for i in range(52)]
```

Download: random/Card-v1.py

Figure 7.15: *A* class *statement defines a new type of object. A method is simply a function defined inside a class. Note how each* def *statement is indented so it becomes part of the body of the class. This module also includes a function named new_deck that is not part of the class.*

The next two methods in Figure 7.15 will use the value of an object's instance variable to return an attribute of a card. We saw earlier that a card's rank is a number between 0 and 12, where a 2 has the lowest rank and aces have the highest rank. Suits are numbers between 0 and 3. There is a simple equation that will figure out a card's rank or suit based on its ID. The rank is just the remainder after dividing _id by 13, and the suit is the truncated result after dividing by 13. Notice how the rank and suit methods simply fetch the value of the object's _id instance variable and plug it into the corresponding equation.

A class that has only the __init__, rank, and suit methods is all that is required to create and use Card objects. Making three instances of the class will lead to the state of affairs shown in Figure 7.14, where there are three small namespaces (one for each instance) plus a namespace for the Card class itself.

The collection of all the methods that work on objects, along with the functions that create new instances, is known as the **application program interface** (or **API**) for a class. The API for the first version of the Card class has three parts: the function that creates a new card and the two methods that return the rank and suit of a card.

We can simplify things for potential users of the Card class if we define a function that returns a complete deck of cards. If a function named new_deck (shown at the bottom of Figure 7.15) is part of the module, programs can call it to get a complete list of all 52 card objects. Programmers do not need to know that card objects have IDs, or to know what ID is associated with each card.

The new API consists of the new_deck function and the rank and suit methods. A card-playing program can shuffle the deck, deal cards, and check ranks and suits just by using the functions defined by the Card class API and without knowing any of the details about how Card objects are represented.

The reason Python programmers use distinctive-looking names that start with underscore characters is to emphasize the fact that these variables have a special purpose. They represent part of the state of an object, and they are not intended to be used outside the class. In object-oriented programming terminology, instance variables should be "hidden" inside each object. In Python the variables are not really hidden—we will see in the tutorial project at the end of this section that a card's ID is accessible from outside the class—but it's a good idea to follow the naming conventions so people who read the code understand what the variables are intended to be used for.

Our new Card module will let us create and use cards, but it is missing two important pieces. If we ask Python to print the value of a Card, it prints a string that's not very informative:

```
>>> Card(0)
<Card object at 0x102d36a10>
```

We can tell from this string that the object belongs to a class named Card, but the number (which is the location in the computer's memory where the object is stored) is not very useful.

To fix this problem we simply have to add a new method named __repr__. The name is short for "representation." Whenever Python needs to show the state of an object, it checks to see if the object's class has a __repr__ method, and if so it calls the method to get a string

```
1    def __repr__(self):
2        return "Card " + str(self._id)
```

Figure 7.16: *A method named   __repr__ is called whenever Python prints an object.*

```
1    def __eq__(self, other):
2        return self._id == other._id
3
4    def __lt__(self, other):
5        return self._id < other._id
```

Figure 7.17: *If c1 and c2 are Card objects the   __eq__   method is called to evaluate c1 == c2 and   __lt__   is called to evaluate c1 < c2.*

that describes the object. By supplying a definition of this method as part of our class we can control what Python prints when it displays a Card object in a shell session.

The version of __repr__ shown in Figure 7.16 is not terribly descriptive, either, but it is an improvement over the default behavior. It creates a string that gives the name of the class and the value of the instance variable _id:

```
>>> Card(0)
Card 0
>>> Card(26)
Card 26
```

The programming project in Section 7.8.1 will show you how to create a more descriptive string like 2♣ or 2♥.

Another improvement to the basic outline is to add two more special methods that allow us to compare Card objects. The method named __eq__ is called if we test to see if two objects are the same, and the method named __lt__ (the name stands for "less than") is called whenever we use Card objects in expressions involving the < operator. If we don't define these methods, Python does not know how to compare two Card objects:

```
>>> Card(0) < Card(26)
TypeError: unorderable types: Card() < Card()
```

But if we define __lt__ as shown in Figure 7.17, Python will call it for us, passing us references to two object instances. The name self will refer to the card on the left side of the <, and other will be a reference to the object on the right side. We simply have to return the result of comparing the two _id variables. This is what we will see after adding the definition of __lt__ to our class:

```
>>> Card(0) < Card(26)
True
>>> Card(33) < Card(26)
False
```

Without a definition of how to compare two Card objects Python would not know how to sort a list of cards. As soon as we provide an implementation of __lt__ as part of our class we can pass a list of objects to sorted and get back an ordered list:

```
>>> sorted( [Card(10), Card(5), Card(7)] )
[Card 5, Card 7, Card 10]
```

The Card class implemented in RandomLab has additional features that are not included in the basic version introduced in this section. The full version prints card names using symbols like ♥ and ♠, it generates a runtime error if a card number is outside the range from 0 to 51, and the __init__ function will make a random card if the constructor is called without an argument. All three of these issues will be addressed in the next section, but first let's do some tests with the basic design.

Object-oriented programming is very well suited to the incremental design process we have been advocating throughout the book. For the project in this section we will start by making the outline of a Card class and testing the basic class structure. Even though this first class has no behaviors defined for its objects, we will learn something about how classes work in Python. We will then add and test methods one at a time, each time making sure the class performs as expected before adding the next method.

Tutorial Project

The project in this section will use the incremental development process to create and test the Card class. You will be working with a file named Card.py. Make sure you start your interactive sessions in the same directory where you save this file.

T92. Create a new file named Card.py and enter the following two lines:
```
class Card:
    pass
```
The word pass is a Python keyword that serves as a placeholder for the body of a class or function that will be filled in later.

T93. Start a new shell session and import your new class:
```
>>> from Card import Card
```
The statement looks redundant, but it means "find the file named Card.py and import the definition of Card from that file."

T94. If you call the built-in function named dir with no arguments Python will show you the contents of its "whiteboard," the set of names defined for the current session:
```
>>> dir()
['Card', '__builtins__', ... '__package__']
```
You should see the name of the class you just imported and the special symbols (names starting with double underscores) defined for every session.

T95. Define a variable named x to be a reference to a Card object:
```
>>> x = Card()
```

The Art of Programming (Part III)

Some of the most important decisions a programmer makes are related to the **API** (application program interface) of new classes: what information to include as part of an object, how to create objects, and the behaviors defined for the objects.

Encapsulation

One of the major benefits of object-oriented programming is the ability to gather all the code that defines an object in one place. In particular, all the functions that access an object's instance variables should be defined as methods. As a result, instance variables can be "hidden" inside the class definition.

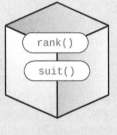

The Card class described in this chapter provides a simple example. From the outside, an instance of this class has two important attributes, its suit and its rank. To get the value of one of these attributes we call a method, for example,

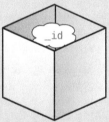

```
>>> c
A♣
>>> c.suit()
0
```

But if we look inside the object, all we see is a single variable that represents the card ID, which is a number between 0 and 51. In this design there isn't any need to have variables to represent the suit and rank; those values are computed as functions of the card's ID.

Encapsulation allows the people who implement the class the freedom to choose more efficient data structures or to make any other changes as long as the object continues to behave the same way. In this case, using a single integer ID makes it easy to compare cards and saves space (not much of an issue for this program, but potentially a problem for larger programs).

◆ Inheritance

Another important benefit of object-oriented programming is the ability to define new classes in terms of existing classes. The technique is not explained in this book, but you can find several examples throughout the PythonLabs modules.

One place where this technique is used is in the definition of the WordQueue class in SpamLab (the lab module for Chapter 6). A WordQueue has a lot in common with a list because it is a linear collection of objects. This declaration tells Python a WordQueue is a special kind of list:

```
class WordQueue(list):
    ...
```

By telling Python a WordQueue is a kind of list, the new class automatically "inherits" methods that access items, print the queue, etc. All we have to do is write the methods that are specific to queues, namely the code that makes sure the queue is always sorted.

T96. Ask Python to show you which class x belongs to:

```
>>> type(x)
<class 'Card.Card'>
```

The result printed above shows that x is an object belonging to a class named Card defined in a module named Card. Congratulations! You have just created your first object using a class you defined yourself.

T97. If you pass x to the dir function Python will show you the names defined in your object's namespace:

```
>>> dir(x)
['__class__', ... '__weakref__']
```

You will see a long list of names that start with double underscores. These are items defined for every object, but other than these names nothing else is defined for your object.

T98. Edit your class definition by entering the definition of the function named __init__ shown on lines 3–4 of Figure 7.15. The name must begin and end with two underscore characters.

T99. Import the new class definition into your shell session. Some IDEs may require you to restart your shell session; others will allow you to import the new class and have it replace the previous definition.

T100. Now when you make a Card object, you need to pass the card number to the constructor because your __init__ function needs a value for the argument it calls n:

```
>>> x = Card(0)
```

T101. If you look at the names defined for this object you should see the card ID created by __init__:

```
>>> dir(x)
['__class__', ... '__weakref__', '_id']
```

T102. Because x has a variable called _id you should be able to access it using a qualified name:

```
>>> x._id
0
```

T103. Create another new Card object and ask to see its ID:

```
>>> y = Card(26)
>>> y._id
26
```

Make sure you understand what Python did when it created these two objects. When asked to evaluate an expression of the form Card(i) Python will call the method named __init__ defined in the class named Card, passing it two parameters: a reference to a new empty namespace that will become the object instance, and the value of i, which is the intended object ID. The body of the function saves the object ID in a local variable in the new namespace.

T104. Edit the definition of your Card class to add the rank and suit methods shown on lines 6–10 of Figure 7.15.

T105. Import the new definition of the Card class and create a new object x:

```
>>> x = Card(51)
```

T106. Because card 51 is the ace of spades, it should have a rank equal to 12 and belong to suit 3 (see Table 7.1 on page 213):

```
>>> x.rank()
12
>>> x.suit()
3
```

T107. Make another card and ask for its rank and suit:

```
>>> y = Card(20)
>>> y.rank()
7
>>> y.suit()
1
```

T108. Try creating some other Card objects and test the rank and suit methods. Do you see how cards 0 to 12 belong to suit 0, cards 13 to 25 belong to suit 1, and so on?

T109. Ask Python to print the values of two of the variables you defined:

```
>>> x
<prng.Card object at 0x1044bf390>
>>> y
<prng.Card object at 0x1044bf350>
```

Because we haven't yet told Python how to print the value of a Card it simply prints the object's address in RAM. Note how each object is stored at a different address.

T110. Add the definition of the __repr__ method (Figure 7.16) to your class. Note there are two underscores at the beginning and end of the name.

T111. Start a new shell session, import your class, and create the two Card objects again:

```
>>> x = Card(51)
>>> y = Card(20)
```

T112. Now when you ask to see the values of x and y you should see strings created by __repr__:

```
>>> x
Card 51
>>> y
Card 20
```

T113. Ask Python to compare your two Card objects:

```
>>> y < x
TypeError: unorderable types: Card() < Card()
```

The error message means Python does not know how to compare two Cards.

T114. Add the definitions of the __eq__ and __lt__ functions from Figure 7.17 to your class, import the class into a new shell session, and create the same two objects again.

T115. Now when you compare two objects Python will use your definition of __lt__ to compare the IDs of the two cards:

```
>>> y < x
True
>>> x < y
False
```

T116. Import the function you wrote to shuffle a deck of cards:

```
>>> from permute import *
```

T117. Make a deck of cards using the new_deck function in your class:

```
>>> d = new_deck()
>>> d
[Card 0, Card 1, Card 2, ... Card 49, Card 50, Card 51]
```

Are all the cards there?

T118. Shuffle the deck and print it:

```
>>> permute(d)
>>> d
[Card 37, Card 43, Card 0, ... Card 14, Card 34, Card 8]
```

You should see all the cards, but in a random order.

T119. Because you told Python how to compare cards you should be able to sort the deck to put it back in order:

```
>>> d.sort()
>>> d
[Card 0, Card 1, Card 2, ... Card 49, Card 50, Card 51]
```

7.8 ◆ Additional Features for the Card Class

This section is an introduction to three issues that confront programmers when they design their own types of objects. They are in this optional section because they are relatively advanced topics that won't be used in projects later in the book. However, they are very important constructs, and the topics covered in Sections 7.8.2 and 7.8.3 are useful in other situations besides object-oriented programming.

7.8.1 ◆ Class Variables

One way to have the __repr__ method include suit symbols in the name of a Card object is to create a dictionary that maps suit numbers to suit symbols. We should do the same thing with ranks, because we want to print A when the rank is 12, K when the rank is 11, etc. If we have two dictionaries, one named suit_sym and another named rank_sym, all we need to do is look up the symbols, and the string we return will just be the concatenation of these two symbols:

```
rank_sym[self.rank()] + suit_sym[self.suit()]
```

The call to self.rank() means "call this card's rank method to get its rank," and similarly self.suit() means "get this card's suit number."

It's easy enough to define the two dictionaries. For suit symbols, we just need to associate a suit number with a one-letter string containing the Unicode character for that suit. The sidebar on escape sequences in the previous chapter (page 151) showed how \uNNNN places Unicode character number NNNN into a string. This assignment statement creates the dictionary that maps suit numbers to suit symbols:

```
suit_sym = {3:'\u2660', 2:'\u2665', 1:'\u2666', 0:'\u2663'}
```

We can do the same thing for ranks, mapping 12, the highest rank, to A, 11 to K, and so on (see the first project in the Tutorial at the end of this section).

The question is, where do we put the definitions? We could simply add two assignment statements to the body of the __repr__ method (Option 1 in Figure 7.18). The problem with this solution is that new dictionaries will be created each time the method is called. Every time we print a representation of a card Python will add two new dictionaries to the object store. If we print the entire deck, the object store will be cluttered with 104 new dictionaries, with one copy of suit_sym and one copy of rank_sym for each card.

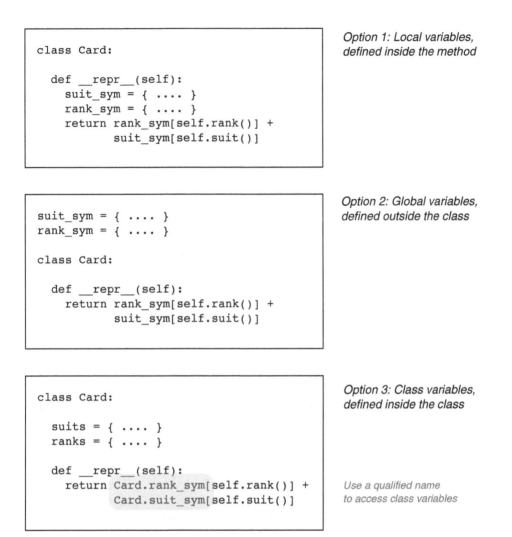

Figure 7.18: *Three ways of defining dictionaries that can be used by methods in the Card class.*

Python will take care of cleaning up the object store when the dictionaries aren't needed anymore, and the overhead of making such small dictionaries is not very great, but using local variables for the two dictionaries is a bad design. In other programs we might need to create much larger objects and the added overhead will become a serious problem.

A second solution is to define the dictionaries as global variables (Option 2 in Figure 7.18). This takes care of the multiple dictionaries problem. When the module is first loaded, the two assignment statements will be executed, and two objects will be created. Every call to __repr__ can access the dictionaries because they are global variables that can be read by any code defined in the same file. While this approach will work, it is not a very satisfying design. Our goal was to encapsulate all the information about Card objects inside the class, and these two dictionaries are defined outside the class. If a user imports the module with

```
from Card import *
```

the dictionaries become global variables in their code also.

The third approach is the one favored in object-oriented programming. By moving the assignment statements that define the objects inside the class, the two variables become part of the class namespace. An important distinction is that they do not belong to any one instance of the class, but instead they are shared by all instances. In object-oriented programming terminology, variables defined to be used by the class as a whole, and that are not part of the objects defined by the class, are known as **class variables**.

Using class variables for the two dictionaries gives us the best of both worlds: only two dictionary objects are created, and they are safely encapsulated inside the class. There is one wrinkle, though: in Python, if a method wants to access the value of a class variable, it has to use a qualified name that includes the class name, a period, and the variable name.

Tutorial Project

♦ Add the following two definitions to your Card class, making sure they are defined inside the class, as shown in Option 3 of Figure 7.18:

```
suit_sym = { 3: '\u2660', 2: '\u2665', 1: '\u2666', 0: '\u2663' }
rank_sym = { 0: '2', 1: '3', 2: '4', 3: '5', 4: '6', 5: '7',
    6: '8', 7: '9', 8: '10', 9: 'J', 10: 'Q', 11: 'K', 12: 'A' }
```

Python is pretty lenient about how you lay out list and dictionary definitions. You can put the second dictionary all on one line in your file, or split it up the way it's shown above.

♦ Change the expression in the body of the __repr__ method so it reads

```
return Card.rank_sym[self.rank()] + Card.suit_sym[self.suit()]
```

Notice how the references to the dictionaries use qualified names: this is how a method refers to a class variable.

♦ Load your updated class definition into an interactive session. Are all the cards in the deck being displayed the way you expect?

```
>>> [Card(i) for i in range(52)]
[2♣, 3♣, 4♣, ... Q♠, K♠, A♠]
```

♦ You can see that class variables are names in the class namespace by calling `dir`:

```
>>> dir(Card)
['__class__', ... 'rank', 'rank_sym', 'suit', 'suit_sym']
```

In addition to all the special names you should see the names of the two class variables and the two methods.

♦ From outside the class you can also refer to a class variable using a qualified name. This statement, typed in the shell window, shows how top-level code can access the value of the suit symbol dictionary inside the Card class:

```
>>> Card.suit_sym
{0: '♣', 1: '♦', 2: '♥', 3: '♠'}
```

7.8.2 ♦ Exceptions

Several of the functions described in previous chapters returned the special object None to indicate something was wrong. For example, the search functions return None if they cannot find an object in a list, and the "spamicity" function of Chapter 6 returns None if a word is not defined in both word lists.

We often want to do something similar when writing the __init__ function for a class. It is a good idea to check to see if the arguments passed to a constructor are valid. The constructor for the Card class, for example, expects to see a number between 0 and 51. If we try to make a card with an ID outside that range we should see an error message.

The problem is there is no way to return any sort of "illegal card" indicator from the __init__ function. Every time we call Card, we will always get back a reference to a new object, regardless of whether the object was initialized properly. It's easy enough to add a print statement that displays an error message, but that won't be very helpful, because the result of the call would still be a new Card object.

The best way to address this problem is to have __init__ generate a runtime error, similar to the way a dictionary generates an error when we look up a value using an undefined key. In Python, these sorts of errors are called **exceptions**, and we say a program **raises** an exception when it detects an error condition.

Raising an exception is simple: we just have to use a raise statement. This example shows how the Card constructor might check to see if n, the requested card number, is outside the range from 0 to 51:

```
if n in range(0,52):
    self._id = n
else:
    raise Exception("Card number must be between 0 and 51")
```

The expression to the right of the word raise creates an object from a builtin class named Exception. In the simplest case, shown here, the argument is an error message that describes the reason the program is raising the exception.

The next question is how to handle an exception. In programming terminology, we want to **catch** the exception. Python programmers use a special type of if statement called a try statement:

```
try:
    c = Card(n)
    print(c)
except:
    ...
```

We can put as many statements as we'd like on the lines following the word try (all indented the same number of spaces). If any of those statements causes a runtime error, Python immediately stops what it is doing and jumps to the statements following the word except. In this example, if n is a valid card number, the card is printed and Python ignores the except clause. If n is not valid, Python never gets to the print statement and instead executes the code in the body of the except clause.

The program in Figure 7.19 shows how the except statement is commonly written. In this form of the statement the Exception object is saved in a variable named e. The code in the body of the except statement can then refer to e, for example, to print the string that was specified when the exception was created.

There is much more to try and except statements than is shown here. The except clause can test for different types of errors, and we can write code that is executed if no errors occur. But the simple form shown here and in Figure 7.19 are sufficient for short programming projects like the ones in this book.

```
1    from Card import Card
2
3    def make_card(n):
4        "Make a Card object"
5        try:
6            return Card(n)
7        except Exception as e:
8            print("invalid card:", e)
9            return None
```

Figure 7.19: *This function will try to create a Card object. If it succeeds, it will return the new card, otherwise it prints an error message that includes a description of why the error occurred and returns None.*

Download: random/makecard.py

Tutorial Project

♦ Import the Card class and other definitions from RandomLab:
```
>>> from PythonLabs.RandomLab import *
```

♦ To see what happens when an exception is raised try to create a card with an illegal ID:
```
>>> c = Card(100)
PythonLabs.RandomLab.RandomError: card number must be between 0 and 51
```

♦ Ask Python to show you the value of c:
```
>>> c
NameError: name 'c' is not defined
```
The fact that c is undefined means Python did not complete the assignment statement because the Card class raised an exception.

♦ Use your IDE to create the class named X shown in Figure 7.20. Load the definition into an interactive session.

♦ Create an object of the class and verify the instance variable is initialized as expected:
```
>>> a = X(10)
>>> a._val
10
```

♦ Call the constructor again, this time passing a negative value:
```
>>> b = X(-10)
n must be positive
```

```
1    class X:
2        def __init__(self, n):
3            "** bad design **"
4            if n < 0:
5                print("n must be positive")
6            else:
7                self._val = n
```

Figure 7.20: *If the constructor for this class is passed a negative number it prints a message but does not raise an exception.*

♦ An error message was printed, as expected, but notice that the constructor still made an object:

```
>>> type(b)
<class '__main__.X'>
```

♦ The variable b refers to an instance of the class, but that instance does not have a val attribute because the assignment statement on line 6 wasn't executed:

```
>>> b._val
AttributeError: 'X' object has no attribute '_val'
```

Do you see why a constructor needs to raise an exception? If it just prints an error message an object is still created and returned.

Edit the class definition so the constructor raises an exception instead of printing a message. Start a new session, load the new definition, and repeat the experiments above. This time Python should not create the object named b because the exception should interrupt the assignment statement before it is complete.

♦ Download the definition of make_card shown in Figure 7.19 or create a new file with your IDE and enter the definition.

♦ Load the function and use it to make some Card objects:

```
>>> make_card(35)
J♥
>>> make_card(55)
invalid card: card number must be between 0 and 51
```

♦ Get ten random integers between 0 and 99 and use them to make a list of cards:

```
>>> a = RandomList(10,100)
>>> a
[18, 33, 48, 40, 0, 71, 24, 81, 13, 67]
>>> [make_card(i) for i in a]
invalid card: card number must be between 0 and 51
invalid card: card number must be between 0 and 51
invalid card: card number must be between 0 and 51
[7♦, 9♥, J♠, 3♠, 2♣, None, K♦, None, 2♦, None]
```

Do you see how make_card caught the exceptions and printed error messages, and how the program kept running after the exceptions were handled?

♦ What do you think would happen if you repeat the previous experiment, but call Card(i) instead of make_card(i)? Run the experiment to test your hypothesis.

7.8.3 ♦ Optional Arguments

It is possible to define a function in such a way that it can be called with a different number of arguments. This technique is especially useful when defining the __init__ method for a class because it gives users some flexibility in how to create an object.

In this section we will see how to modify __init__ in the Card class so a new card object can be created either by specifying the card ID or by specifying its rank and suit symbols. With the new version of the constructor, either of these expressions will create a new card representing the jack of hearts:

```
>>> c = Card(35)
>>> c = Card('J', '\u2665')
```

The first step is to write a def statement that specifies default values for one of the arguments. The bmi function in Figure 7.21 shows how this is done. Notice how the third argument listed in the def statement looks like an assignment. This definition is telling Python that when bmi is called the user must supply values for the first two arguments. If the call has only two arguments, the string 'metric' will be used as the value of units, but if the call includes a third argument, that value will be used instead.

Given this definition, we can compute a body mass index by passing a height in meters and a weight in kilograms:

```
>>> bmi(1.8, 80)
24.691358024691358
```

But we can also specify a height in inches and a weight in pounds if we pass a third argument that tells the function to treat the numbers as standard U.S. units:

```
>>> bmi(72, 180, 'standard')
24.409722222222225
```

A version of the __init__ function for the Card class that allows users to specify either a card number or a rank and suit is shown in Figure 7.22. This new definition illustrates all three Python constructs described in this section:

- It uses class variables to define suit and rank symbols.

- It allows the constructor to be called with either one or two arguments.

- It raises an exception if the arguments are not valid.

Two new dictionaries, suit_num and rank_num, are the inverse of the dictionaries introduced earlier. Before we saw how

```
suit_sym[i]
```

could be used to look up the string representing one of the suits. The new dictionary allows us to do the opposite:

```
suit_num[s]
```

evaluates to a number between 0 and 3 if s is one of the four symbols used to display suits. When two arguments are passed to the constructor, it first checks to make sure the two strings are valid suit and rank strings, and if so, converts them into a card number.

```
1  def bmi(height, weight, units = 'metric'):
2      "Compute a Body Mass Index value"
3      if units == 'metric':
4          return (weight / height**2)
5      elif units == 'standard':
6          return (weight / height**2) * 703
7      else:
8          return None
```

Figure 7.21: *This version of the body mass index calculator lets users pass height and weight in either metric or standard U.S. units.*

```
1   suit_sym = { 3: '\u2660', 2: '\u2665', 1: '\u2666', 0: '\u2663' }
2   rank_sym = { 0: '2', 1: '3', 2: '4', 3: '5', 4: '6', 5: '7',
3       6: '8', 7: '9', 8: '10', 9: 'J', 10: 'Q', 11: 'K', 12: 'A' }
4
5   suit_num = dict(zip(suit_sym.values(), suit_sym.keys()))
6   rank_num = dict(zip(rank_sym.values(), rank_sym.keys()))
7
8   def __init__(self, arg1, arg2 = None):
9       "Create a new card object from a card ID or rank and suit name"
10      if type(arg1) == int:
11          if arg1 in range(0,52):
12              self._id = arg1
13          else:
14              raise Exception("Card number must be between 0 and 51")
15      elif arg1 in Card.rank_num and arg2 in Card.suit_num:
16          self._id = Card.rank_num[arg1] + 13 * Card.suit_num[arg2]
17      else:
18          raise Exception("Expected Card(n) or Card('rank', 'suit')")
```

Figure 7.22: *This version of the function that initializes new Card objects allows users to specify either one argument, which is expected to be a card number, or two, in which case they should be strings representing the card's rank and suit.*

Tutorial Project

You can download the complete definition of the Card class in a file named `Card.py` from the book website, or you can add the code shown in Figure 7.22 to your own copy of the file.

♦ Load the new definition of the Card class into an interactive session and use it to make some Card objects:

```
>>> c1 = Card('2', '\u2663')
>>> c2 = Card('Q', '\u2660')
```

♦ Verify the card IDs have the expected values:

```
>>> c1._id
0
>>> c2._id
49
```

♦ Type a few expressions using these cards to verify they work like other Card objects:

```
>>> c2
Q♠
>>> c1 == Card(0)
True
>>> c2.rank()
10
>>> c1 < c2
True
```

♦ Print the two dictionaries that map suit symbols to numbers, and vice versa:

```
>>> Card.suit_sym
{0: '♣', 1: '♦', 2: '♥', 3: '♠'}
>>> Card.suit_num
{'♦': 1, '♥': 2, '♣': 0, '♠': 3}
```

Do you see how the suit_num dictionary is the inverse of suit_sym?

♦ Type expressions that will raise exceptions:

```
>>> Card(99)
Exception: Card number must be between 0 and 51
>>> Card('two of clubs')
Exception: Expected Card(n) or Card('rank', 'suit')
>>> Card('jack', 'hearts')
Exception: Expected Card(n) or Card('rank', 'suit')
```

7.9 Summary

Making lists of random values is a subtle problem. It's hard for humans to do, and it's not quite so easy for machines, either.

In this chapter we saw how to make a long list of numbers that, for short stretches in the middle of the list, appear to be random. Given the right combination of parameters, the equation that defines how to add a new value to the list will make sure each new number appears to be unrelated to the previous one, and that every number between 0 and $n - 1$ eventually appears in a list of n numbers.

But even though the list has all n numbers, and they appear to be in a random order, there can be some hidden patterns. For example, when we took a closer look at one of the sequences we studied, we saw that it alternated between even and odd numbers. Even more subtle patterns might be hidden in the data. We explored two methods for testing randomness by drawing pictures based on values produced by the random number generator. These informal methods can help us tell, at a glance, whether there is a hidden bias or pattern.

Random number generators are among the most important and widely used algorithms in computer science. They are used in games, of course, but there are many other application areas where random values play a key role. We used random sequences in Chapters 4 and 5 to test searching and sorting algorithms. Many scientific algorithms also use random numbers. For example, some algorithms that reconstruct the evolutionary history of a set of genes use a function based on taking a random sample of all possible family trees to find the one that is most likely given the similarities between the genes. Later in this book, Chapter 12 describes a type of problem known as optimization, where the goal is to find the optimal solution to a problem, and one common approach also uses random samples. E-Commerce, Internet banking, and other network traffic depend on encryption algorithms to turn a piece of text into what appears to be a random sequence of letters; many of the important concepts used to design effective random number generators are also used in encryption algorithms.

This chapter also provided an opportunity to introduce some new Python programming constructs. We saw that Python source files correspond to modules, and that when a file is loaded by an import statement, Python executes the statements in the file as it reads the module. Anything defined in the module, whether it is a class or function definition or a

---

Concepts and Terminology Introduced in This Chapter

random	A sequence of items is random if the values are independent of one another
pseudorandom	A pseudorandom sequence is one produced by an algorithm; pseudorandom sequences are not truly random, but small subsequences may appear to be random
PRNG	Pseudorandom number generator, an algorithm that creates a sequence of pseudorandom values
mod operator	An operator that computes the remainder of a division; in Python, the remainder after dividing x by y is written x % y
permutation	A reordering of the items in a list
uniform distribution	The result of generating random values where each value is equally likely
histogram	Also known as a bar chart; a visual display that shows the number of times various events occur
namespace	A collection of names, often associated with a module
class	A keyword in Python used to start the definition of a new type of object
instance variable	A variable that helps define the state of an object

---

variable created by assignment statements, becomes part of the namespace of the module. We can then access these items using the qualified name syntax, in which we write the module name, a period, and the item name.

Namespaces provide the foundation for object-oriented programming in Python. When we call a constructor, Python creates a new empty namespace, and the __init__ function defined in the body of the class is invoked, allowing it to initialize the instance variables that define the state of the object.

Methods are simply functions defined in the body of a class. When we call a method, Python saves a reference to the object in the first parameter (which we typically call self), so the method can access the object's state. For the project in this chapter we simply used objects to represent playing cards, which are simple objects that do not change after they are created, but in future chapters we will see objects that are updated by method calls.

Exercises

7.1. What does Python print as the value of the following expressions?

    a) 19 % 5

    b) 21 % 7

    c) ((21 * 7) + 16) % 31

    d) ((99 * 81) + 337) % 1000

7.2. Suppose a list a is initialized to be the list [0], and then Python executes this for statement with one of the statements below in the body of the loop:

```
for i in range(10):
```

In each case, show the value of a after the loop terminates:

a) `a.append(i ** 2)`

b) `a.append(a[-1] + 10)`

c) `a.append((a[-1] + 3) % 10 )`

7.3. The schedule in Section 7.1 for events that occur every seven hours had a period of 12, i.e., every number between 0 and 11 appears once in the schedule before it repeats. Can you find other intervals besides seven hours that also lead to a period of 12?

7.4. Redo the schedule for events that occur every seven hours using a 24-hour clock. Is the period still 12? Or is it now 24?

7.5. What are some intervals that lead to schedules with a period of less than 24 when a 24-hour clock is used?

7.6. What are some of intervals besides every seven hours that lead to a full schedule with 24 times and every time between 0 and 23 occurring exactly once?

7.7. What do the answers to the previous problem have in common? Is there a common attribute for intervals that lead to schedules with a period of 24 as opposed to those that lead to a period of less than 24?

7.8. Create a dot plot for numbers between 0 and 999:

```
>>> view_dotplot(1000)
```

Recall from Exercise T19 on page 26 that Python's randint function can be used to get a random integer between a specified upper and lower bound. These two statements import the function from its library and make a random value between 0 and 999:

```
>>> from random import randint
>>> randint(0, 999)
```

Use randint to make a dot plot of 3,000 points, where the x and y coordinates are all random values between 0 and 999. Do the points appear to be spread uniformly all over the canvas?

7.9. Use Python to run an interactive experiment that counts how often different cards appear in poker hands. Make a histogram with 13 bins. Create a deck of cards, and then execute a for loop 1,000 times. On each iteration, shuffle the deck, make a hand with the first five cards, and update the histogram bins for the ranks of those five cards.

7.10. If you took the challenge at the beginning of Section 7.1 and entered a series of random digits in a file, make a histogram from your numbers. One way to make a histogram is to load the numbers into a spreadsheet application and use its "chart" command. Were you able to make a uniform distribution? Or are some numbers a lot more frequent than others?

7.11. ◆ Here are some values of a, c, and m that should generate much better pseudorandom sequences than the ones we used in this chapter:[3]

a	c	m
1255	6173	29282
171	11213	53125
421	54773	259200

Repeat some of the experiments in this chapter to evaluate the quality of the pseudorandom sequences defined by these combinations. Use the resulting values to make histograms (not just of values from 1 to 6, but for other ranges, too) and dot plots.

---

[3]From W. H. Press, et al., *Numerical Recipes in C: The Art of Scientific Computing*, Second Edition, Cambridge University Press, 1992.

7.12. Is it possible for a call to permute to have no effect? That is, for a list a to have the same order before and after a call to permute(a)? Explain why or why not.

7.13. Write a Python expression that uses list comprehension to make a list containing the four aces:

[A♣, A♦, A♥, A♠]

7.14. Tutorial exercise T102 showed that it is possible to figure out a card's ID by asking for the value of the _id instance variable:

```
>>> c = Card(35)
>>> c._id
35
```

Do you think Python will allow you to change a card's ID? What would happen if you typed this statement after defining c as shown above?

```
>>> c._id = 38
```

Print the card again. Does this experiment help convince you it's a good idea to try to keep instance variable names hidden inside a class definition?

7.15. An import statement typed into an interactive session creates a new name on the "whiteboard." For example, if you type

```
>>> import math
```

the name math is defined, just as if you had typed an assignment such as

```
>>> math = 3.14159
```

Can you think of an experiment that will test the claim that an import statement is a type of assignment? Is there anything special about a name defined by an import compared to a name defined by an assignment statement?

Programming Projects

7.16. Write your own version of prng_sequence. Your function should take four parameters: n, the number of items to include in the sequence, and a, c, and m, the parameters that define the sequence. The first number in the sequence should be 0, but the remaining numbers should be defined by the equation we used in this chapter, namely, $x_{i+1} = (a \times x_i + c) \bmod m$.

7.17. Modify your prng_sequence function so the first item in the sequence is defined by the system clock.

7.18. Modify the function named roll so it takes an argument that specifies the number of sides on the die. For example, call roll(6) to roll a 6-side die, or roll(20) to play a game with a 20-side die.

7.19. ♦ Modify the roll function from the previous problem so the number of sides is 6 by default. Calling roll() with no argument will return a random value between 1 and 6, but roll(20) will return a value between 1 and 20.

7.20. Add a method named random to your prng module. A call of the form random(i,j) should return a random number between i and j.

7.21. Write a function named points that will return the number of points in a bridge hand. The simplest version will simply count 4 points for an ace, 3 points for a king, 2 points for a queen, and 1 point for a jack. If you want to make a more complicated version, add distribution points: 3 points if there are no cards in a suit, 2 points if there is one card in a suit, and 1 point if there are two cards.

7.22. Write a function named print_hand that will print a bridge hand on four lines. The function should take a single argument, which you can assume will be a list of 13 Card objects. The function should print all the clubs on the first line, diamonds on the second line, hearts on the third line, and spades on the last line. The cards should be printed in decreasing order on each line. Here is an example (assuming deck is a list with all 52 Card objects):

```
>>> permute(deck); h = deck[0:13]; print(h)
[J♦, A♥, Q♦, 8♥, K♣, A♣, Q♣, 8♠, K♦, 10♦, Q♠, 5♠, 6♣]
>>> print_hand(h)
A♣  K♣  Q♣  6♣
K♦  Q♦  J♦  10♦
A♥  8♥
Q♠  8♠  5♠
```

7.23. Write a version of the print_hand function (see the previous problem) that displays a bridge hand the way it typically appears in a newspaper column. Suit symbols are printed only once, at the front of each line, and the suits are printed in reverse order, so the hand shown above would be printed as

```
♠ Q 8 5
♥ A 8
♦ K Q J 10
♣ A K Q 6
```

7.24. Define a function named new_pinochle_deck that will create a list of Card objects representing the cards used in pinochle. This game is played with a special deck of 48 cards that has only the 9, 10, J, Q, K, and A of each suit, but it has two of each of those cards.

```
>>> new_pinochle_deck()
[9♣, 9♣, 10♣, 10♣, ... K♠, K♠, A♠, A♠]
```

7.25. Define a class named Point that will represent a point on a graph. To create a Point object, pass the x and y coordinates to the constructor:

```
>>> p1 = Point(1,1)
>>> p2 = Point(4,5)
```

Include the following methods in your class (the examples refer to the two points p1 and p2 shown above):

- The __repr__ method should display the point in standard mathematical notation:

  ```
  >>> p1
  (1,1)
  ```

- A method named dist should compute the distance between two points:

  ```
  >>> p1.dist(p2)
  5.0
  ```

- A method named polar should return a pair of values corresponding to the polar coordinates of the point:

  ```
  >>> p1.polar()
  (1.4142135623730951, 0.7853981633974483)
  ```

  The polar coordinates of a point (x, y) are a pair of numbers (r, θ) where $r = \sqrt{(x^2 + y^2)}$ and $\theta = \tan^{-1} y/x$ (Python's math library has a function named atan that computes $\tan^{-1}$).

7.26. ♦ The formula for polar coordinates in the previous problem is valid only if the x coordinate of a point is greater than 0. Modify the polar method so it returns the correct value if x is negative or 0.

7.27. ◆ Define a class named Sequence that will represent DNA sequences. To create a new Sequence, call the constructor with a string of letters that are either A, C, G, or T:

```
>>> s = Sequence('GATTACA')
```

The constructor should save the string in an instance variable named _seq. Your class should define the __repr__ method so it returns the value of _seq when a user wants to print a sequence:

```
>>> s
GATTACA
```

Here are some suggestions for additional methods to add to your class. Before you start to write any code, experiment with Python's slice operator. Because _seq is a string, _seq[i:j] represents the characters from locations i through j in _seq. This operation will be very useful when accessing substrings of _seq when you implement the methods described below.

- Implement a method named mutate that will change a single letter in the sequence. For example, this call will change the letter at location 4 to a T:

  ```
  >>> s.mutate(4,'T')
  >>> s
  GATTTCA
  ```

  One way to implement mutate is to have it replace _seq with a new string that has the first i letters in _seq (i.e., _seq[:i]) plus the new letter plus the rest of the letters in _seq (i.e. _seq[i+1:]).

- Define a method named insert. A call to s.insert(i,t) should insert the letters in string t before location i:

  ```
  >>> s = Sequence('AAAA')
  >>> s.insert(2,'TT')
  >>> s
  AATTAA
  ```

- Define a method named delete. A call to s.delete(i,n) should remove the n letters starting at location i:

  ```
  >>> s
  AATTAA
  >>> s.delete(3,2)
  >>> s
  AATA
  ```

- Define a method named subseq. A call to s.subseq(i,j) should create a new Sequence object using the letters in locations i through j-1 of s.

  ```
  >>> t = s.subseq(1,3)
  >>> t
  AT
  >>> type(t)
  <class 'Sequence'>
  ```

7.28. ◆ Add statements to the __init__ function for your Card class so it raises an exception if the card ID passed as an argument is not a number between 0 and 51.

7.29. ◆ Modify the __init__ function for your Card class so users can pass either a card ID or two strings, such as 'jack' and 'hearts'.

Chapter 8

Bit by Bit

Binary codes and algorithms for text compression and error detection

Computer systems use a variety of technologies to store and transmit data. Processor and memory chips are based on semiconductor technology; disks are made from magnetic materials spread over the surface of a thin, circular plate; removable disks have a reflective surface with tiny pockmarks that are detected by lasers; digital cameras and cell phones store digital information in "flash" memories that retain their data even when the device is turned off; networks transfer information in the form of light waves over fiber-optic cables.

As diverse as these technologies are, they have one thing in common that allows them to be used to store or transmit information: they are physical devices that can be set to one of two different states. A semiconductor memory cell will be in a state where we can measure one of two different voltage levels, or a section of a DVD will be smooth or have an indentation, which can be detected by a laser scanning the surface of the disc.

Exactly how the two states are managed physically is in the realm of computer engineering, or, as programmers like to say, is "a hardware problem." The role of computer science is to figure out how to represent information in a form that can be stored or transmitted using these technologies.

The starting point is to use two symbols to stand for the two different states. Traditionally, the symbols are 0 and 1, the digits of the binary number system. Engineers designing a memory system decide what voltage levels will be used inside memory chips, and we can assign the symbol 1 to a level of 3.5 volts and the symbol 0 to a level of 0 volts.

Computer scientists figure out how to represent text and other information as a sequence of binary symbols and devise algorithms that manipulate these symbols. A variety of real-world objects, things like pieces of text from a book, images captured by a camera, or sounds

---

Codes (Secret and Otherwise)

In everyday usage the word "code" is commonly associated with secrecy or privacy, as in "diplomatic code" or "personal access code."

In computer science, however, a code is simply a sequence of *0*s and *1*s used to represent data inside a machine or in a message being sent over a network.

A **coding scheme** defines codes for a particular set of items. A good example is ASCII, which is a common scheme for representing text. To transmit a message, each letter is translated into a code, a process known as **encoding**. At the other end of the network, a **decoding** operation translates codes back into text. There is no attempt to keep the message secret, so anyone with a table of ASCII codes can recover the original string.

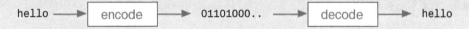

The process of turning a piece of text into a secret message is **encryption**. An encryption algorithm uses a password (usually called a **key**) to scramble the message so it looks like a random string of *1*s and *0*s. The only way to recover the original message is to pass the same key to a **decryption** algorithm.

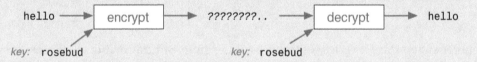

Programmers also use the word "code" to simply mean "program," as in "the Python code is in this file" or "the compiler generated the object code."

---

recorded in a studio, can be described by a sequence of 0s and 1s, and it is these sequences that are stored or transmitted.

The process of transforming information into a sequence of 0s and 1s is known as **encoding**. The projects in this chapter explore algorithms that work with encoded text. We will begin by looking at basic methods for encoding characters as sequences of binary digits, and see how strings in Python are basically sequences of encoded characters.

Encoding text is a surprisingly complex topic. While it might seem like a very straightforward process—just look in a table that defines codes to find the sequence of 0s and 1s for each character and store the result in memory—there are many difficult and subtle issues. To take just one example, how should the encoding scheme deal with accents? If a text has a letter in its basic form and also a form that includes an accent, such as *e* and *é*, do we need two different codes? Or should we have one code for the basic letter *e*, and a second code for the accent, so that when *é* appears in a text it is encoded with two codes? How will this decision affect a sorting algorithm when it compares letters? In French, putting an accent on a letter does not have any effect on the ordering of words, so *école* comes before *elegant* in a dictionary. When sorting French words the scheme that encodes accents separately from letters might make life easier as the sorting algorithm could just ignore accent symbols. But in Swedish, å and a are considered two different letters. What looks like an accent symbol—the little "o" above an "a"—to an English speaker is not a separate piece in Swedish, it's part of the letter itself. Words starting with å follow words starting with z in a dictionary, so an algorithm that sorts Swedish words would prefer an encoding that uses different codes for å and a.

Going in the opposite direction, from the coded form back to the original data, is a process known as **decoding**. For text, decoding is usually just a matter of reverse translation, of finding the letters or symbols that correspond to each binary code. But as we will see, even this process can be tricky, as the decoder needs to know what rules the encoder followed when the text was first turned into a sequence of bits.

Errors occur with surprising frequency, both when data is transmitted over a network or is simply sitting in memory. No matter what sort of technology is used, data is often lost or changed in transmission, or the storage media can be corrupted. One way to deal with potential errors is to store extra information along with the data, and then to use this extra information to see if an error has occurred. In Section 8.4 we will look at algorithms that figure out the necessary extra information and use it to detect errors.

Another issue related to binary representation is data compression. In some cases, data can be compressed by coming up with an alternative representation that captures all the essential information. A good example is audio compression. MP3 files are approximately $\frac{1}{10}$ the size of the original digital recordings, but in many cases the music sounds almost as good as the original. With text, however, we can't afford to lose any information. Changing a single letter or substituting a single punctuation mark can greatly modify the meaning. Although we can't replace the original text with an approximate copy, there are ways to reduce the size of the encoding without losing any information. In Sections 8.5 and 8.6 we will look at a simple but effective algorithm that generates alternative codes for letters, resulting in an encoding that is shorter yet retains all the information from the original text.

8.1 Binary Codes

The 0s and 1s in a binary encoding are commonly referred to as **bits**. In this context the word "bit" is an abbreviation for "binary digit." Because the devices that store information or transmit it from one system to another are binary (two-state) devices, a bit is the smallest unit of information that can be stored or transmitted.

One of the fundamental relationships in computer science is between the number of bits in a code and the number of items that can be encoded. The relationship is defined by the formula $n = 2^k$, which says that n, the number of different codes we can construct, is equal to 2^k, where k is the number of bits in the code.

To see how this formula is derived, start with the fact that there are two possible 1-bit codes, 0 and 1. With two bits, there are four codes: 00, 01, 10, and 11. When written in this order, one can see a pattern: the first two codes begin with 0, and the second two begin with 1.

Let's use this same pattern to write out all the 3-bit codes (Figure 8.1). Write the four 2-bit codes on a piece of paper, and then write them again, so there are two groups of codes.

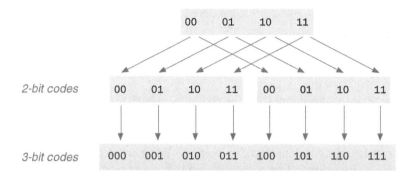

Figure 8.1: *To create a 3-bit code, make two copies of the 2-bit code and put a 0 bit in front of one copy and a 1 bit in front of the other copy.*

Put a 0 in front of each code in the first group, and a 1 in front of each code in the second group, and you will have all eight possible 3-bit codes.

The general rule is that we can write twice as many codes using $n + 1$ binary digits as we can with n digits, because we write each n-bit code twice, and then write a 0 in front of half the codes and a 1 in front of the other half. There are 2^2 2-bit codes, $2 \times 2^2 = 2^3$ 3-bit codes, and, more generally, 2^n n-bit codes.

As a practical example of this relationship, the ASCII code described in the next section is a 7-bit code, which means there are $2^7 = 128$ different patterns. Each pattern can be used to encode a different character. For example, the letter A corresponds to the pattern 10000001, B is 10000010, and so on. 128 patterns gives us enough codes for ordinary English text, as we can use 26 of the codes for lowercase letters, another 26 for uppercase letters, and still have several codes left over for punctuation marks and other symbols.

The inverse equation is also an important formula. Often we are given a set of objects, and we want to know how many bits will be required to encode each item in the set with a different sequence of bits. Because $n = 2^k$, it follows that $k = \log_2 n$, where n is the size of the set. For example, if we want to encode a piece of text, and we know the text contains only 26 uppercase letters and 6 punctuation marks, for a total of 32 characters, we could design a code that uses $\log_2 32 = 5$ bits per character.

When the logarithm is not an integer, we just "round up" to the next higher integer. The more formal way to express the relationship between the size of a set, n, and k, the number of bits required to assign a unique code to each item in the set, is

$$k = \lceil \log_2 n \rceil$$

(recall from Chapter 5 the notation $\lceil x \rceil$, pronounced the "ceiling of x," means "the smallest integer greater than x").

Binary codes can be used to represent more than just numbers and characters. Any type of data based on a finite set of values can be encoded so it can be stored in a file and analyzed by a computer. As an example, suppose a research group is collecting data through a survey, and the questionnaire has places for respondents to fill out their age, the state they live in,

Alabama	000000	A's	00000
Alaska	000001	Angels	00001
...		...	
Wyoming	110001	Yankees	11101

0011011	010111	10110
age	state	team

Figure 8.2: *Responses from a survey can be encoded in binary by using $\lceil \log_2 n \rceil$ bits for a field that has n alternatives. Each response can be encoded in 18 bits: 7 bits for numbers up to 127, 6 bits to represent one of the 50 U.S. states, and 5 bits for one of 30 different baseball teams.*

and their favorite baseball team. Age is a number, so we can write the number in base 2 to represent a respondent's age. If the researchers assume everybody who fills out the survey will be less than 128 years old, they only have to use $\lceil \log_2 127 \rceil = 7$ bits to encode the age of the respondent.

To encode the response for the state a person lives in, the researchers can use the fact there are 50 U.S. states, which means $\lceil \log_2 50 \rceil = 6$ bits are needed to encode a state name. One way to assign codes is to go in alphabetical order, so Alabama is 000000, Alaska is 000001, and so on. The same sort of reasoning would lead to 5-bit codes for each of the 30 major league baseball teams (Figure 8.2).

The projects for this chapter begin with experiments on binary codes. The BitLab module defines a new class called Code. One way to use the constructor for this class is to have it create a sequence of 1s and 0s corresponding to the pattern of bits used to encode a number. For example, to create the binary code for the number 23 we simply write

```
>>> Code(23)
10111
```

If we want a code that has a specific length, the constructor lets us pass a code length as a second parameter. To see the same code, but as an 8-bit number, the expression is

```
>>> Code(23, 8)
00010111
```

The code is the same pattern as before, but it has extra 0s at the front (the equivalent of writing *02* instead of *2* when entering a month in a form).

When we are working on a project that uses codes, a convenient way to manage them is to put them in a dictionary. For example, suppose we want to encode DNA sequences. These sequences are commonly represented as strings that have only the letters *A*, *C*, *G*, and *T*. Because there are only four letters, the dictionary would associate each letter with a different 2-bit code.

Figure 8.3 shows a function that will create such a dictionary. The function expects us to pass it a string of characters, and it will return a dictionary that has a unique binary code for each character. Here is a call that will make the code for the DNA letters:

```
>>> make_codes('ACGT')
{'G': 10, 'T': 11, 'C': 01, 'A': 00}
```

```
1    from PythonLabs.BitLab import Code
2    from math import ceil, log2
3
4    def make_codes(seq):
5        "Create a list of unique binary codes for each item in seq"
6        n = ceil(log2(len(seq)))          # required number of bits
7        codes = {}
8        for i in range(len(seq)):         # associate seq[i] with
9            codes[seq[i]] = Code(i,n)     #    the code for i
10       return codes
```

Download: encoding/makecodes.py

Figure 8.3: *A call to* make_codes(s) *(where* s *is either a string or a list of items) will create a dictionary that associates every item in* s *with a different binary code.*

Recall that when Python prints a dictionary, it can print the key-value pairs in any order it wants. However, if you look at each item in this dictionary you will see that letters were assigned codes in the order they were given in the string: *A* is 00, *C* is 01, and so on.

The name seq used for the argument passed to make_codes is a hint the function will make a set of codes for any kind of Python sequence. The len function is used to figure out how many items are passed to the function (line 6). Because len works for both strings and lists, and because we can iterate over both strings and lists with a for statement, we can pass either of those types of objects to make_codes and get back a dictionary containing binary Code objects:

```
>>> make_codes('AEIOU')
{'U': 100, 'I': 010, 'E': 001, 'O': 011, 'A': 000}
>>> make_codes(['do', 're', 'mi'])
{'do': 00, 'mi': 10, 're': 01}
```

Tutorial Project

Include the BitLab module when you start a new shell session for the projects in this chapter:

```
>>> from PythonLabs.BitLab import *
```

Some of the experiments in this section use the make_codes function in Figure 8.3. Either create a new file with your IDE and enter the function definition or download the file from the book website.

T1. Type this statement to see the binary representation of the number 12:

```
>>> Code(12)
1100
```

T2. This for statement will print the binary codes for the numbers from 0 to 3:

```
>>> for i in range(4):
...     print(i, "=", Code(i))
...
0 = 0
1 = 1
2 = 10
3 = 11
```

T3. As shown above, by default the constructor for Code objects uses the minimum number of bits needed to encode an integer. Repeat the for loop but make a 2-bit representation of each number:

```
>>> for i in range(4):
...     print(i, "=", Code(i,2))
...
0 = 00
1 = 01
2 = 10
3 = 11
```

T4. Repeat the exercise again, but this time print eight numbers, and ask for 3-bit codes:

```
>>> for i in range(8):
...     print(i, "=", Code(i,3))
...
0 = 000
1 = 001
...
7 = 111
```

By looking at the bit patterns printed in the previous exercise, can you see why there are twice as many 3-bit numbers as there are 2-bit numbers? Look at the last 2 bits on each line. Do you see how all four 2-bit patterns occur in the first four lines, and again in the last four lines?

This same pattern will occur if you ask Python to print all 16 4-bit patterns, and then all 32 5-bit patterns. Does this exercise help convince you that the number of codes grows exponentially with the number of bits?

T5. The formula for determining how many bits are needed to give each item in a set a unique code is $\lceil \log_2 n \rceil$. Ask Python to compute the number of bits needed for a set of four items:

```
>>> from math import ceil, log2
>>> ceil(log2(4))
2
```

T6. For a set of five items, we have to "round up" to use one more bit:

```
>>> log2(5)
2.321928094887362
>>> ceil(log2(5))
3
```

T7. Make a list of four color names:

```
>>> a = ['yellow', 'green', 'white', 'black']
```

T8. Use make_codes to create a binary code for each color:

```
>>> make_codes(a)
{'black': 11, 'white': 10, 'green': 01, 'yellow': 00}
```

Note that each possible 2-bit binary number was used to make the codes.

T9. Add a fifth color to the end of the list:

>>> *a.append('steel')*

T10. Make a new set of codes for the extended list:

>>> *make_codes(a)*
{'steel': 100, 'black': 011, 'white': 010, 'green': 001, 'yellow': 000}

Because every 2-bit pattern was needed for a list of four colors, we have to make 3-bit patterns for the list of five colors.

T11. How many bits do you think it will take to give each of 18 different colors a unique code? First try answering this question using a call to log2 and ceil, and then verify your answer by making a list of 18 color names and passing it to make_codes:

>>> *a*
['plum', 'sienna', ... 'sea green', 'seashell']
>>> *make_codes(a)*

T12. Is the code for the first item in your set of 18 colors the binary representation of the number 0? Is the last code the binary representation of the number 17?

8.2 Codes for Characters

Processor chips generally work with several bits of data at once. When data is transferred from memory to the processor, or vice versa, it is done in groups of 32 or 64 bits at a time, called **words**. A single word can hold a piece of data such as a number, a set of characters, or a reference to an object. A word can also hold the binary representation of an instruction that tells the processor what to do for a single step in an algorithm. Words are often divided into smaller units, called **bytes**. In a modern system there are eight bits in a byte, so a single word will hold 4 or 8 bytes.

In the early days of commercial computing, transferring data between computers was rare, and networks were nonexistent, so each system was able to use its own technique for encoding data. The size of a word varied greatly from computer to computer, and it was not uncommon to find systems with 12-bit, 18-bit, or even 35-bit words. Not surprisingly, the number of bits in a byte also varied from system to system; common sizes were 6, 7, or 8 bits.

As people started sharing data across multiple computer systems, the need for a standard encoding scheme became apparent. One of the first such standards was the American Standard Code for Information Interchange, or **ASCII**, which was formed by a committee representing several American computer manufacturers. The ASCII code specified which pattern of bits to use for encoding text. ASCII used 7 bits for each character, with codes for upper and lowercase letters and the symbols commonly found on a QWERTY keyboard.

Eventually, computer designers settled on 32 bits as a typical word size. Later, as technology improved, the standard word size increased to 64 bits. An even number of 8-bit bytes will fit in either 32 or 64 bits, so 8 bits became the standard size for a byte. Because it was convenient to store each character code in its own byte when writing data to a file, text files were collections of 8-bit codes, where the first bit was always 0 and the remaining bits held an ASCII code. This naturally led to suggestions for extending ASCII to an 8-bit code. Eventually ISO, the International Standards Organization, defined a scheme called Latin-1 that includes codes for accented letters and currency symbols used in European languages.

ASCII

The American Standard Code for Information Interchange (ASCII) was developed in the 1960s, and formally adopted as a standard in 1968. It has codes for the symbols found on a QWERTY keyboard in the United States, a few of which are shown here.

					@		`	
	0100000	0	0110000	@	1000000	`	1100000
!	0100001	1	0110001	A	1000001	a	1100001
"	0100010	2	0110010	B	1000010	b	1100010
#	0100011	3	0110011	C	1000011	c	1100011
$	0100100	4	0110100	D	1000100	d	1100100

In the late 1980s, researchers began to consider how to design an encoding scheme that would be able to represent a much wider collection of characters, including Arabic and Hebrew alphabets and the ideograms used in Chinese, Japanese, and Korean. Their effort led to the definition of Unicode, which is now the most widely used scheme for representing text.

Modern operating systems, including versions of Microsoft Windows and Mac OS X, use Unicode as the default format for encoding text files. E-mail applications and web browsers are also based on Unicode, which makes it possible to transmit files from one system to another while preserving all the letters and symbols that were written by the person who created the text.

As of version 3.0, Python uses Unicode to represent the characters in string objects. That means we can create strings that have not only the characters found on a keyboard, but also any symbol defined as part of the Unicode standard. This assignment statement creates a string that has two characters commonly used in English text plus a Greek letter:

```
>>> formula = '2πr'
```

The obvious question at this point is "how do we include a character like π when we type a string?" Whether or not a symbol can be entered directly from the keyboard depends on several factors, including the terminal emulator or IDE being used, the operating system, and language settings configured by the user. For example, Mac OS X users with a U.S. keyboard can press the Option key while typing "p" to type a π symbol and can type the assignment statement exactly as it appears above.

Another approach is to write a \u escape sequence as part of the string. This is the technique used in previous chapters to include suit symbols in the names of cards. If we look for the Greek letter π in a table of Unicode symbols, we will see its 4-digit hexadecimal (base 16) code number is 03C0, so this is another way to write the assignment statement that defines the formula:

```
>>> formula = '2\u03C0r'
```

A third option is to use a built-in function named chr that converts a decimal code number into a character. The chr function was part of earlier versions of Python and is still part of the language. Because the decimal form of the code number for π is 960 this is another way of writing the assignment statement, using chr and the string concatenation operator:

```
>>> formula = '2' + chr(960) + 'r'
```

No matter which technique we use to include a symbol as part of a string, a Unicode character is treated just like any other character. If we ask Python how long the string is, it will tell us there are three characters:

```
>>> len(formula)
3
```

If we look at the middle character in the string we will see that single Greek letter:

```
>>> formula[1]
'π'
```

Note that Python does not actually have a separate data type for characters. If we take a closer look at these expressions that access a single item in a string object we will see that characters are just strings that happen to be only one character long. If we save a reference to the last letter in our test string this is what will see:

```
>>> ch = formula[2]
>>> ch
'r'
>>> type(ch)
<class 'str'>
>>> len(ch)
1
```

As the examples above show, Python hides all the details of how the individual characters are represented inside the computer. We can think of a string as a list of numbers, where each number is a Unicode character ID. Python will take care of allocating the space in memory, figuring out the binary encoding of each symbol, etc.

Unicode Characters in Strings

Three ways of creating the string '2πr' in Python are shown at right. If your IDE provides a way of typing characters directly, or if you copy and paste a symbol from a web page or other source, you can enter the symbol directly in your IDE.

```
>>> formula = '2πr'
```

```
>>> formula = '2\u03C0r'
```

```
>>> formula = '2' + chr(960) + 'r'
```

The other two techniques are based on character codes. The \u escape syntax expects the 4-digit hexadecimal (base 16) code for a character, but the chr built-in function expects a decimal number. Often a website that lists Unicode symbols will show both the decimal and hexadecimal forms.

◆ **UTF-8**

One might expect the binary encoding for Unicode, which has over 110,000 symbols, to require 17 bits for each code. But a clever scheme named UTF-8 allows the size of the code to vary, so the most common characters require only 8 bits.

The code uses 8-bit groups called **octets**. If the first bit in a code is 0 it means the entire code fits in a single octet, and the remaining 7 bits are used to hold the ASCII code. This is the octet for the letter *a*:

0	1	1	0	0	0	0	1	*one octet*

If the first bit is a 1, the code continues with the next octet, and the code requires 16 bits total. Here is the code for å (an *a* with a circle over it):

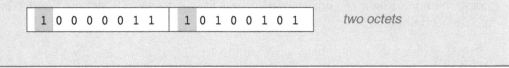

1	0	0	0	0	0	1	1	1	0	1	0	0	1	0	1	*two octets*

If for some reason we need to know the ID of a character, we can ask for its "ordinal value" using the built-in function named `ord`:

```
>>> ord(formula[0])
50
>>> ord(formula[1])
960
```

Note that `chr` and `ord` are inverse functions: one maps numbers to characters, and the other maps characters to numbers. But remember that characters are really just one-letter strings, so when you call `ord` you are passing it a string object that has only one character in it.

Because Unicode supports such a wide variety of symbols—over 110,000 in the latest version of the standard—you might guess that it takes more bits to represent a single character than it does in ASCII.

The group that developed the Unicode standard came up with a scheme, called UTF-8, that allows just one byte to be used for all the symbols that are in both ASCII and Unicode. The idea is that if the first bit in a byte is 0, the remaining seven bits can simply hold an ASCII code. If the first bit is a 1, however, the code continues into the byte next to it. Letters from Cyrillic, Chinese, Japanese, Arabic, Hebrew, and other alphabets require anywhere from 2 to 4 bytes.

One place where we do need to be concerned with how characters are mapped into sequences of binary digits is when we write a string out to a file. The `str` class has a method named `encode` that will generate the sequence of bytes that encode the characters in the string. When we call `encode`, we can pass it the name of an encoding scheme, like UTF-8 or Latin-1, and the result will be a sequence of bytes that can be written to an output file:

```
>>> b = formula.encode('utf-8')
```

If we want to see how these might be represented as binary patterns that will be written on the disk, we can make an 8-bit Code object for each byte. This `for` loop displays the binary encoding of the bytes:

```
>>> for x in b:
...     print(Code(x,8))
...
00110010
11001111
10000000
01110010
```

Binary numbers with 7 or 8 bits are hard to read, so many tables that show ASCII codes also give the hexadecimal version of the codes. BitLab has a class named Hex that represents hexadecimal codes. It works just like the Code class, except the symbols used to represent codes are from the base 16 number system. Here is an example of a number encoded both in binary and hexadecimal:

```
>>> Code(42)
101010
>>> Hex(42)
2A
```

Tutorial Project

Before you start the projects in this section, use your web browser to find a reference page that has a table of ASCII character codes. Try to find a table that lists codes in decimal, binary, and hexadecimal. The Wikipedia entry for ASCII is a good source, or you can do a web search for "ASCII" to find others.

T13. Make a string that has a combination of letters, digits, and punctuation marks:
```
>>> s = "The 19th Century (1801-1900)."
```

T14. Notice that Python includes the spaces and parentheses as part of the string. To verify this, count the number of characters between the quotes, and then ask Python to print the length of the string:
```
>>> len(s)
29
```
Does your count agree with Python's count?

T15. This expression will return the ordinal value (character ID) for the first character in s:
```
>>> ord(s[0])
84
```

T16. Look for the letter *T* in the ASCII table. Does it have code number 84?

T17. To see the binary representation of the first letter in the string type this expression:
```
>>> Code(ord(s[0]))
1010100
```
Is this the binary code for *T* shown in the table you found on the Internet?

T18. Repeat the exercise, but make a hexadecimal code:
```
>>> Hex(ord(s[0]))
54
```
Is this the hexadecimal code for *T* in your table?

T19. This for loop will print the binary, decimal, and hexadecimal codes for every letter in the string:

```
>>> for ch in s:
...     print(Code(ord(ch),7), ord(ch), Hex(ord(ch)))
...
1010100 84 54
1101000 104 68
...
```

Look up these codes in the ASCII table you found. If your table has 8-bit binary codes, you can print 8-bit codes by changing the 7 to an 8 in the body of the for loop.

Can you find each of the characters in s, including the spaces and the punctuation marks? Did you notice there are different codes for upper and lowercase letters, for example, the letters T and t?

T20. Make some strings with Unicode symbols:

```
>>> s1 = 'We paid ' + chr(163) + '3.50 for a pint of ale.'
>>> s2 = 'The Greek letter \u03A9 is the symbol for resistance.'
>>> s3 = '42 ' + chr(247) + ' 6 = 7'
```

What do you see when you ask Python for the value of s1, s2, and s3?

The following optional exercises look at the binary encoding of UTF-8 characters, showing how letters that were not part of the ASCII character set encoded using two or more bytes.

♦ Type a statement that creates a 3-letter string containing one accented letter:

```
>>> boat = 'b' + chr(229) + 't'
>>> boat
'båt'
```

♦ This for statement will print each byte in the UTF-8 encoding of the string:

```
>>> b = boat.encode('utf-8')
>>> for x in boat.encode('utf-8'):
...     print(Code(x,8))
...
01100010
11000011
10100101
01110100
```

The middle letter å is part of the Latin-1 character set; these characters have codes between 128 and 255 and can all be encoded with 2 bytes. Notice how the encoding takes a total of 4 bytes, and the bit patterns for the middle bytes both start with a 1 to indicate they are part of a multi-byte encoding.

♦ Repeat the previous two exercises with a string that contains a symbol that is not part of ASCII or Latin-1:

```
>>> card = 'J\u2665'
```

How many bytes are in the encoding of the heart symbol?

8.3 Error Detection

No matter what technology is used to store or transmit character codes, there is always a chance that errors will occur. Noise can interfere with network communications, or a bump might disrupt the motion of a disk. Magnetic tapes, and even CDs and DVDs, can lose information over time, and new data can be stored in a flash memory card only a certain number of times before the card begins to lose data.

There are a variety of ways of encoding text, music, and images in order to deal with potential errors. The simplest techniques implement some form of **error detection**. The idea is to add extra descriptive information as part of the encoding, so the code contains bits for the data plus additional bits for error detection. When the data is retrieved, an algorithm can compare the description with the data to determine whether anything has changed. If an error is detected, the receiver can send a message back to the sender, to request that a message be sent again.

In this section we will look at the simplest form of error detection. It will work for any type of data, but we will use 7-bit ASCII character codes in our experiments. The descriptions use terminology from computer networking to describe how the scheme works: a **sender** will create a message, adding error detection information, and a **receiver** will examine the message to see if errors occurred.

This simple scheme is based on a concept called **parity**. If a binary code has an even number of 1 bits we say it has "even parity," and if it has an odd number of 1s the code has "odd parity." For example, the sequence 0101 has even parity, because two of the four bits are 1s, but 0111 has odd parity because it has three 1s.

We can use parity to implement error detection by having the sender attach an extra bit to a code before transmitting it. If the sender starts with a 7-bit character code the extended code will have eight bits after the parity bit has been added. The sender can choose whether to attach a 1 or a 0 to the end of the original code. The idea is to attach the bit that will make the total number of 1s in the new 8-bit code an even number.

Here is an example. If we pass a single character in a call to Code we will get back the binary pattern for that letter. Since we're doing experiments with ASCII characters we will make a 7-bit code:

```
>>> c = Code('X', 7)
>>> c
1011000
```

A method named add_parity_bit will extend the code with an additional bit. There are three 1s in this code, so the parity bit should be a 1 in order to give the extended 8-bit code even parity:

```
>>> c.add_parity_bit()
>>> c
10110001
```

Notice how the code is one bit longer, and the bit that was attached at the end is a 1.

At the other end of the communication channel, the receiver will count the number of 1s in the 8-bit code, and if the count is an even number, the receiver assumes no bits were changed during the transmission. The original 7-bit code can be extracted by detaching the parity bit from the end of the 8-bit extended code (Figure 8.4).

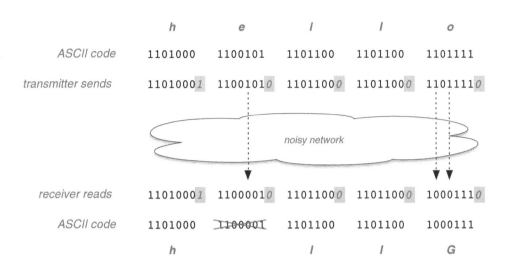

Figure 8.4: *When a message is transmitted the sender adds a parity bit to each 7-bit character to make an 8-bit code. The receiver counts the number of 1s in the entire code. If the code has odd parity an error has occurred, otherwise the character is taken from the first 7 bits.*

If an error occurs during transmission one of the bits will change: either a 1 will turn into a 0, or a 0 will turn into a 1. Either way, the total number of 1s in the code seen by the receiver will not be an even number, and the 8-bit packet can be marked as an error.

However, this scheme only works if there is a single error within any 8-bit code. As the last column in Figure 8.4 shows, the receiver can be misled if two errors occur. Again it doesn't matter if a 1 changes to a 0 or vice versa; if two bits change the 8-bit code seen by the receiver will have an even number of 1s, and the receiver can't tell that two of the bits were not the same as the ones originally transmitted.

In the experiments for this section, we will create messages like the one shown in Figure 8.4 and then see what happens as errors are deliberately added. The BitLab module defines two functions, encode and decode, that translate a string into a list of Code objects and vice versa. Here is a simple example using a string with three characters:

```
>>> msg = encode("cat")
>>> msg
[1100011, 1100001, 1110100]
```

If we decode this list of code objects we should get back the original string:

```
>>> decode(msg)
'cat'
```

An optional second argument to encode will tell it to add a parity bit to each code:

```
>>> msg = encode("cat", with_parity = True)
>>> msg
[11000110, 11000011, 11101000]
```

Notice how each code now has 8 bits, and that each code has an even number of 1s.

To simulate a noisy communication channel, a function named `garbled` will make a copy of a list of codes and make changes to random bits. When we call this function we pass it a message and a number *n* that specifies how many errors we want to introduce. Using the example message created above:

```
>>> rec = garbled(msg, 1)
```

Printing both the original and the garbled copy will let us find the erroneous bit:

```
>>> print(msg); print(rec)
[11000110, 11000011, 11101000]
[11000110, 11000011, 11111000]
```

In this example the error is in the 4th bit from the left in the last code.

When we try to decode a garbled message, we'll still get a string, but any character that has a parity error will show up as a bullet symbol instead of a valid ASCII code:

```
>>> decode(rec)
'ca•'
```

Tutorial Project

The encode and decode functions used in this section are defined in BitLab, so make sure you import the module before doing these exercises.

T21. Make a code object for the letter *A* and look at the pattern of bits:
```
>>> c1 = Code('A', 7)
>>> c1
1000001
```

T22. The code for *A* has two 1s, so it already has even parity. The parity bit for this letter will be a 0:
```
>>> c1.add_parity_bit()
>>> c1
10000010
```
Note how there are now eight bits and the newly attached parity bit is a 0.

T23. Repeat the experiment, but with a letter that has an odd number of 1s:
```
>>> c2 = Code('C', 7)
>>> c2
1000011
>>> c2.add_parity_bit()
>>> c2
10000111
```
Do you see why the method added a 1 as the parity bit?

T24. To see how encode works, pass it a short string and verify the result is a Code object for each letter in the string:
```
>>> encode('hello')
[1101000, 1100101, 1101100, 1101100, 1101111]
```

T25. Call encode with a longer string. This time, pass the optional argument that tells it to add parity bits to all the codes, and save the result in a variable called `msg`:
```
>>> msg = encode('hello, world', with_parity = True)
```

T26. If you want, look at a few of the codes to make sure they have parity bits:

```
>>> msg[0:3]
[11010001, 11001010, 11011000]
```

T27. Make a copy of the message, telling `garbled` to make a single bit change somewhere:

```
>>> rec = garbled(msg, 1)
```

T28. If we decode the garbled message there should be a bullet symbol somewhere:

```
>>> decode(rec)
'h•llo, world'
```

Because the change was made to a random bit you will probably see a different outcome.

T29. Ask Python to show you the two messages:

```
>>> print(msg); print(rec)
[11010001, 11001010, 11011000, ... 11001001]
[11010001, 11000010, 11011000, ... 11001001]
```

Can you see where a single bit changed? In this example it's in the second code.

T30. Make a short message that has only one letter:

```
>>> msg2 = encode('X', with_parity = True)
```

T31. Add two errors to the message and save the result:

```
>>> rec2 = garbled(msg2, 2)
```

T32. These print statements will print the codes and the corresponding strings on separate lines:

```
>>> print(msg2, decode(msg2)); print(rec2, decode(rec2))
[10110001] X
[11100001] p
```

Notice how two bits changed. Because the altered code has even parity, it appears to be valid, so `decode` turned it into a letter, not a bullet symbol.

T33. Repeat the previous two exercises several more times. You should never see a bullet symbol that shows there was an error in the received message.

T34. Garble the long message again, but this time add ten errors:

```
>>> rec = garbled(msg, 10)
```

Look at the bytes in the received message, and print the decoded message. Did the errors go in ten different codes, *i.e.*, are there ten bullet symbols? If not, can you explain what happened?

T35. Try this experiment again, but introduce 30 errors. Did you find characters where three errors were introduced? Four? What happened to these characters? Is there a general rule for when a bullet symbol is in the decoded message?

8.4 Parity Bits and the Exclusive OR

We're now going to take a closer look at how a transmitter figures out whether to add a 0 or a 1 to the end of a character in order to make an even parity code. There are several clever algorithms to do this with only a few basic steps, but in this section we'll do a simple iteration that looks at all the bits in a code.

One approach is to write a `for` loop that iterates over the bits of a word to count the number of 1s. Once we have the sum in a variable named n, we can do a quick check to see if n is odd by finding the remainder after dividing n by 2. If the remainder is 1 then we know n is odd; if the remainder is 0 n is even.

The algorithm we are going to implement also iterates over each bit in a code, but it does not need to do any additions and it does not need to perform a division to figure out the remainder. Our function will simulate operations in the logic circuits of a communications controller, using a type of logical operation known as an **exclusive OR**. In logic, the expression "*a* or *b*" is true if either *a* or *b* is true. The exclusive OR is similar, except it is false if *a* and *b* are both true. In other words, the exclusive OR means "*a* or *b* but not both."

Both types of OR, as well as other logical operations, can be performed by the Code objects defined in BitLab. When we are working with bits, the convention is to use the symbols 1 and 0 instead of *true* and *false*, but otherwise the operations are the same. For example, the ampersand symbol is used for the logical AND function, so 1 & 1 = 1 (Figure 8.5). The exclusive OR function is designated by a caret operator (^), so we have 1 ^ 0 = 1 and 0 ^ 1 = 1 but 1 ^ 1 = 0.

To put the exclusive OR to work in computing the parity of a code, we'll use a variable named p to represent the parity bit. The first step, shown on line 5 of the function (Figure 8.6) initializes p so it is a 1-bit code containing a 0 bit. The for statement on line 6 tells Python to execute the loop once for each bit in x, the Code object passed as an argument to the function. On each iteration the variable named bit is set to a 1-bit code taken from x, and inside the loop we update p by computing the exclusive OR of p and the current bit. When the loop terminates, the current value of p is the parity bit.

To see why this function definition works, consider what happens each time the bit in the code is a 0. If p is currently 0, it stays 0 because 0 ^ 0 = 0. If p is currently 1, it will stay 1 because 1 ^ 0 = 1. Thus whenever the loop sees a 0 bit in x the parity bit does not change.

But now consider what happens with the bit from the code is a 1. If p is currently 0, it will change to 1, because 0 ^ 1 = 1. Similarly, if p is a 1, it will change to 0 because 1 ^ 1 = 0. So whenever the loop sees a 1 bit in x the parity bit "flips" to the opposite state.

To summarize, the parity bit p stays the same whenever there is a 0 in the code and changes value for every 1 bit in the code. If there is an even number of 1s, the parity will always end up back at 0: the first 1 changes it to 1, but the next 1 will change it back, and at the end of the loop an even number of changes will leave p set to 0. If there is an odd number of 1s, the first 1 will change the parity to 1, and after that it flips back and forth, and the final value will be 1.

The parity function of Figure 8.6 can be used as a helper function by both the transmitter and the receiver. When a transmitter is encoding a string, it should create a 7-bit code for a

		AND	OR	XOR
a	b	a & b	a | b	a ^ b
0	0	0	0	0
0	1	0	1	1
1	0	0	1	1
1	1	1	1	0

Figure 8.5: *Operators defined for binary codes are similar to Python's Boolean operators. The three operators shown in this table are AND (ampersand, &), OR (vertical bar, |), and exclusive OR (caret, ^).*

```
1    from PythonLabs.BitLab import Code
2
3    def parity(x):
4        "Return the parity bit (a 1-bit Code object) for code x"
5        p = Code(0,1)       # initialize p to 0
6        for bit in x:
7            p = p ^ bit     # ^ means "exclusive OR"
8        return p
```

Download: encoding/parity.py

Figure 8.6: *This function uses the exclusive OR operator to compute the parity of a binary code.*

letter, call parity to get back a 1 or a 0, and then extend the code with that bit. For example, when 01100001, the code for letter *a*, is passed to parity, the result is a 1. Extending the code with a 1 will give it an even number of 1s, which is what we want:

```
>>> c = Code('a')
>>> c.extend(parity(c))
>>> c
11000011
```

On the other end, a receiver will read an 8-bit code. It calls parity to make sure there is an even number of 1s. If the result of a call is a 1, it means the 8-bit code has an odd number of 1s and an error occurred somewhere. But if the call to parity returns 0 the code is valid, and the receiver can use the ASCII code in the first 7 bits:

```
>>> c[0:7]
1100001
>>> decode(c[0:7])
'a'
```

Tutorial Project

T36. Make a Code object for the letter *x*:
```
>>> c = Code('x')
```

T37. The code should have 7 bits:
```
>>> c
1111000
```

T38. When we access individual bits, the bits are numbered from left to right. Type this expression to get the value of the first bit in c:
```
>>> c[0]
1
```

T39. Recall that in Python a negative index can be used to access items at the end of a list. Type this expression to get the value of the last bit in c:
```
>>> c[-1]
0
```

T40. Use a `for` loop to iterate over the bits, printing them one at a time on the console:
```
>>> for bit in c:
...     print(bit)
...
```
How many lines of output did you get? Were the bits printed in order?

The next exercises involve expressions with logical operators. Check your results using the truth tables for the logic functions shown in Figure 8.5.

T41. The OR of the 1 at the front of c and the 0 at the end is 1:
```
>>> c
1111000
>>> c[0] | c[-1]
1
```

T42. The Exclusive OR of these two bits is also 1:
```
>>> c[0] ^ c[-1]
1
```

T43. Since the first two bits are both 1 the AND of these two bits should be 1:
```
>>> c
1111000
>>> c[0] & c[1]
1
```

T44. The difference between OR and Exclusive OR is that the Exclusive OR of the two 1s at the beginning of c is 0:
```
>>> c[0] | c[1]
1
>>> c[0] ^ c[1]
0
```

The remaining projects in this section use the `parity` function shown in Figure 8.6. Either download the function from the book website or use your IDE to create a file named `parity.py` and type in the definition exactly as its shown (don't forget the `import` statement on line 1).

T45. Print the value of c (which is the ASCII code for *x*) again:
```
>>> c
1111000
```

T46. From the exercises in the previous section it should be clear that the parity bit for this code is 0. Call your new `parity` function to see if it returns 0:
```
>>> parity(c)
0
```

T47. The ASCII code for *y* has an add number of 1s. Verify your function returns 1 for this code:
```
>>> Code('y')
1111001
>>> parity(Code('y'))
1
```

If you're not sure how the algorithm works add a `print` statement to the body of the loop. Re-run the examples shown above. You should see that the temporary variable p changes from 0 to 1 or vice versa only when the current bit is a 1.

8.5 Huffman Trees

The goal for a text compression algorithm is to rewrite a piece of text in a new form that requires fewer bits. One way to do this, and the approach we will explore in this chapter, is to devise an encoding scheme that uses fewer bits for some of the characters. We will use a technique known as *Huffman encoding*, named after David A. Huffman (1925–1999), who originally described the method in 1952.

A Huffman code relies on the fact that some characters are more common than others. In ordinary English text, the letters *e*, *t*, and *a* appear much more often than *q*, *j*, and *z*. An encoding scheme that uses fewer bits for the most common letters might be able to encode a piece of text with a smaller total number of bits than one that uses the same number of bits for every letter.

To test this idea we are going to use Huffman's algorithm to develop a special purpose code for Hawaiian words. The Hawaiian alphabet has a total of thirteen letters: five vowels, seven consonants, and a symbol called the *okina* that sometimes appears between two vowels. An encoding scheme that uses the same number of bits to code each letter in this alphabet would require $\lceil \log_2 13 \rceil = 4$ bits per letter. Our goal is to devise a coding scheme that uses fewer bits for the common letters and more bits for the uncommon letters. After we build the code, we will test it with a few Hawaiian words to see if in fact the new code uses fewer bits.

Huffman's algorithm uses a data structure known as a **priority queue** to manage information about characters. The WordQueue class in SpamLab (the lab module for Chapter 6) was a special type of priority queue designed specifically for that project; in this chapter we will use a more general type of queue that is commonly used in a wide variety of situations.

The Hawaiian Alphabet

If you've ever traveled in Hawaii you may have noticed that geographical names, names of cities, and other Hawaiian words did not have all the letters of the English alphabet.

Hawaiian words are spelled with the same five vowels (A, E, I, O, U) but use only seven consonants (H, K, L, M, N, P, W).

A 13th symbol, called the *okina*, is used between two vowels when they are to be pronounced as separate syllables. For example, *a'a*, one of the words for lava, is pronounced "ah-ah".

The table at right (from the American Cryptogram Association) shows the frequency of each letter. A is by far the most common, making up 26% of all the letters in Hawaiian words. W is least common, used just 1% of the time.

'	0.068
A	0.262
E	0.072
H	0.045
I	0.084
K	0.105
L	0.044
M	0.032
N	0.083
O	0.107
P	0.030
U	0.059
W	0.009

Structurally, a priority queue is just like a list: it's a linear container with references to other objects. What distinguishes a priority queue from a regular list is that the objects in the priority queue are always sorted. Each time we add a new item to a priority queue, the item is automatically saved in a location that preserves the order.

The priority queues we will be using for the project in this chapter are instances of a class named PriorityQueue. After importing the class from BitLab, we can make a new, empty, queue with this assignment statement:

```
>>> pq = PriorityQueue()
```

The method that adds items to a priority queue is named insert. We saw earlier that lists also have a method with this name. If a is a list, the call a.insert(i,x) will place the item x in location i of the list. But because priority queues are always sorted, we can't specify the location for an item. We just pass a reference to an object we want to insert, and the method will figure out where it goes. This example adds a few strings to the new queue created above:

```
>>> pq.insert('grape')
>>> pq.insert('banana')
>>> pq.insert('papaya')
>>> pq.insert('mango')
```

If we look at the queue we will see the strings are in alphabetical order:

```
>>> pq
[banana, grape, mango, papaya]
```

Recall from earlier chapters that we remove an item from a list by calling a method named pop. Our new data structure is called a "queue" because when items are removed they are taken from the front of the line. So when we call pop for a priority queue object, we don't specify the location, and we always get back the item that was removed from the front of the queue. This example shows how a call to pop removes strings in alphabetical order from the queue:

```
>>> pq.pop()
'banana'
>>> pq.pop()
'grape'
>>> pq
[mango, papaya]
```

Another new data structure used in this project is a **binary tree**. A tree consists of **nodes** that are connected to each other by a set of branches (Figure 8.7). Trees, like lists and queues, are containers that hold references to other objects. When we add an object to a tree, we will make a node to refer to the object and then insert the node into the tree somewhere by attaching it to an existing node.

Every tree has one special node called the **root**. Computer scientists prefer to draw trees with the root at the top of the picture, as shown in Figure 8.7. Nodes at the bottom of a tree that do not have any other nodes below them are known as **leaves**; the leaves of the tree in the figure are shaded gray. The other nodes (including the root) are referred to as **interior nodes**.

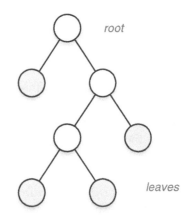

Figure 8.7: *A binary tree is a collection of* ***nodes***. *Lines connecting nodes define relationships similar to those in a family tree. At the top of the tree is a single node, the* ***root***, *that has no predecessors. Nodes that have no descendants are called* ***leaves***.

The BitLab class we will use to build binary trees is named Node. The leaf nodes in Huffman's algorithm contain two pieces of information: the name of a letter from the alphabet and the frequency of that letter. For example, the letter *M* has a frequency of 0.032 in Hawaiian words (meaning 3.2% of the letters in random Hawaiian text should be the letter *M*). Here is how we call the Node constructor to make a node for this letter:

```
>>> leaf = Node('M', 0.032)
```

If we ask Python to print the object this is what we'll see:

```
>>> leaf
( M: 0.032 )
```

This notation tells us the node has a label of *M* and a frequency of 0.032. The parentheses surrounding the label and frequency are intended to look like the sides of a circle in a drawing of a tree.

An interior node is also created by a call to Node, but in this case we pass references to two existing nodes to the constructor. The result of the call will be a new node object that has the existing nodes as its children. Interior nodes also have a frequency associated with them. In this case, the node's frequency is the sum of the frequencies of its children.

As an example of how to create an interior node, suppose we have two existing nodes named t0 and t1, created by these assignment statements:

```
>>> t0 = Node('W', 0.009)
>>> t1 = Node('P', 0.030)
```

This statement will make a new interior node with t0 and t1 as its descendants:

```
>>> t2 = Node(t0, t1)
```

The string printed by Python to describe the new node is more complicated, but if you look closely, you'll see three nodes are included in the description:

```
>>> t2
( 0.039 ( W: 0.009 ) ( P: 0.030 ) )
```

The outer set of parentheses identify the new interior node, and next to the opening parenthesis is the frequency of the new node, which is the sum of the frequencies of t1 and t2.
The two descendants are shown as they were before, printed one after the other, but inside
the outer parentheses to indicate they are now below this new interior node.

One way to envision how Huffman's algorithm works is to imagine a set of tinker toys,
where the circular pieces correspond to nodes, and sticks are used to connect nodes to
each other. The algorithm starts with a collection of unconnected circular pieces that will
eventually be the leaves of the tree. Each step of the algorithm will find two pieces of
the tree that were built on a previous step and connect them to each other, as shown in
Figure 8.8.

Each new piece of the tree will be an interior node that has the existing parts as its
children. Because the only pieces available initially are leaf nodes, the first step will connect
two leaves. Later steps might connect a leaf to an interior node, or connect two interior
nodes. Eventually all the nodes will be connected and the final product will be a tree that
looks like the one in Figure 8.7.

Now that we have the two main pieces of the Huffman tree algorithm—a type of object
to represent nodes of a binary tree and a queue that will keep the nodes in order—the steps
in the algorithm are very simple. First we make leaf nodes for every symbol in the alphabet
and put the resulting nodes into a priority queue. Then, as long as there are two or more
nodes in the queue:

- Remove the first two nodes from the queue.
- Create a new interior node using these two nodes.
- Insert the new node back into the queue.

Because each iteration combines two nodes into a single new node the queue grows shorter
on each round. When the queue has been reduced to a single node this node will be the
root of the final Huffman tree.

Figure 8.9 shows the initial steps in the construction of a small tree for the five vowels,
using their relative frequency in Hawaiian words. The first step initializes the priority queue
with a new node for each letter. Note the letters are sorted, with the least frequent letters
at the front of the queue.

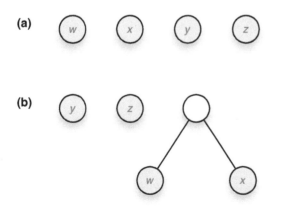

Figure 8.8: *Building a Huffman tree is
analogous to making a tree shape with
tinker toys. **(a)** At the start of the
algorithm, all the circle pieces that will
be leaves of the tree are not connected
to anything. **(b)** Each step will take
two existing pieces and connect them
to a new piece. The new piece will be
the common ancestor of the two
existing pieces.*

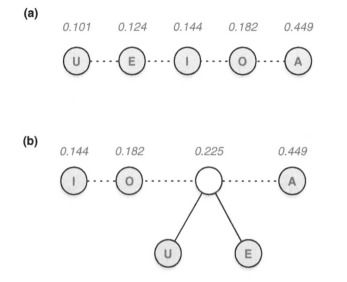

(a)

0.101	0.124	0.144	0.182	0.449

U ···· E ···· I ···· O ···· A

A	0.449
E	0.124
I	0.144
O	0.182
U	0.101

(b)

0.144	0.182	0.225	0.449

I ··· O ········· ○ ········ A

U   E

Figure 8.9: *The table shows the relative frequency of the five vowels in the Hawaiian alphabet. (a) The first step in Huffman's algorithm initializes a priority queue with one leaf node for each letter. (b) The first iteration makes an interior node for the two least frequent letters and inserts the node back into the queue.*

The second row in the figure shows what happens on the first iteration. The two letters at the front of the queue, which are the two least frequently used vowels, are removed from the queue. They are connected by a new interior node, and the new node is put back in the queue. Because the frequency of an interior node is the sum of the frequencies in nodes below it, this new node has a frequency of 0.225. When it is inserted back into the queue it will be placed between the leaf nodes for O and A.

The two important things to remember about this algorithm are (1) the nodes in the queue are ordered according to their frequency and (2) on each round, the queue grows shorter, so the algorithm is guaranteed to terminate.

As you work on the project in this section, you will be able to view a drawing of the current state of the priority queue. After you initialize the queue, call a method named `view_queue` to draw the queue on the PythonLabs canvas. Initially each queue entry will be a leaf node, but at each iteration you will see new interior nodes appearing in the queue.

The frequency data for letters used in Hawaiian words is in a set of data files included as part of PythonLabs. To read one of these files call a function named `read_frequencies`, which is defined in BitLab. The function reads the data in a file and returns it in the form of a dictionary that associates letters with their frequencies. This call reads the vowel frequencies:

```
>>> vf = read_frequencies(path_to_data('hvfreq.txt'))
```

Recall that `path_to_data` is the PythonLabs function that returns a string containing the path to one of the data files included with the lab modules. If you want to create your own frequency data, you can save it in a text file in your project directory and pass the name of your file to `read_frequencies` (the Lab Manual has details about the file format.)

Tutorial Project

If you start a new shell session don't forget to import the classes used for this project:

```
>>> from PythonLabs.BitLab import *
```

T48. Make a new priority queue object and save it in a variable named pq:

```
>>> pq = PriorityQueue()
```

T49. Ask Python to print the queue:

```
>>> pq
[]
```

As expected, the new queue is empty.

T50. Priority queues, like lists in Python, can contain any type of object, as long as the objects can be compared (if they can't be compared to one another there is no way to keep them in order). Try adding a couple of strings to your new queue:

```
>>> pq.insert('lemon')
>>> pq.insert('grape')
>>> pq
[grape, lemon]
```

Note how the insert method found the right place for the second string. The queue figures out where to insert an object and automatically puts it in a place that makes sure the queue remains sorted.

T51. Add a few more strings to your queue:

```
>>> for x in ['kiwi', 'strawberry', 'pear']:
...     pq.insert(x)
...
```

T52. Check to make sure the queue now has five items, and that they are stored in the queue in alphabetical order:

```
>>> len(pq)
5
>>> pq
[grape, kiwi, lemon, pear, strawberry]
```

T53. Even though the queue looks like a list, it is not a list. If you try to call methods that work for lists you will get an error message:

```
>>> pq.append('ugli')
AttributeError: 'PriorityQueue' object has no attribute 'append'

>>> pq.reverse()
AttributeError: 'PriorityQueue' object has no attribute 'reverse'
```

This makes sense because if we were allowed to apply these other operations the order of the items in the queue might change, and the items would no longer be stored according to their priority.

T54. Type these expressions to remove the first two strings from your queue:

```
>>> s = pq.pop()
>>> t = pq.pop()
```

T55. Print the two variables to confirm they have references to the two strings that were at the front of the queue:

```
>>> print(s,t)
grape kiwi
```

T56. Ask Python to print the queue, and double-check to make sure there are three items left after removing the first two:
```
>>> pq
[lemon, pear, strawberry]
>>> len(pq)
3
```

T57. To make a new leaf node (a node that corresponds to a single letter) call Node and pass it a letter and its frequency:
```
>>> t0 = Node('A', 0.2)
>>> t1 = Node('B', 0.3)
```

T58. Print the two new nodes to see how they are displayed in a shell window:
```
>>> t0
( A: 0.200 )
>>> t1
( B: 0.300 )
```

T59. Create a new interior node by passing the two leaves in a call to the Node constructor:
```
>>> t2 = Node(t0, t1)
>>> t2
( 0.500 ( A: 0.200 ) ( B: 0.300 ) )
```
The outer parentheses show that this is a node with a frequency of 0.5, which is the sum of the frequencies of the two nodes passed as arguments. The two descendants of this interior node are shown following the frequency.

The next set of exercises will use the letter frequency data for vowels in Hawaiian words to make the Huffman tree shown in Figure 8.9. The frequency data is in a file that is installed as part of PythonLabs.

T60. Use read_frequencies to make a dictionary that maps vowels to their frequencies:
```
>>> f = read_frequencies(path_to_data('hvfreq.txt'))
```

T61. If you print f (the name stands for "frequency") this is what you will see:
```
>>> f
{'A': 0.449, 'I': 0.144, 'E': 0.124, 'U': 0.101, 'O': 0.182}
```
(Python may print the letters in a different order but the frequencies should be as shown).

T62. A function named init_queue will initialize a priority queue with a node object for each item in the dictionary:
```
>>> pq = init_queue(f)
>>> pq
[( U: 0.101 ), ( E: 0.124 ), ... ( A: 0.449 )]
```
Does this look accurate to you? Are the items in the queue Node objects? Are they sorted? How many are there?

T63. Call view_queue to draw the queue on the PythonLabs canvas:
```
>>> view_queue(pq)
```

T64. Type this expression to remove the first node from the queue and save it in a variable named n1:
```
>>> n1 = pq.pop()
```
On the canvas, notice how the first item in the queue was moved out of the queue and the queue now has four nodes.

T65. Call pop again, and save the node in a second variable named n2:

>>> `n2 = pq.pop()`

The queue should now be down to three nodes.

T66. Call the Node constructor to make a new parent of the two nodes you just removed:

>>> `n3 = Node(n1,n2)`

Look at the new node on the canvas. Do you see how its frequency is the sum of the frequencies of its children?

T67. Insert the new interior node back in the queue:

>>> `pq.insert(n3)`

Is the new interior node in the correct location in the middle of the queue?

T68. To continue building the tree, repeat the previous steps. It will be easier to repeat the operations if you put them all on a single line:

>>> `n1 = pq.pop(); n2 = pq.pop(); pq.insert(Node(n1,n2))`

T69. Repeat the combined statement above until there is only one node left in the queue. This single node represents the complete tree for all five vowels.

T70. If you want to watch the steps for the construction of the Huffman tree for the entire alphabet, get the set of frequencies from a data file named `hafreq.txt` (note the file name starts with "ha", for "Hawaiian alphabet").

>>> `f = read_frequencies(path_to_data('hafreq.txt'))`
>>> `f`
`{'A': 0.262, 'E': 0.072, "'": 0.068, ... 'W': 0.009}`

Now repeat the steps that initialize the queue, draw it on the canvas, and combine the nodes.

8.6 Huffman Codes

After we have built a tree using Huffman's algorithm, we can use the structure of the tree to define codes for each letter.

To create a Huffman code, start by attaching labels to the lines that connect interior nodes to their children. At every interior node, including the root, the line going down to the left child should be labeled with a 0, and the line going down to the right child should be labeled with a 1. The labels on the connections in the final tree for the Hawaiian vowels are shown in Figure 8.10.

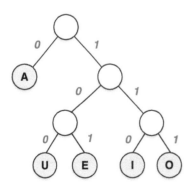

Figure 8.10: *The Huffman tree for the Hawaiian vowels, with labels attached to each connection.*

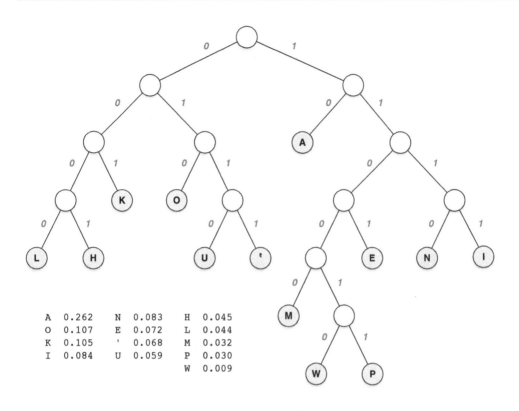

Figure 8.11: *Huffman tree for the Hawaiian alphabet. The frequencies in this table are the same as the ones on page 269, but here they are sorted by frequency.*

The labels in the tree can now be used to define codes for letters. Each letter of the alphabet used to make the tree appears somewhere in the tree as a leaf node, because the tree construction algorithm initialized the priority queue with one leaf for each letter, and all the leaves were eventually added to the tree. The code for a letter is defined by the labels on the path from the root to the leaf for that letter. For example, the three connections from the root to the leaf for the letter *U* in Figure 8.10 are labeled 1, 0, and 0, so the code for *U* is 100. The path to *O* goes down the other side of the tree, and the bits on this path give 111 as the code for *O*.

The Huffman tree for the full Hawaiian alphabet is shown in Figure 8.11. An important thing to notice about the tree is that the most common letters are closest to the root, so the codes for these letters use fewer bits. It's easy to see why the least frequently used letters are farthest from the root: they correspond to nodes that were toward the front of the queue when it was initialized, and were thus the first to be attached to interior nodes. Now that these letters are below an interior node, the path to these leaves will be longer than paths to letters at the end of the queue.

As an example of how infrequent letters end up farther from the root, the two letters that were initially at the front of the queue for the complete Hawaiian alphabet were *W* and *P*. The first iteration attached these two letters to a new interior node and put the new node back in the queue. This new combined node still had a low frequency (0.039), so it remained near the front of the queue, where it was picked up and combined with another node, moving the letters even farther down the tree.

```
1    from PythonLabs.BitLab import Node, read_frequencies, init_queue
2
3    def build_tree(filename):
4        "Build a Huffman tree using letters and frequencies in a file."
5        pq = init_queue(read_frequencies(filename))
6        while len(pq) > 1:
7            n1 = pq.pop()
8            n2 = pq.pop()
9            pq.insert(Node(n1,n2))
10       return pq[0]
```

Download: encoding/buildtree.py

Figure 8.12: *The* build_tree *function creates a Huffman tree based on letter frequencies in a data file. The return value will be a reference to the root of the tree.*

The experiments in this section will use a Huffman tree to encode Hawaiian words. The goal is to see if shorter codes for more common letters will lead to a shorter message, that is, if Huffman codes do in fact compress the text. Words with lots of *A*s will require fewer bits, because the code for *A* is the shortest code. But will the benefit of using a shorter code for *A* be negated by the fact that codes for *W* and *P* are longer, so there is no overall benefit from using a Huffman code?

The steps used to create a Huffman tree have been collected into a single function named build_tree, shown in Figure 8.12. If we pass the function the name of a file containing letters and their frequencies, we will get back a reference to the root node of the Huffman tree. This example shows how to create the tree from the full Hawaiian alphabet (Figure 8.11):

```
>>> tree = build_tree(path_to_data("hafreq.txt"))
```

A function named assign_codes will iterate over the tree to collect the codes for all the leaf nodes and put them in a dictionary:

```
>>> hc = assign_codes(tree)
>>> hc
{'A': 10, 'E': 1101, "'": 0111, ... 'W': 110010}
```

If you want to see the code for a letter, just look it up in the dictionary:

```
>>> hc['H']
0001
```

Given codes for individual letters, encoding a complete word is simply a matter of writing the bits for each letter in one long string. For example, the encoding of *ALOHA* is

```
100000010000110
```

The first two bits are the code for *A*, the next four bits are the code for *L*, the next three the code for *O*, then four bits for *H*, and finally two bits for the last *A*.

Decoding a string of bits in order to determine the original word is also defined in terms of the tree. The idea is to use the sequence of bits in a code to choose a direction at each step on a journey that starts at the root of the tree.

As an example, suppose we want to decode the string 1101. The first bit is 1, so the first step goes down the right path from the root. From this node, the second step is also a 1, so we go down the right side again. The third step is a 0, so we go left, and the last step is a 1, so we go right again. Now we're at a leaf node, and the label there is *E*, so we're done: the code 1101 is for the letter *E*.

To decode a longer bit string that corresponds to a complete word, we just need to repeat the process described in the previous paragraph. Each time we reach a leaf node we can write the letter at that node, and then go back to the root and continue decoding with the next bit in the input.

Suppose we receive a message containing the bits 10011110. Start walking from the root and after two steps (10) we're at the letter *A*. Go back to the root and start a new path from the third bit (the first two have already been decoded). This path takes us to the okina, or glottal stop symbol, after four steps (0111). Go back to the root again, and the final two bits take us again to *A*. There are no more bits to decode, so this bit string contained the encoding of the word *A'A*.

Earlier in the chapter we saw how to call a BitLab function named encode to generate a binary encoding of a string. The analogous function for encoding with a Huffman tree is named huffman_encode. When we call it, we need to give it the string to encode and the tree that defines the encoding:

```
>>> huffman_encode('ALOHA', tree)
100000010000110
```

To go in the opposite direction, to convert a message that was encoded using a Huffman tree back into a string of letters, pass the tree to a function named huffman_decode:

```
>>> huffman_decode(msg, tree)
'ALOHA'
```

For test data, the BitLab module includes two files of Hawaiian words. One file contains plain text that can be passed to huffman_encode, and the other contains encoded words that can be passed to huffman_decode. To read one of the files, call a function named read_test_data and pass it an argument to tell you which data set you want: 'words' to get a list of unencoded strings, or 'codes' to get a list of Code objects.

Tutorial Project

T71. Make the tree shown in Figure 8.11 by passing the file with the full Hawaiian alphabet to build_tree:

```
>>> tree = build_tree(path_to_data("hafreq.txt"))
```

T72. Call the assign_codes function to create the binary encoding for each letter:

```
>>> hc = assign_codes(tree)
```

T73. Looking at the tree in Figure 8.11, can you figure out the binary encoding for the letters *U* and *P*?

T74. Check your answers by looking up the codes for the same letters:

```
>>> hc['U']
0110
>>> hc['P']
110011
```

Were you able to figure out a code by tracing a path from the root to the leaf for a letter? Try some more examples on your own until you are sure you understand how Huffman codes are defined.

T75. Read the data set that contains a list of Hawaiian words:

```
>>> words = read_test_data('words')
```

T76. Print some of the words in the list, for example,

```
>>> words[1]
"ALI'I"
>>> words[26]
'NENE'
```

T77. Using the codes in hc write the binary code for the word you selected. Check your answer by passing the word to huffman_encode:

```
>>> huffman_encode(words[1], tree)
100000111101111111
>>> huffman_encode(words[26], tree)
1110110111101101
```

T78. Try some more tests on your own. Print some strings in words, encode them by hand, and then pass them to huffman_encode to check your results.

When you are confident you understand how words are encoded in binary, you are ready for the next set of projects, which will decode binary strings to recover the original text.

T79. According to the tree in Figure 8.11, which letter is encoded by the bits 010? By 110011?

T80. You can check your answer to the previous problem by calling a function named print_codes, which will print the dictionary in order, with codes followed by letters:

```
>>> print_codes(hc)
0000 :  L
0001 :  H
...
110011 :  P
```

T81. Read the data set that contains a list of codes:

```
>>> codes = read_test_data('codes')
```

T82. Ask Python to show you the first few items in the list:

```
>>> codes[0:4]
[0110100, 0010100, 00111110, 1100110100]
```

The codes are sorted so the shortest codes are at the front of the list.

T83. Try decoding the first code in the list. You can either use the tree in Figure 8.11 or the list displayed by calling print_codes, whichever is easier. Check your result by passing the code to huffman_decode:

```
>>> huffman_decode(codes[0], tree)
'UA'
```

T84. Try decoding some more codes by repeating the previous two exercises with different items from the codes list.

The idea behind Huffman codes is that a word should be encoded with fewer bits when the encoding scheme has a short bit pattern for more common letters. In the next set of exercises we'll do some tests to see whether this idea works for Hawaiian words.

T85. The alphabet we are using has 13 letters. This expression will tell us how many bits it would take to encode each letter if we use the same number of bits for each letter:

```
>>> from math import ceil
>>> ceil(log2(13))
4
```

T86. The length of the code for a word would be the product of the number of bits per letter and the length of the word:

```
>>> 4 * len("ALOHA")
20
```

So it would take 20 bits to encode *ALOHA* using four bits per codes.

T87. To count the number of bits used by a Huffman code we can use `len` to compute the length of an encoded message:

```
>>> huffman_encode('ALOHA', tree)
100000010000110
>>> len(huffman_encode('ALOHA', tree))
15
```

The Huffman code for this test word saves five bits. That's not surprising, since the word has two *A*s, and the Huffman code for *A* is the shortest code.

T88. We're going to run this same test a few more times, so let's define a function that will print the number of bits in each encoding:

```
>>> def test(s, t):
...     print(len(s)*4, len(huffman_encode(s, t)))
...
```

T89. Pass a string and the tree to this new method:

```
>>> test('MAHALO', tree)
24 20
```

The first number printed is the number of bits for a 4-bit code and the second is the number of bits for the Huffman code.

T90. Test a word that has one of the infrequent letters:

```
>>> test('WIKI', tree)
16 17
```

Even though the code requires six bits for the *W*, the *K* is encoded with three bits, so the total number of bits is almost the same.

T91. Try a few more tests:

```
>>> test("KAKA", tree)
>>> test("POHAKU", tree)
>>> test("HUMUHUMUNUKUNUKUAPUA'A", tree)
```

♦ What do you think will happen if an error occurs in a message encoded with a Huffman code? Encode a long word or place name, change one of the bits, and decode the result:

```
>>> msg = huffman_encode("HONOLULU", tree)
>>> msg.flip(8)
>>> huffman_decode(msg,tree)
'HOAAOLULU'
```

Can you explain what happened to this message?

8.7 Summary

The general topic for this chapter was *data representation*: how to encode information as a sequence of 1s and 0s so it can be stored in a computer's memory, saved in a file, or transmitted over a network.

We focused on techniques for encoding text. The simplest method uses a scheme where each letter in an alphabet has the same number of bits, and each letter has a different pattern. The number of bits to use depends on the size of the alphabet. For example, to encode the 26 letters used for English words, we need 5 bits for each letter. That's because there are 2^5 different 5-bit numbers, and with $2^5 = 32$ patterns there are enough to assign a different sequence of 1s and 0s to each letter.

The general formula for figuring out how many bits are needed to encode a set of items is $\lceil \log_2 n \rceil$, where n is the number of items, and the notation $\lceil x \rceil$ means "round x up to the next integer." So to include upper and lowercase letters, digits, and punctuation marks in a coding scheme, we would simply add up the number of symbols, compute the base-2 logarithm, and round up. The same idea works for any type of data; an example used in Section 8.1 was a coding scheme for answers on a questionnaire, where there are a fixed number of choices for each response.

We also looked at algorithms that work with binary encodings. An error detection algorithm examines a bit pattern to see if anything has changed since the data was originally encoded. The simplest algorithms are based on the idea of *parity*, which is just a count of the number of 1 bits in a piece of data. When the data is first encoded, an extra bit, called the parity bit, is added to the end of the code. The bit that is added is chosen so the total number of 1s is an even number. Then, if any one of the bits changes, the total will be an odd number, and we know something has gone wrong. More sophisticated algorithms can detect more than one change, and in some cases can even figure out which bits changed and correct the error.

Another type of algorithm that works with encoded information is a data compression algorithm. We looked at a simple technique that is based on the frequency of the letters in an alphabet. The idea is to use only a few bits to encode common letters, and more bits for uncommon letters, and as a result a complete piece of text will (usually) require a smaller number of bits to encode.

Exercises

The first four questions refer to the survey that uses the encoding scheme shown in Figure 8.2.

8.1. Suppose another team is added to the list of baseball teams. Could team IDs still be encoded with 5 bits? What if two teams were added? Three teams?

8.2. Suppose the survey is modified to ask which country a person lives in, instead of which state. How many bits would be needed to encode a country ID? (There are 192 member states in the United Nations.)

8.3. If the survey is expanded to include other sports, how many different team IDs could be encoded if the code used 9 bits for a person's favorite team?

8.4. Design an encoding scheme for your own set of localities and teams. Use city names, provinces, *etc.* near where you live, and use names of teams for any sport you like.

Concepts and Terminology Introduced in This Chapter

encoding	A method for representing a piece of data as a sequence of bits (binary digits 0 and 1)
decoding	The opposite of encoding, a process for figuring out what is represented by a sequence of bits
error detection	An algorithm that decides whether a binary code was altered since it was first created
parity	A simple technique for error detection, attaches an extra bit to a code
text compression	A process that reduces the number of bits needed to encode a piece of data
binary tree	A data structure that represents relationships between objects; each node in a binary tree has at most two descendants
Huffman tree	A binary tree where each leaf node corresponds to a character, and the path from the root to a leaf defines the code for that character

---

8.5. What are the ASCII codes of the following characters?

```
Q   w   e   r   t   !
```

Note: you can use Python to check your answer:
```
>>> for ch in 'Qwert!':
...     print(ord(ch))
...
```

8.6. What characters are encoded by the following binary numbers (based on the ASCII coding scheme)?

(a)	1111000	(d)	1101111	(g)	0101000
(b)	0101110	(e)	1100100	(h)	0110111
(c)	1100011	(f)	1100101	(i)	0101001

8.7. Extend each of the codes shown in the previous problem with a parity bit that results in an 8-bit code with even parity (i.e., an even number of 1 bits).

8.8. The codes shown below were originally made by a transmitter that added an even parity bit. Which codes have a single-bit error somewhere in the code?

(a)	11100111	(d)	11100001	(g)	11000111
(b)	11111011	(e)	11100100	(h)	11001010
(c)	11100100	(f)	11010000	(i)	01000010

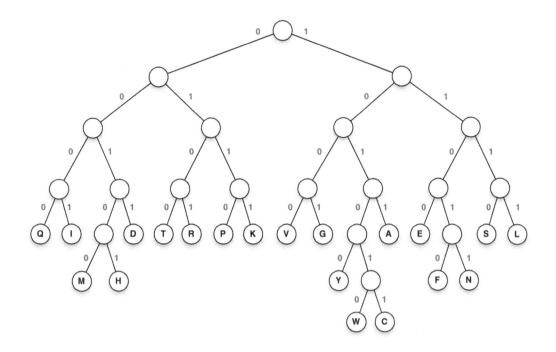

The next set of questions refer to the Huffman tree in the figure above. The tree defines a code for the 20 letters used to abbreviate amino acids, based on their frequency in protein molecules. If you want to check your answers, you can use BitLab to build a Huffman tree from the letter frequencies in a file named aafreq.txt.

8.9. Which letters have the shortest codes?

8.10. Which letters have the longest codes?

8.11. What is the code for the letter *M*? For *V*?

8.12. Use the tree to encode the following sequences of letters:

    a) *MGF*

    b) *GARW*

    c) *VAEYNK*

8.13. Which letter is encoded by the bits 1001? By 101011?

8.14. Use the tree to decode the following strings of bits (*note:* the decoded strings will all be common English words):

    a) 10110011

    b) 011011001001

    c) 01101100101101011110

8.15. DNA sequences can be written as strings made up of the letters *A*, *T*, *C*, and *G*. Describe the code you would get if you built a Huffman tree for these characters, assuming each letter has the same probability (0.25).

8.16. ♦ Find a website with a Hawaiian-English dictionary, or another source of Hawaiian words. Predict which words should be longer if encoded with a Huffman code than with a standard 4-bit code. Check your predictions using the Huffman code functions in BitLab.

8.17. ◆ Would it be possible to use parity bits with Huffman codes? Does it make sense to add parity bits to each character? What about at the end of a word or the end of a message? Explain how you might set up a "communication protocol" so a sender attaches parity bits and a receiver uses them to check words or messages.

8.18. ◆ An exercise on page 281 at the end of Section 8.6 showed what would happen if an error changes a bit in a message encoded with a Huffman code. Repeat that exercise a few times, each time changing a different bit near the start of the message.

    a) Can an error corrupt only one letter? Or is it always the case that several letters will change?

    b) It seems like even when several letters are changed the decoding algorithm eventually gets back in sync again and starts returning the proper letters again. Can you explain why?

Programming Projects

Some of the following projects ask you to implement your own version of functions or methods that are already defined in BitLab. Your programs can import definitions of classes and variables from BitLab to use as helper functions, but the challenge is to write your own top level functions.

Note: You can use the index operator to look at a single bit in a Code object, and you can also iterate over the bits in a Code. It may look like these bits are integers but they are actually 1-bit code objects:

```
>>> c = Code('x')
>>> c
1111000
>>> c[0]
1
>>> type(c[0])
<class 'PythonLabs.BitLab.Code'>
```

BitLab defines two constants named Zero and One to use in expressions that compare codes. For example, if you want to see if the first bit in c is a 1, use this expression:

```
>>> c[0] == One
True
```

8.19. Define a function named print_hex that will print the ASCII codes in a string using hexadecimal notation:

```
>>> print_hex("Hello")
48 65 6C 6C 6F
```

8.20. Write a function named non_ascii_count that will return the number of characters that are not ASCII characters:

```
>>> formula = '2\u03C0r'
>>> formula
'2πr'
>>> non_ascii_count(formula)
1
```

8.21. Define a function named pop_count that returns the number of 1 bits in a binary code:

```
>>> c = Code('A')
>>> c
1000001
>>> pop_count(c)
2
```

8.22. Define a function named add_parity that will extend a code with either a 0 or a 1 in order to make the total number of 1s an even number:

```
>>> c = Code('x')
>>> c
1111000
>>> add_parity(c)
>>> c
11110000
```

You can use the parity function of Figure 8.6 or the pop_count function from the previous problem as a helper function.

8.23. Write a function named check_parity. The input should be an 8-bit code. If the total number of 1s in the code is even, return the ASCII character encoded in the first 7 bits of the code, otherwise return a bullet symbol:

```
>>> c
11110000
>>> check_parity(c)
'x'
>>> c.flip(0)
>>> c
01110000
>>> check_parity(c)
'•'
```

Note: You can use a slice to access parts of a code. For example, c[1:4] is a 3-bit code created from bits 1 through 3 of c. Also, see the note for Problem 8.24 to see how to convert a code to a character.

8.24. Write your own versions of the encode and decode functions. If s is a string, a call to encode(s) should return a list of code objects, with one code for each character in s. Call decode to return a string created from each code in a list:

```
>>> a = encode("aloha")
>>> a
[1100001, 1101100, 1101111, 1101000, 1100001]
>>> decode(a)
'aloha'
```

Note: You can use the char method of a Code object to convert it into a character:

```
>>> c = Code('z')
>>> c
1111010
>>> c.char()
'z'
```

8.25. Write a function named compression that will compare the lengths of two encodings of a word, one using a 4-bit code and one using a Huffman code. The function will be similar to the test function defined in Problem T88, except instead of printing the two lengths, it should return the difference of the lengths. For example if tree refers to the Huffman tree for the full Hawaiian alphabet:

```
>>> compression('HANALEI', tree)
4
>>> compression('POIPU', tree)
-3
```

The result is negative in the second example because the Huffman code is 3 bits longer (not surprising as the word has 2 *P*s and the code for *P* has 6 bits).

8.26. ♦ Carry out an experiment that determines the average compression for English words. To do this experiment you will have to do the following:

- Find a website that lists frequencies of letters in English, and create a text file using this data (see the Hawaiian alphabet data files to figure out the data format).

- Modify the compression function from the previous problem so it uses 5 bits per character (because English words use 26 letters).

- Write a program that scans a document (or uses random words from the PythonLabs word list) and computes the average compression over all words.

8.27. ♦ Problem 5.27 in Chapter 5 described a recursive algorithm for generating a list of strings that corresponded to Gray codes. Write a new version of the function graycode that generates list of Code objects instead of a list of strings:

```
>>> graycode(1)
[0, 1]
>>> graycode(2)
[00, 01, 11, 10]
>>> graycode(3)
[000, 001, 011, 010, 110, 111, 101, 100]
```

Note: You can concatenate two codes using the + operator. If c is a Code object, Zero + c is a new code with 0 in front of c.

A **tree traversal** is an algorithm that "visits" every node in a tree to perform some operation, for example printing some information associated with the node (see the sidebar at the top of the next page). Traversal algorithms use the following methods from the Node class to get information about a tree node. If n is a reference to a node,

- n.is_leaf() returns True if n is a leaf node.

- n.char() returns the character stored at node n (or None if n is not a leaf).

- n.left_child() returns a reference to the node at the root of the left subtree below n (or None if n is a leaf).

- n.right_child() is the same as n.left_child() except it returns a reference to the right subtree.

8.28. ♦ Write a recursive function named depth_first that takes a reference to a tree node as an argument. The function should do a depth-first traversal of the tree to print the characters at the leaf nodes. For example, if tree refers to the root of the Huffman tree for the full Hawaiian alphabet, this is what you should see from a depth-first traversal:

```
>>> depth_first(tree)
L
H
K
...
N
I
```

8.29. ♦ Modify your depth_first function so it takes a list as an optional second argument. If the function is passed a list, the characters at leaf nodes should not be printed but instead should be appended to the list. Return the updated list as the result of the call:

```
>>> depth_first(tree, [])
['L', 'H', 'K', 'O', 'U', "'", 'A', 'M', 'W', 'P', 'E', 'N', 'I']
```

Tree Traversal Algorithms

Traversing a tree is analogous to iterating over a list: the goal is to "visit" every node in order to carry out an operation. In a **depth-first** traversal the algorithm walks down the tree as deep as it can and then works its way back up. In a **breadth-first** traversal the nodes closest to the root are visited first.

Suppose the goal is to print the value at each leaf node in a Huffman tree. A recursive algorithm to do a depth-first traversal of a tree t is to first check to see if t is a leaf node. If it is, print the character at that node, otherwise traverse the left subtree below t and then traverse the right subtree below t. For the tree shown at right, the depth-first traversal will print the nodes in the following order: G, T, C, A.

To do a breadth-first traversal, initialize a list where the root of the tree is the only item in the list. Then use a `while` loop that does the following as long as the list is not empty: remove the first item from the list; if it is a leaf, print the character at the node, otherwise add the left subtree and the right subtree to the end of the list. A breadth-first traversal of the same tree will print A, G, T, C.

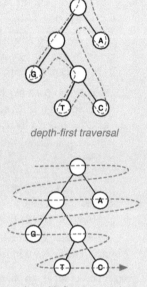

depth-first traversal

breadth-first traversal

8.30. ♦ A breadth-first traversal visits nodes in a tree from top to bottom, so that nodes closest to the root are visited first. Write a function named `breadth_first` that will print the characters it finds at leaf nodes. Here is a breadth-first traversal of the tree for the full Hawaiian alphabet:

```
>>> breadth_first(tree)
A
K
O
...
W
P
```

8.31. ♦ Modify your definition of `breadth_first` so that if it is passed a list as the second argument it will append the characters at leaf nodes to the list instead of printing them:

```
>>> breadth_first(tree, [])
['A', 'K', 'O', 'L', 'H', 'U', "'", 'E', 'N', 'I', 'M', 'W', 'P']
```

Chapter 9

The War of the Words

Computer architecture and machine level programming

From the time the earliest computing machines were proposed, progress in computing went hand in hand with progress in technology. The first machines capable of carrying out the steps of an algorithm were designed in the early 1800s, and they were based on the technology that was available at the time, including basic mechanical devices like gears, levers, pulleys, and rotating axles.

Improved materials and better manufacturing techniques allowed steady improvement in calculators and computers. The introduction of the vacuum tube, which was basically an "electronic switch" that could take the place of older mechanical switches, was a major step. ENIAC, the first electronic machine, was put into service in 1946. The use of electronic technology allowed it to carry out arithmetic operations 1,000 times faster than its electromechanical predecessors.

It was another innovation from this era that made it possible for later computers to evolve into the machines we use today. This innovation was more than simply an improvement in technology. It was a conceptual breakthrough in how computers were organized.

In early computers, the program that controlled the sequences of operations carried out by a machine was external to the machine itself. The data processed by the machine was stored internally, encoded in the form of decimal or binary digits, using techniques similar to those described in the previous chapter. The programs that controlled the machines, however, were typically written on punch cards or paper tapes, and the operations were read one at a time as the machine worked its way through the algorithm. To program the ENIAC, cables were plugged into a panel on the front of the machine (Figure 9.1). These cables connected different parts of the processor, so that output from one component was fed as input to another component.

Figure 9.1: *The ENIAC was programmed by plugging cables into a panel on the front of the machine.*

The major innovation that changed computer science and enabled the phenomenal growth in computing since the 1940s was the idea that the steps in a program could be stored inside the computer along with the data. Computer scientists use the word "architecture" to describe how a set of processors, memories, and I/O devices are interconnected. This new design eventually became known as the **von Neumann architecture**, after John von Neumann (1903–1957), the Hungarian-born American mathematician who wrote the first paper that proposed storing both programs and data in a computer's main memory unit (Figure 9.2).

It would be hard to overstate just how important this idea has been in the development of computer science. It has had a major impact on almost every area, including computer technology, programming languages, software engineering, and theoretical computer science. In terms of computer technology, modern systems would not be able to execute programs anywhere nearly as quickly if programs were still stored externally. Computer chips today can execute billions of arithmetic operations each second, but all that computing power would be wasted if processors had to read instructions from an external source that couldn't provide commands quickly enough to keep the processor running at full speed.

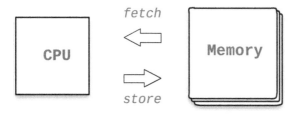

Figure 9.2: *In a von Neumann (stored-program) computer, the memory holds the encoded forms of both instructions and data.*

When the program is in the same memory as the data, processors can access instructions just as quickly and easily as data.

The idea of a stored-program computer opened up the possibility of implementing algorithms in programming languages like Python, Java, or C++ to describe the steps in an algorithm. The first stored-program computers were programmed by writing out extremely detailed step-by-step instructions, where each step corresponded to one of the operations the machine was capable of executing. This type of program is called a "machine language" program because the operations were defined in terms of what the machine could do. Now, however, we write programs in "high-level languages" that describe operations in more abstract mathematical terms. Because programs are saved in plain text files, text editors, compilers, and other applications can read the programs as data. Most languages use a **compiler** to translate programs from a high-level language into a machine-executable form, often doing a variety of transformations and optimizations.

In this chapter we will explore the idea that programs are, like numbers and strings, a form of data that can be encoded in binary and stored in memory. The projects are based on an old computer game named **Corewar**. The game is essentially a computerized version of Battleship. Two programs are loaded into a computer's memory, and they take turns executing instructions. The idea is for a program to lob a "bomb" at the other program that will cause it to stop executing. The bombs are the encoded form of the machine language instruction that tells the machine to halt. A program will pick up a bomb and store it somewhere else in memory, hoping to write it over the code of the other program, so that when the second program executes the instruction it halts and loses the game.

The games are run inside a specially constructed machine that is organized along the lines suggested by von Neumann, where a processor is connected to a memory that holds both instructions and data. To the first program, the bomb is simply a piece of data that can be moved around from one place to another in memory. To the second program, however, the bomb is a machine instruction that tells the program to halt. The fact that instructions and data are both encoded and stored in the same memory unit is the central concept explored in the projects in this chapter.

Core Memory

The "core" in Corewar comes from the main type of technology used to implement computer memories from the 1950s through the 1970s.

A core was a small "donut" made of a material that could be magnetized in a clockwise or counterclockwise direction. Wires threaded through a core could read the direction or set the field to either direction.

The word "core" is still used as a synonym for "main memory" in a computer system. An in-core algorithm is one that carries out all its steps on data that has been read into memory, while an out-of-core algorithm can work on data that is too large to read into memory all at once. A "core dump" is a file created by writing the entire contents of memory out to a disk or CD.

Computer architects refer to an item that can be stored in a single location in memory as a "word." In a von Neumann architecture, a word can be either an instruction or a simple piece of data like a number or a character. The title of this chapter summarizes what is going on inside the machine during a game of Corewar: two collections of words are throwing data at each other until one of them is forced to halt.

9.1 Hello, MARS

The computer used to play Corewar is named MARS. In order to play a game, we don't need to build a MARS system out of processor and RAM chips, but instead we can write a program in Python that mimics the actions of a real system. A program that emulates the actions of a computer is known as a **virtual machine**, or VM.

MARS and other processors based on the von Neumann architecture execute programs according to what is known as the **fetch-decode-execute** cycle. The cycle starts when the machine reads an instruction from memory. This is the "fetch" portion of the cycle.

When the instruction arrives, the processor decodes it in order to figure out what to do. Often this phase performs additional memory operations; for example, if the instruction tells the machine to add two numbers, the decode phase is when the machine reads the values from memory. After all the necessary information has been gathered, the machine carries out the final phase of the cycle and does the actual instruction execution. When the operation is complete the cycle repeats with the next instruction in the program.

In our virtual machine simulator, memory is simply a list. The items in the list will be a new type of object, called a Word, that will represent either a single MARS machine language instruction or a single piece of integer data. When the virtual machine fetches a word from memory, it specifies the list index of the item it wants. For example, if the memory is named mem, and the processor wants to execute the instruction in the first location in memory, it retrieves the word from mem[0]. When computer architects talk about reading or writing items in memory, they use the term **address** to refer to a memory location, so they would say "The processor fetches the word from memory address 0."

Virtual Worlds

The word "virtual" was originally used in computer architecture to describe something that was not really present but was implemented instead by software.

For example, some old computers had 64KB of **physical memory**, meaning there were 65,536 locations to store information in the RAM chips. Using special techniques implemented in the operating system, a program could behave as if there was several megabytes of memory. This extended memory was known as the **virtual memory**.

The idea of a **virtual machine** is still widely used in computer science. A familiar example is the Java Virtual Machine, or JVM. Web designers write "applets" in Java, and when users connect to a website, the applet is downloaded and executed by a JVM that is part of a web browser.

```
1    ; test_hello.txt -- a "hello, world" program for MARS
2
3    x        DAT #7
4    y        DAT #4
5    hello    ADD x, y      ; add the contents of x to y
6             DAT #0        ; a dat instruction halts the machine
7             end hello
```

Figure 9.3: *A "hello, world" program in Redcode, the assembly language of the MARS virtual machine. This program executes one instruction (the ADD on the line labeled hello) and then halts.*

To run a MARS program, we will use a text editor to create a file that has a single MARS instruction on each line. But instead of trying to figure out what the binary encoding of each instruction might be, we can use a symbolic notation known as **assembly language**. An assembly language program is a plain text file that has one line for each machine instruction, along with comments and a few other lines that tell the computer how to execute the program. The assembly language for our virtual machine is known as Redcode. The name of the machine, "MARS," is an acronym for Memory Array Redcode Simulator.

Because Corewar programs don't do any I/O there are no instructions for reading or writing data. Any data used by Redcode program must also be included in the file with the program. When we're ready to run the program, we will save the file and then load the instructions into the MARS virtual machine. As the simulator runs the Redcode program it will change the contents of memory. When the machine halts we can examine the memory to see what the final result was.

Our first project with MARS will be the machine language equivalent of a "hello, world" program. This program doesn't print a message, like a normal "hello, world" program, but it serves the same general purpose: it's a trivial program that illustrates how to use the machine. The MARS version of "hello, world" will just add two numbers and then halt. The Redcode instructions for this program are shown in Figure 9.3.

The main thing to remember when looking at an assembly language program is that each instruction in the program will be translated into a word object that will be stored in memory. The program of Figure 9.3 has four instructions, and when we test it, the program will be loaded into memory locations 0 through 3. The first two lines define the data values that will be added, the third has the instruction that performs the addition, and the fourth is the word that halts the program.

At the front of each line there can be a **label**, which is like a variable name or method name in a Python program. Labels are typically simple names, written in lowercase, that remind us what is stored in memory at that location. If a line does not have a label, spaces must be included at the front of the line so the machine knows there is no label.

The most important part of each line is the **opcode**, which is short for "operation code." It is the opcode that specifies what the processor should do when it fetches this word. The two kinds of instructions used in this trivial little program are ADD and DAT. ADD, as you might expect, tells the machine to add two numbers, and DAT indicates that the word holds a piece of data.

Following the opcode there will be one or two **operands**. ADD instructions have two operands. When the machine executes an instruction of the form ADD x, y it goes out to memory to find the values in locations specified by x and y, adds them, and stores the result back in memory at location y. Notice that this instruction updates memory, and that the old value in location y will be replaced by the sum.

For our programs, each DAT instruction will have just one operand, which will be a piece of data used by the program. MARS programs can only operate on integer data, but the numbers can be positive or negative.

Line 7 in Figure 9.3 is a special statement. It just has the word end followed by a label. This line does not contain a machine instruction; it just tells MARS that when the program is loaded into memory the first instruction to execute is the one on the line labeled hello.

For the first set of experiments with MARS programs we will make a small "test machine" (called MiniMARS) to run our test programs. These machines have a tiny memory unit that is just big enough to hold a single program. You can think of it as a USB "memory stick" containing a machine language program being inserted into the side of a MARS processor. To make one of these test machines, pass the name of a text file that has the program to the constructor for the MiniMARS class. This statement will create a test machine for the "hello, world" program:

```
>>> m = MiniMARS(path_to_data('test_hello.txt'))
```

We now have a small MARS computer, in the form of a MiniMARS object with the name m. If we ask Python to print the value of m we will get a one-line summary of the state of the machine:

```
>>> m
<PythonLabs.MARSLab.MiniMARS name: Hello PC: 2 status: ready>
```

The description tells us what type of object we have and the name of the program loaded into its memory. The acronym PC stands for **program counter**. When we turn the machine on and tell it to start executing instructions, this is the address of the first instruction to execute. Finally, the ready status means the machine is in standby mode, waiting for us to tell it to execute its program.

The method that translates a Redcode source program into a binary format that can be stored in memory is called an **assembler**. One of the main jobs of an assembler is to replace symbolic labels like x and y with addresses so the processor knows where to find the data. We can look at how the assembler translated the instructions in our program by calling a method named dump. Here's what the "hello, world" program looks like:

```
>>> m.dump()
  00: DAT #0 #7
  01: DAT #0 #4
 >02: ADD -2 -1
  03: DAT #0 #0
```

The > in front of line 2 is a reminder that this is the instruction that will be executed when the program starts.

Notice how the labels in the ADD instruction on line 2 have been translated into numbers. Because the word labeled x in the Redcode program is two places before the ADD, the assembler replaced the x with -2. Similarly, the word labeled y is one location in front of the ADD, so y was replaced by -1. In later examples we might see positive numbers as operands; in these cases the number refers to data that is later in the program.

The important thing to remember is that when MARS executes the ADD instruction, the numbers following the opcode are not the data to be added; rather, they are used to create addresses where the data will be found. Because ADD -2 -1 is in memory location 2, this instruction tells MARS to get the words stored in locations 0 and 1, add them, and store the result back in memory at location 1.

Tutorial Project

The MARS virtual machine and other Python functions you will use for this project are in a module named MARSLab.. Start a Python session and import this module:

```
>>> from PythonLabs.MARSLab import *
```

T1. The "hello, world" program is in a file named test_hello.txt. Make a test machine for this program:

```
>>> m = MiniMARS(path_to_data('test_hello.txt'))
```

This statement asked Python to make a MiniMARS test machine object and save a reference to it in the variable named m.

T2. Print the value of m to confirm you have a new MiniMARS object:

```
>>> m
<PythonLabs.MARSLab.MiniMARS name: Hello PC: 2 status: ready>
```

T3. Call the dump method to get a "core dump" showing the complete contents of the memory of this machine:

```
>>> m.dump()
  00: DAT #0 #7
  01: DAT #0 #4
 >02: ADD -2 -1
  03: DAT #0 #0
```

The number at the front of each line is a memory address. Following each address is a machine instruction created by the assembler and stored at that address when the program was loaded into memory.

You may have noticed the dump method prints two operands for the DAT instructions. That's because the internal machine language always has two operands, even though some Redcode instructions just have one. In our programs the first operand on a DAT instruction will always be #0 and we can safely ignore it. The data we use will always be the second item printed after DAT.

T4. Call a method named step to tell the machine to fetch and execute one instruction:

```
>>> m.step()
ADD -2 -1
```

The value returned by step is the instruction that was just executed. As expected, the machine executed the ADD instruction in location 2.

What we want to do now is see if the ADD instruction did in fact add the contents of the two locations specified in the instruction. The ADD instruction should have added the contents of the memory cell two places in front of the instruction to the memory cell one place in front of the instruction. Because the instruction is in memory address 2, the contents of cell 0 should have been added to cell 1.

T5. Call the dump method again:

```
>>> m.dump()
  00: DAT #0 #7
  01: DAT #0 #11
  02: ADD -2 -1
>03: DAT #0 #0
```

Do you see how the item in memory location 1 is now the sum of 7 and 4?

Every time we call step the machine executes a single instruction. It also updates the program counter so it refers to the next instruction to execute. You may have noticed the > is now in front of instruction 3. When we call step again to continue the program the machine will fetch and execute the DAT instruction in memory location 3.

T6. Print the state of the machine again:

```
>>> m
<PythonLabs.MARSLab.MiniMARS name: Hello PC: 3 status: continue>
```

Two things have changed: the PC is now 3, and the status has changed to continue. The new status means the MARS program has not halted yet, and we can call step to execute the next instruction.

T7. Call step again:

```
>>> m.step()
DAT #0 #0
```

As expected, the machine executed the DAT instruction in location 3.

T8. Print the machine state again:

```
>>> m
<PythonLabs.MARSLab.MiniMARS name: Hello PC: 0 status: halt>
```

The status has changed to halt because the DAT instruction terminated the program.

T9. If we try to call step again we'll get an error message because there are no more steps to execute:

```
>>> m.step()
'machine halted'
```

9.2 The Temperature on MARS

As a way to introduce some of the other operations that can be used in Redcode programs, let's look at a program that converts temperatures from Fahrenheit to Celsius. When we did this calculation in Python, we wrote it as a single expression:

```
c = (f - 32) * 5 / 9
```

If we want to do this same computation with MARS, we need to write a Redcode program that has DAT instructions for the variables f and c and a sequence of machine instructions that carry out the steps in the calculation. When the program halts we'll look in the memory cell designated to hold the value of c to find the converted temperature value.

MARS is a very simple machine, with a total of only 11 instructions. There are ADD and SUB instructions, to do addition and subtraction, but there are no multiply or divide instructions. We can still do multiplication and division, though. We just have to implement the algorithms that do these operations in the form of a sequence of MARS instructions.

To compute the value of the mathematical expression x times y we store those two numbers in memory cells which we will label, appropriately enough, x and y. We also set initialize another memory cell with a 0, and then add x to it y times.

This algorithm can be implemented using three Redcode instructions, as shown in Figure 9.4. The instruction on the line labeled mult is one we've seen already. It just tells the machine to add the contents of the cell named x to a cell named acc, which has been initialized with a DAT #0 instruction. The label acc is short for "accumulator," which is an old name for a part of a machine that was used to hold temporary results.

The instruction on the next line subtracts 1 from the memory cell that holds the value of y. What is new about this instruction is the # symbol in front of the 1. The # tells MARS to subtract the number 1, instead of using the 1 to create an address to fetch a piece of data from memory. An operand with a # in front is called an **immediate** operand, and it means "use this number as the data" instead of "use this number as an address to fetch the data from."

The third instruction is what is known as a **branch** instruction, or a "jump." This instruction tells MARS to look at the value in y, and if it is not 0, to go back to the cell labeled mult and continue execution there. The JMN opcode stands for "JuMp if Not zero." The two operands are the label on an instruction to jump to and the name of a memory location to check to see if it is 0.

Taken together the three instructions define an iteration, in this case a simple loop that is executed over and over until y counts down to 0. The number of loop executions depends on the value initially stored in memory location y. As an example, if x contains DAT #6, and y contains DAT #7, this loop will be executed seven times, each time adding 6 to acc. In other words, it will compute 6×7.

A complete MARS program that can be used to test the multiplication algorithm is shown in Figure 9.5. This small program includes the cells that hold the values to multiply (lines 5 and 6) and the loop that implements the multiplication (lines 7 through 10). When the loop terminates, the program continues on to line 11, which is the halt statement that ends the computation.

```
1    acc      DAT #0
2    mult     ADD x, acc
3             SUB #1, y
4             JMN mult, y
```

Figure 9.4: *To multiply x by y, initialize an accumulator with 0 and then add x to it y times.*

```
1    ;redcode
2    ;name mult
3    ;strategy    multiplication demo: compute acc = x * y
4
5    x           DAT #7              ; multiplicand
6    y           DAT #6              ; multiplier
7    acc         DAT #0              ; accumulator -- result goes here
8    mult        ADD x, acc          ; add x to acc
9                SUB #1, y           ; subtract 1 from y
10               JMN mult, y         ; repeat if y is not 0
11               DAT #0              ; algorithm halts here
12               end mult
```

Download: mars/test_mult.txt

Figure 9.5: *A program that tests the multiplication algorithm. When executed by a MiniMARS virtual machine the program will compute 6×7 and halt.*

The division step required by a program that computes $(f - 32) \times 5/9$ is implemented by a loop similar to the multiply loop. The algorithm for dividing x by y is to count how many times y can be subtracted from x. Because we want to divide the value in acc by 9, the division loop repeatedly subtracts 9 from acc and keeps track of the number of times the loop is executed by adding 1 to a counter each time through the loop.

The complete code for the MARS program that converts temperatures is in a file named test_celsius.txt (Figure 9.6). The input temperature is placed in the memory location labeled fahr. When the program halts, the output will be in the location labeled cels. The program begins execution on line 8, which has an instruction to subtract 32 from the input temperature. Next, we want to multiply this value by 5. Notice how the accumulator and counter (lines 18 and 19) are initially set to 0 and 5, respectively, when the program is loaded. The loop that starts at the instruction labeled mult is iterated five times, and after the last iteration, the product of (fahr-32)*5 will be in the cell labeled acc.

The division loop begins with the statement labeled div. At this point the cell labeled acc holds the value we want to divide, so the loop repeatedly subtracts 9 from acc. The test for the end of this second iteration is a little trickier. The instruction SLT #0, acc means "skip the next instruction if $0 < acc$" (SLT is an abbreviation for "Skip Less Than"). So as long as acc has a value greater than 0, the machine skips over the halt instruction, adds 1 to cels, and jumps back to the top of the loop at the instruction labeled div. As soon as acc has a value less than 0, the machine does not do the skip, but instead executes the DAT #0 instruction and halts.

Tutorial Project

T10. Make a test machine with the multiplication program of Figure 9.5:

```
>>> m = MiniMARS(path_to_data("test_mult.txt"))
```

```
 1    ;redcode
 2    ;name Celsius
 3    ;strategy   cels = (fahr - 32) * 5 / 9
 4
 5    fahr    DAT #80             ; input temperature
 6    cels    DAT #0              ; store result here
 7    ftmp    DAT #0              ; save fahr-32 here
 8    start   MOV fahr, ftmp      ; (1) subtract 32
 9            SUB #32, ftmp
10    mult    ADD ftmp, acc       ; (2) multiply by 5
11            SUB #1, count
12            JMN mult, count
13    div     SUB #9, acc         ; (3) divide by 9
14            SLT #0, acc
15            DAT #0              ; stop here when division done
16            ADD #1, cels
17            JMP div
18    acc     DAT #0              ; accumulator
19    count   DAT #5              ; counter
20            end start
```

Download: mars/test_celsius.txt

Figure 9.6: *A Redcode program to convert a temperature value from Fahrenheit to Celsius. The input temperature is the value of the DAT instruction with the label* fahr. *After the program runs, the DAT instruction labeled* cels *will hold the output temperature.*

T11. Get the machine's status by asking Python to print the value of m:

```
>>> m
<PythonLabs.MARSLab.MiniMARS name: mult PC: 3 status: ready>
```

The PC is 3, meaning the machine will start with the instruction in location 3. Note this 3 means "memory cell 3", not line 3 in the text file shown in the figure, because the comments on lines 1 through 4 are not loaded into memory. The first instruction will be the ADD instruction on the line labeled mult.

T12. Use the dump method to look at the machine language. We can pass a starting and ending location as arguments and the method will show us only those cells in memory. This command prints the DAT instructions in memory locations 0 through 2:

```
>>> m.dump(0,2)
  00: DAT #0 #7
  01: DAT #0 #6
  02: DAT #0 #0
```

These lines show the two values that will be multiplied are 7 and 6, and the result is initialized to 0.

T13. Call the step method three times, so that each instruction in the loop is executed once:

```
>>> m.step()
ADD -3 -1
>>> m.step()
SUB #1 -3
>>> m.step()
JMN -2 -4
```

The lines printed by step are machine language instructions, where labels have been turned into addresses. ADD -3 -1 means "add the contents of the word three locations before this instruction to the word one location before this instruction." Do you see what the SUB instruction is supposed to do? And how the JMN instruction means "jump back two instructions if the value in the word four locations back is not zero"?

T14. Call dump again to look at the memory cells that hold the data:

```
>>> m.dump(0,2)
00: DAT #0 #7
01: DAT #0 #5
02: DAT #0 #7
```

Can you see how the number 7 was added to acc (memory location 2), and how the counter (memory location 1) decreased by 1?

T15. Instead of calling step three times, we can call another method named run. run will call step for us, and return either when the program stops or it has called step the specified number of times:

```
>>> m.run(3)
3
```

The return value is the number of instructions that were executed.

T16. Call dump again:

```
>>> m.dump(0,2)
00: DAT #0 #7
01: DAT #0 #4
02: DAT #0 #14
```

Once again 7 was added to acc, and 1 was subtracted from count.

T17. Because there are three instructions in the loop, running the machine for twelve steps will have it complete four more iterations of the loop:

```
>>> m.run(12)
12
```

T18. Print the first three words again:

```
>>> m.dump(0,2)
00: DAT #0 #7
01: DAT #0 #0
02: DAT #0 #42
```

Notice how y, the word in location 1, has counted all the way down to 0. Because the loop ran six times, the number 7 was added to the accumulator that many times, and the result of 7×6 is now in location 2.

T19. Print the machine's status:

```
>>> m
<PythonLabs.MARSLab.MiniMARS name: mult PC: 6 status: continue>
```

Because y is 0 the JMN instruction did not tell the machine to go back to the top of the loop. Instead, the program is going to continue with the instruction in location 6.

```
1   ;redcode
2   ;name div
3   ;strategy    division demo -- compute 20 / 9
4
5   x         DAT #20
6   y         DAT #9
7   count     DAT #0              ; result goes here
8   div       SUB y, x
9             SLT #0, x
10            DAT #0              ; stop here when division done
11            ADD #1, count
12            JMP div
13            end div
```

Download: mars/test_div.txt

Figure 9.7: *A test for the division algorithm (see the optional exercises at the end of this section).*

T20. Use the step method to execute a single instruction:

>>> *m.step()*
DAT #0 #0

Sure enough, the DAT instruction in location 6 was executed.

T21. Because a DAT instruction tells the machine to stop the status should change to halt:

>>> *m*
<PythonLabs.MARSLab.MiniMARS name: mult PC: 0 status: halt>

T22. Make a test machine that has the temperature conversion program loaded into its memory:

>>> *m = MiniMARS(path_to_data("test_celsius.txt"))*

T23. Look at the first two locations in memory. The first is the input temperature, and the second is where the output temperature will be placed when the program is done:

>>> *m.dump(0,1)*
00: DAT #0 #80
01: DAT #0 #0

T24. Run the program:

>>> *m.run()*
124

The return value means the machine executed 124 instructions before it halted.

T25. Look at the first two memory locations again:

>>> *m.dump(0,1)*
00: DAT #0 #80
01: DAT #0 #26

Does the second location hold the Celsius equivalent of 80°F?

You can now move on to the next section, which will explain how MARS runs two or more programs that compete against each other in a game of Corewar. If you want to explore the division algorithm you can work on the following exercises.

♦ Make a machine with the test program for the division algorithm (Figure 9.7):

```
>>> m = MiniMARS(path_to_data("test_div.txt"))
```

♦ Use dump to look at the first three memory locations:

```
>>> m.dump(0,2)
  00: DAT #0 #20
  01: DAT #0 #9
  02: DAT #0 #0
```

♦ When the program runs, it should compute $20 \div 9$ and store the result in memory location 2. Run the program and look at memory after the program halts. Are the results correct?

♦ Download a copy of the program and try some experiments with other data values. For example, to divide 17 by 5, edit the first two lines so they look like this:

```
x          DAT #17
y          DAT #5
```

Save the file, make a MiniMARS object with your file, and run it.

♦ There is a bug in this code: it gives the wrong answer for $x \div y$ when y divides x evenly, that is, when the remainder is 0. Try the program with $x = 20$ and $y = 10$. Can you see what the problem is?

9.3 Corewar

In the classic board game Battleship, players choose where to place a set of ships on a rectangular grid that cannot be seen by the other player. After both players are set up, they take turns calling out board locations, defined by a row and column. For example, player A might call out "C-5." If player B has a ship on square C-5, it is hit. When a ship takes enough hits it sinks.

In Corewar, our computer-based version of Battleship, two Redcode programs are assembled and loaded into the MARS machine's memory at random locations. Our VM is set up to have a default memory size of 4096 words, so programs can be loaded into any location between 0 and 4095. Neither program is given any information about where the other program is located.

When the machine starts, it will alternate instructions from the different programs, first executing a single instruction from Program A, then an instruction from Program B, then A again, and so on. The idea is for a program to lob a "bomb" onto a random location in memory, hoping to hit the other program. A bomb is simply a DAT instruction—if the other program tries to fetch and execute the DAT, instead of the instruction that was originally there, it will halt because a DAT serves as a halt instruction in Redcode. When a program runs into a halt instruction it loses and the other program is declared the winner. If both programs are still running after a predefined number of rounds the game ends in a draw.

A very simple Corewar program, named Dwarf,[1] illustrates one strategy for playing the game. The program is very small: it has one DAT instruction (the bomb) and a three-instruction loop (Figure 9.8). The DAT instruction in the location labeled vache[2] serves two purposes in this program: it is the address where the bomb should be thrown, and, because it is a DAT instruction, it's also the bomb itself. The main idea is to add 4 to this address each time through the loop, so the bomb is stored in every fourth location in memory.

---

[1]Corewar programs often have names inspired by Dungeons and Dragons® or *Lord of the Rings*.
[2]Monty Python fans: do an Internet search for "fetchez la vache."

Figure 9.8: *The Dwarf program stores DAT instructions in every fourth memory cell.*

```
1   ;redcode
2   ;name Dwarf
3
4   vache   DAT #0
5   dwarf   ADD #4, vache
6           MOV vache, @vache
7           JMP dwarf
8           end dwarf
```

Download: `mars/dwarf.txt`

The first and third instructions in the loop are straightforward: the ADD instruction adds 4 to the address where the bomb will go, and the JMP branches back to the top of the loop. The MOV, which stores the bomb at the specified address, illustrates a new programming technique, called **indirect addressing**. Notice that there is an @ symbol before the second operand. Without the @, an instruction like MOV x, y means "copy x to location y." But when the operand includes @, as in MOV x, @y, it means "copy x to the location pointed to by y." In other words, y is not the destination, it holds the *address* of the destination. In this program the word labeled bomb holds the address where we want to throw the DAT instruction. By adding 4 to it each time through the loop we end up storing DAT instructions every four locations in memory.

As the programs start getting more complex, and especially when there are two programs in memory at once, it becomes harder and harder to follow the execution of the programs by simply calling dump to print the contents of memory. The MARSLab module includes another method for viewing the contents of memory. This method, called MARS.view, opens the PythonLabs canvas and draws a series of gray boxes, one per memory cell. Because there are 4096 memory cells, they can't all be displayed on a single line, so they are split into a set of rows. The window shown in Figure 9.9 has 48 cells per row. The top row has boxes for memory cells 0 through 47, the next row for cells 48 through 95, and so on.

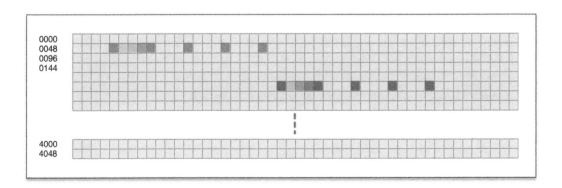

Figure 9.9: *Visual display of MARS memory, showing the progress of a Corewar tournament with two Dwarf programs.*

When the MARS virtual machine is running a set of programs the view will be updated. Each time the machine fetches a word from memory or writes a word into memory, the corresponding box will change color. The viewer uses different colors for each program to make it easier to watch the progress of the different programs.

Figure 9.9 shows what you might see during a tournament being played by two Dwarf programs. A cluster of four colored cells in a row indicates where the instructions are being executed. Each time MARS fetches an instruction it colors one of these cells. As the programs iterate over the three instructions in the body of the loop these cells will be changing color so you can watch the progress of each program. The animation also changes the color of a cell where a bomb is dropped. As the programs run you will see how bombs are placed in every fourth cell.

Figure 9.9 also helps explain how MARS deals with two issues related to memory addresses. The pointer used in indirect addressing to specify where a bomb is written out to memory is initially set to 0, and then on successive iterations it is set to 4, 8, 12, and so on. As you can see from the pattern of colored squares in the figure, MARS interprets the pointer to be a distance from the start of the program. Just as the number -2 in an instruction like JMP -2 tells the machine to go to an address relative to the current instruction, a value in a pointer tells the machine to use an address relative to where the pointer is located.

This is a fussy little detail that you will need to understand if you want to write your own Redcode programs, but for the "big picture" it's enough to know that the Dwarf program always tosses bombs at locations that start at the same address the program is loaded into. The leftmost colored cell for each program in Figure 9.9 is the location of the pointer. The first bomb went into a memory cell four places after this one, and each new bomb goes into a location four places past the previous one.

The other issue is, what happens when a pointer value reaches the end of memory? With a 4096-word memory, the addresses run from 0 to 4095. If a Dwarf program is loaded at location 4000, will it only be able to put bombs in locations 4004, 4008, and so on, up to location 4092, before it runs out of room? If that were the case, the contest wouldn't be very fair. The program loaded closer to address 0 would always have an advantage. To deal with this, MARS memory references "wrap around" to the start of memory again. If you watch the display, you will eventually see that bombs reach the end of memory and then start over again in the low addresses.

Tutorial Project

T26. Create the graphics window that will display the state of the MARS machine by calling the MARS.view method:

```
>>> MARS.view()
```

You should see a grid of gray cells similar to the one in Figure 9.9.

T27. Call MARS.load to have the machine load a Redcode program into a random location in memory. This command will load the Dwarf program:

```
>>> MARS.load(path_to_data('dwarf.txt'))
```

T28. MARS.status will show you where the programs was placed:

```
>>> MARS.status()
Dwarf     : 1809..1812  PC: [ *1810 ]
```

The machine status shows the program was loaded into locations 1809 through 1812 (you will see different values since the programs are placed in random locations). The program counter shows that the next instruction to execute will be the second instruction in the program, the start of the loop.

T29. A method named run will execute the program for a specified number of rounds:

```
>>> MARS.run(10, single=True)
```

Passing single=True tells the machine we're not running a contest but just testing a single program. As the program runs, the squares in the graphics window will start filling in.

T30. To slow down the display specify a value for a variable named delay. This command will pause the machine for 1/10 second between each instruction:

```
>>> Canvas.delay = 0.1
```

T31. Continue the program for another 100 instructions:

```
>>> MARS.run(100, single=True)
```

Can you see how the program is in a three-instruction loop? Bombs are being dropped on every fourth location, starting at the end of the set of instructions for the program and moving toward higher addresses. If the program runs long enough, you should see bombs "wrap around" from high addresses to low addresses.

The next part of the experiment will run a contest between two copies of the program. This time we'll specify where to put them so we can see what happens when one program wins.

T32. Call MARS.reset to clear the machine:

```
>>> MARS.reset()
```

T33. Load two copies of Dwarf into the machine at addresses 1960 and 1990:

```
>>> MARS.load(path_to_data('dwarf.txt'), 1960)
>>> MARS.load(path_to_data('dwarf.txt'), 1990)
```

The second argument passed to load is the location where we want the program to go. You should see two programs in the middle of the display.

T34. Ask to see the machine's status:

```
>>> MARS.status()
Dwarf     : 1960..1963  PC: [ *1961 ]
Dwarf     : 1990..1993  PC: [ *1991 ]
```

There are now two programs, and each has its own program counter (PC).

T35. Tell the machine to run for 20 rounds:

```
>>> MARS.run(20)
```

The machine executed 40 instructions, 20 from each program.

T36. Look at the pattern of bombs stored by the program on the left (red cells). Do you see how the next bomb will land in the middle of the blue program?

T37. Run for 10 more instructions:

```
>>> MARS.run(10)
'halted'
```

Notice how the return value from the call to run is the string 'halted'. That means one of the two programs has stopped because it tried to execute a DAT instruction.

T38. Look at the display. Do you see how the location where the bomb hit the blue program has turned black? This is the location where the blue program fetched the DAT instruction that caused it to halt.

T39. If you try to continue the program again the machine won't do anything; it just returns
'halted' right away.

T40. Call the status method:

```
>>> MARS.status()
Dwarf    : 1960..1963  PC: [ *1963 ]
Dwarf    : 1990..1993  PC: [ ]
```

The output confirms what we see on the display: the second program does not have a PC
value, so it cannot fetch an instruction, and the other program has won the game.

T41. If you're still not sure how the Dwarf program works, and you'd like to do some more experiments with it, create a test machine. You can pass a second parameter to the MiniMARS
constructor to make a memory with extra words:

```
>>> m = MiniMARS(path_to_data("dwarf.txt"), 20)
```

Using the techniques shown in the previous section, call dump to look at the contents of
memory, run the machine for a few steps, and then look at memory again.

9.4 Self-Referential Code

The MARS projects up to this point have demonstrated the main idea of the von Neumann
architecture. The two major components in a computer are the processor and memory. The
memory holds encoded forms of both instructions and data, and programs are executed
through a fetch-decode-execute cycle.

One of the implications of this design is that programs can be treated as data. Programs
can make copies of other programs—or even themselves—and modify code by writing new
instructions. The project in this section will explore this idea in a little more depth.

The Dwarf program tries to force the other program in memory to halt by writing DAT #0
over the code of the other program. But the Dwarf is not limited to writing DAT instructions; it could write any MARS instruction. One possibility is to write JMP 0 to every fourth
location. Recall that the operand in a JMP instruction is the distance, relative to the location of the JMP, to the next instruction in the program. The Dwarf itself implements its
loop through a JMP -2 that means "go back two instructions." So JMP 0 means "go to the
instruction in the same location as this instruction." In other words, any program executing
this instruction would be caught in an infinite loop! It's a small loop, containing only one
instruction, but it's a loop nonetheless.

In more abstract terms, a program that writes JMP 0 instead of DAT #0 is trying to "stun"
the other program instead of killing it. The other program will still remain in the game,
but it can't make any progress: it will be stuck in the location of the JMP instruction. This
strategy might be effective in certain situations that will be explained in the next section.

The fact that a program can use any instruction as a type of data opens up another Corewar strategy: a program can copy itself, and then jump to the location of the copy and
continue execution there. For example, suppose the Dwarf program has executed 1024 iterations of its loop and the other program is still running. Because the memory size in our
MARS virtual machine is 4096 words, and the Dwarf drops a bomb on every fourth location,
the next 1024 iterations will just write DATs over the same locations. One way to get the
program to write DATs to different locations is to copy the four instructions in the Dwarf
code to a new location in memory and then jump to this new location. If the difference

Figure 9.10: *A program named Imp copies itself.* **(a)** *When it is first loaded it occupies only one memory cell.* **(b)** *After two rounds the program has copied itself twice and is ready to make a new copy.*

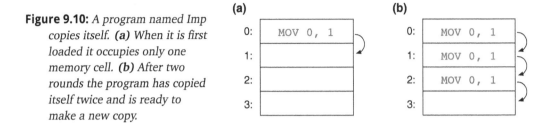

between the address of the copy and the address of the original is not a multiple of four, the copy will start writing DATs over new locations.

A very simple program that copies itself is named Imp (Figure 9.10). This program has just one instruction: MOV 0, 1. An instruction of the form MOV x, y means "copy the contents of location x to location y." Here again, the 0 and the 1 are interpreted as memory locations relative to the address of the instruction. The 0 means "the address of this instruction" and the 1 means "the address following this instruction." When the machine executes the MOV, the instruction copies itself one location ahead in memory. When MARS goes to fetch the next instruction for this program, it sees the newly made copy, and the cycle repeats.

This tiny little one-instruction program moves relentlessly through the entire memory of the machine. Unless it is stopped or modified by the other program, after 4096 cycles the entire MARS memory will be filled with MOV 0, 1 instructions. The natural question is, how effective is this strategy? What will happen when the IMP writes itself over the code of the other program?

The project in this section will be to run a contest between an Imp and a Dwarf. Before you start, make some predictions about what might happen in this game:

- What will happen if the Imp writes its code over the Dwarf?

- Can the Imp ever cause the Dwarf to halt?

- What are the odds of the Dwarf being able to stop the Imp by writing a DAT instruction over it?

Tutorial Project

T42. Load the Imp program into a small test machine with ten words:
```
>>> m = MiniMARS(path_to_data("imp.txt"), 10)
```

T43. Call dump to look at the memory:
```
>>> m.dump()
>00: MOV 0 1
  01: DAT #0 #0
  02: DAT #0 #0
  . . .
```

As you can see, the program is only one instruction long, and it is loaded into location 0.

T44. Tell the machine to execute one instruction, and print the memory again:

```
>>> m.step()
MOV 0 1
>>> m.dump()
  00:  MOV 0 1
 >01:  MOV 0 1
  02:  DAT #0 #0
  ...
```

Do you see how the Imp duplicated itself by simply moving a copy of what is in location 0 to location 1?

T45. Step the machine a few more times, and print the memory again. Each time the machine executed the Imp instruction it made a new copy one location further down in memory.

What do you think will happen if the Imp ever reaches the end of memory? Make a prediction and then test your hypothesis by running this small experiment.

T46. Reset the MARS machine and load a copy of the Imp into a location near the end of the machine's memory:

```
>>> MARS.reset()
>>> MARS.load(path_to_data('imp.txt'), 4090)
```

Because the program is only one instruction long you should see a single red cell in the lower right corner of the display.

T47. Run the program in single-player mode for ten instructions:

```
>>> MARS.run(20, single=True)
```

Did you see how the program "wrapped around" and continued at location 0?

T48. Reset MARS and set up a contest between an Imp and a Dwarf:

```
>>> MARS.reset()
>>> MARS.load(path_to_data('imp.txt'))
>>> MARS.load(path_to_data('dwarf.txt'))
```

T49. Run the game for 1,000 steps:

```
>>> MARS.run(1000)
```

As these programs run it should be easy to see how the Imp moves through memory. If the contest stops before the Imp catches up to the Dwarf, type MARS.run to run for another 1,000 cycles.

Because you can control where programs are loaded you can design your own experiments. Here are some ideas.

T50. What happens if you load the Imp into a location one space before the first location of the Dwarf?

T51. Repeat the previous experiment, but move the Dwarf one location farther away from the Imp. What happened that time?

T52. Can you set up an experiment where the Dwarf wins?

9.5 ♦ Clones

In the previous section we saw how a Redcode program can make a copy of its instructions, and then branch to the copy so that it starts executing from a new location. We can take this idea one step further—we can also tell MARS we want to execute both copies of the program. Not only is the program executing instructions from the new location, but it also continues to execute the original instructions. It's as if the program was able to clone itself, so that two identical copies are running.

The MARS instruction that activates the second copy of the program is named SPL, which stands for "split." When the machine executes an instruction of the form SPL x, it starts running a new program located at memory address x. But it also keeps the current program running, and the instruction immediately following the SPL is also executed.

It is not necessary for the program at label x to be an exact copy of the original. Corewar programs often have separate little pieces that each do a specialized task. A piece can be activated by executing an SPL instruction that tells the machine to start the secondary task while also keeping the original program going.

In modern computing terminology, we would say that the SPL instructions starts a new **thread**. Writing a program that has two or more threads is more difficult than programming with a single thread, but it can be very useful for complex applications. For example, web browsers typically have multiple threads, with threads for the different windows and for pieces of code that manage menus and operations inside browser windows.

Monitoring a program on the visual display when it executes an SPL instruction is similar to playing an old video arcade game. In the game named Centipede® a worm-like creature winds its way down the screen. When a player shoots the centipede, it splits in two, and each piece becomes a new, shorter worm. When you watch a MARS program that contains SPL instructions, you will see a new thread appear in the display. For example, if you were to run a version of Dwarf that copies itself and then uses SPL to start the copy, you would soon see two places on the screen running the three-instruction main loop of the Dwarf program. To let you know that both copies are from the same player, both will have the same color.

A simple program that demonstrates how to use a split instruction is shown in Figure 9.11. This short program is an example of an "Imp cannon." The SPL instruction on the first line launches a new thread that starts execution at the line labeled imp. This part of the program is a simple Imp, and it will start replicating itself through memory. The original thread continues at the JMP instruction, which goes back to the top of the program where

Figure 9.11: The "Imp cannon" strategy fills the
MARS machine with multiple Imps.

```
1   ;redcode
2   ;name threads
3
4   start   SPL imp
5           JMP start
6   imp     MOV 0, 1
7           end start
```

Download: mars/test_threads.txt

it spawns yet another thread. This short two-instruction loop keeps generating Imps that spread through memory, one after another.

Making new threads of execution comes at a cost, both in real applications and in MARS programs. In MARS, the new thread alternates steps with the old thread. Suppose program A executes an SPL, breaking itself into two threads, but program B still has a single thread. The machine will execute one instruction from A, then one from B, then one from the second thread in A, then back to B, then back to the first thread in A again. In other words, the machine cycles are still split evenly between A and B, but within program A, the machine will switch back and forth between the two threads. Even though A has two threads, the programs running in those threads proceed half as quickly as the single program before the split.

What this means in terms of Corewar strategy is that there is an advantage in breaking into multiple threads. A program is still alive as long as any of its threads is still running. But this advantage is offset by the fact that each thread is running more slowly, and as a result it will be easier for the other program to knock it out.

In the previous section there was a discussion of whether it made sense for a program to try to "stun" an opponent by writing JMP 0 instead of killing it by writing DAT #0. One place where this strategy might be advantageous is if a program expects to compete against an opponent that starts several threads. If a thread is killed, the surviving threads will all speed up slightly because there is one less thread to share that program's machine cycles. But if a thread is stunned, it still consumes its share of machine cycles. All of the opponent's threads still run at the same slow speed, and it may eventually be easier to kill them all off.

Tutorial Project

♦ Reset your MARS machine and make sure it is on the canvas, then load a single copy of the test_threads program into location 0:

```
>>> MARS.reset()
>>> MARS.view()
>>> MARS.load(path_to_data('test_threads.txt'), 0)
```

♦ Get the machine's status:

```
>>> MARS.status()
threads   :   0..2     PC: [ *0 ]
```

The program counter (PC) shows there is one thread, and that the next instruction for this thread is in location 0.

♦ Execute one instruction in single-player mode:

```
>>> MARS.run(1, single=True)
```

♦ Now get the status again:

```
>>> MARS.status()
threads   :   0..2     PC: [ *1 2 ]
```

There is a lot of new information here. First, notice there are two threads. The next instruction to be executed for the first thread is in location 1. That's the JMP instruction that will be executed by the original thread. The new thread is starting at location 2, which is where the Imp is located.

♦ Run another instruction and look at the machine status:

```
>>> MARS.run(1, single=True)
>>> MARS.status()
threads  :   0..2    PC: [ 0 *2 ]
```

The new status shows that the first thread executed the JMP, so its next instruction will come from location 0 again. The asterisk has moved to the second thread to show that the next instruction for this program will come from that thread.

♦ Run one more instruction and look at the status:

```
>>> MARS.run(1, single=True)
>>> MARS.status()
threads  :   0..2    PC: [ *0 3 ]
```

The Imp has moved to location 3, and the next instruction will come from the main thread, at the top of the loop.

♦ Run the program for 20 steps and print the status again:

```
>>> MARS.run(20, single=True)
>>> MARS.status()
threads  :   0..2    PC: [ 0 *8 7 5 3 ]
```

Do you see how the main thread keeps spinning off copies of the Imp, and how those copies are steadily moving through the machine?

♦ Run the program for 1,000 steps:

```
>>> MARS.run(1000, single=True)
```

As you are watching the program, do you see how it seems to slow down the longer it runs? Each new thread is taking a portion of the instruction cycles.

A program named Mice uses the strategy of cloning itself, making so many copies that it would be hard for the opponent to stop them all. After the program copies itself, it uses an SPL instruction to activate the copy. Now two threads are copying themselves, and each of those will eventually execute an SPL, meaning there will be four copies of the program running.

♦ Reset the machine, load Mice, and run it in single-player mode:

```
>>> MARS.reset()
>>> MARS.load(path_to_data('mice.txt'))
>>> MARS.run(25, single=True)
```

25 instructions is enough for the program to duplicate itself somewhere else in memory.

♦ Run another 50 steps:

```
>>> MARS.run(50, single=True)
```

Do you see how each copy is now duplicating itself?

♦ Run for 1,000 steps:

```
>>> MARS.run(1000, single=True)
```

Are there mice all over the place?

♦ To run a contest between Mice and Dwarf, load them both into random locations and run the machine for a 1,000 or more cycles:

```
>>> MARS.reset()
>>> MARS.load(path_to_data('mice.txt'))
>>> MARS.load(path_to_data('dwarf.txt'))
>>> MARS.run(1000)
```

Is the Dwarf ever hitting one of the mice? Can it win this contest?

Concepts and Terminology Introduced in This Chapter

von Neumann architecture	A design for a computer system in which programs are encoded in binary and stored in memory
processor	The part of a computer system that contains the logic circuits that carry out steps of an algorithm
memory	The component that stores data and instructions
virtual machine	A piece of software that simulates the actions of a processor and memory
assembly language	A programming language where each statement is a single instruction that can be carried out by a processor
opcode	The part of an assembly language statement that tells the processor which operation to perform
operand	Part of an assembly language statement that specifies data to be used in an instruction or where in memory to store a result
thread	A sequence of instructions that implements a single program or method

---

9.6 Summary

The important idea introduced in this chapter is the concept of a stored-program computer. Proposed by mathematician John von Neumann in 1945, the plan of connecting a processor to a memory that holds both instructions and data is the dominant architecture used in practically every computer system today. Modern systems have a variety of different kinds of memories, sophisticated "multi-level cache" systems to improve performance, and separate caches for instructions and data, but these are all methods for efficiently implementing the basic plan of the stored-program computer.

Being able to encode programs as data has had far-reaching implications beyond the area of computer engineering. In order to store a program in memory, it has to be encoded in the form of binary symbols. The first computer programs were written in a binary machine language, but over the years more abstract notations were developed. An assembly language, like the Redcode language we used in this chapter, has symbolic names like ADD and MOV for machine operations, and allows us to assign names to memory locations. Modern programmers use higher level (more abstract) programming languages. Compilers, debuggers, and other applications used for software development are all based on this notion that programs are data that can be operated on by other programs.

The projects in this chapter played a game called Corewar, a computer-based version of Battleship, a classic board game where players try to guess the location of their opponent's ships. In Corewar, the contestants are programs, and the goal is for one program to halt the execution of the other program. The game depends on the fact that a program can overwrite instructions in the other program, copy itself to a new location in memory, or use any of a variety of other strategies derived from having instructions and data in the same memory.

Our Corewar contests were played on a virtual machine named MARS, a simulator written in Python that mimics the actions of a real processor. The simulator allows us to test programs in isolation, executing the program one step at a time and watching what each instruction does. We can also load two programs into the machine and let them run until one program stops or until a predetermined number of instruction cycles have been executed.

The technique of using software to implement virtual machines is widely used in computing. One familiar example is the Java Virtual Machine, or JVM, which is a standard part of almost every web browser. When you connect to a website that contains an "applet" written in Java, the web server sends your browser the encoded form of the JVM instructions for the applet, allowing the program to run on your computer as the browser executes the JVM instructions. A growing new technology known as "cloud computing" is also based on the notion of a software description of a real machine. Computer administrators can configure a machine as if it were a real piece of hardware, but then upload the machine specification to a server on a network. When someone wants to use the machine the server finds a place to execute programs on an actual piece of hardware that matches the description of the virtual machine.

The advanced section on "clones" introduced the idea of a thread. The SPL instruction splits MARS programs into two pieces that appear to run in parallel. In reality, however, the virtual machine divides its time among the different threads in a round-robin manner, running one instruction at a time. Threads are also widely used in real-world programming, especially in complicated applications where threads execute separate tasks.

The fact that applications are often organized as a group of threads that all cooperate to solve a complex problem brings up an interesting question: can we build a computer system with more than one processor, so that all the threads can run at the same time on different processors? New generations of processor chips do in fact use "multi-core" technology that is able to speed up these sorts of applications. Systems known as "multi-CPU clusters" are built from hundreds or even thousands of processors, and are being used to solve very large problems in science and other areas that require high-performance computing.

About Corewar

The game of Corewar was first described by A. K. Dewdney in a series of columns written for *Scientific American* in 1984. The MARS virtual machine we used for our projects is based on a 1988 standard, but the most widely used virtual machines today are based on a newer standard that was published in 1994.

If you would like to try your hand at writing your own Corewar programs, you can find more information on the web at http://corewar.co.uk. There you will find tutorials, articles on the history of the game, and implementations of the newer virtual machine that you can download and install on your own computer.

The PythonLabs data directory has several more Redcode programs with names like "ferret," "piper," and "plague." Call path_to_data('piper.txt') to see where "piper" was installed on your system and you will find these other files in the same directory.

Exercises

The first questions are about the multiplication and division algorithms presented in Section 9.2. The multiplication algorithm computes $x \times y$ by repeatedly adding the contents of a word labeled x to an "accumulator." The division algorithm computes $x \div y$ by counting how many times y can be subtracted from x.

9.1. How many iterations does the multiplication algorithm make if it is asked to compute 2×8?

9.2. How many iterations are required to compute 8×2?

9.3. Will the algorithm give the correct result when $x = 0$? Explain what the program named test_mult would do if the instruction labeled x is changed from DAT #7 to DAT #0.

9.4. Repeat the previous problem, but explain what would happen if $y = 0$. In other words, what would the algorithm do if the instruction labeled y is changed from DAT #6 to DAT #0.

9.5. Will the multiplication algorithm work if either input is negative? For example, would the algorithm give the correct result if it is asked to multiply -3×4? What about 3×-4?

9.6. What would the division algorithm do if it is asked to compute $7 \div 0$?

9.7. When the bombs thrown by the Dwarf program reach the end of memory, they "wrap around" and start going into low addresses. Will the bombs ever overwrite the Dwarf's own instructions, so it accidentally "shoots itself in the foot" and causes itself to halt?

9.8. What would happen if the ADD instruction in a Dwarf program is changed from ADD #4, bomb to ADD #2, bomb? Would this change your answer to the previous question?

9.9. ◆ Explain what would happen if a program executes the instruction SPL 0.

Programming Projects

9.10. Modify the temperature conversion program (celsius.txt) so it converts temperatures in the opposite direction, from Celsius to Fahrenheit.

9.11. Fix the bug that causes the division program (test_div) to give the wrong answer when x is a multiple of y. Hint: Add a test that checks to see if $x - y = 0$.

9.12. ◆ Modify the Dwarf program (dwarf.txt) so it uses the strategy described in Section 9.4 (page 307), where the program throws 1,024 bombs and then copies itself to a new location an odd number of words away.

9.13. ◆ Implement the Sieve of Eratosthenes in Redcode.

- The program should find all prime numbers less than $n = 1,000$.

- You can put the "worksheet" in 1,000 cells that immediately follow the last instruction in your pgoram.

- Write a loop that initializes the list with the numbers from 2 to 999. Use a DAT #0 instruction as the placeholder value. This loop will be easier to write if you use the DJN (decrement and jump if non-zero) instruction. Read about this instruction and see examples of how it is used in the Lab Manual.

- For the first version of the main loop, remove multiples of 2, then 3, then 4, and so on, up to multiples of 31.

- Write a more efficient version that will find the first non-zero location in the list, and then remove multiples of that value. For example, after removing multiples of 5, the next non-zero value is 7. The loop should store DAT #0 in list location 14, 21, and so on.

Chapter 10

I'd Like to Have an Argument, Please

A program that understands English (or does it?)

In 1950, Alan Turing (1912–1954), one of the founders of modern computer science, published a paper with the title "Computing Machinery and Intelligence." Electronic computers were just starting to be used outside of math and science, and they were being adopted by businesses and other organizations. There was widespread interest in this new technology, and people began to wonder just what these machines were capable of doing. The topic of Turing's paper was the nature of human intelligence, specifically whether a machine could ever be able to handle problems routinely solved by humans.

To answer the question of whether or not a computer was intelligent, Turing proposed a simple criterion, based on language. He argued that if a computer could carry on a conversation with a person, without the person ever suspecting they were interacting with a machine, then the computer should be considered intelligent.

Building a robot that looks and sounds like a human is, by itself, a daunting prospect. Turing suspected that if a person talked with a robot, no matter how well the robot could converse in English, the person would be biased and would never accept the machine as being intelligent.

To separate the question of intelligence from questions of looking and sounding human, Turing proposed a simple "thought experiment." In this experiment, which is now known as the **Turing Test**, a human judge is asked to carry on a conversation using a computer terminal. In modern terminology, we would say the judge would use a chat application, running on their laptop or desktop personal computer. The judge chats with two other participants over a local network connection (Figure 10.1). The other two participants will

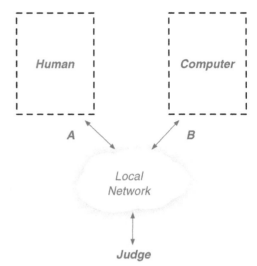

Figure 10.1: *In the Turing Test, three participants are using "chat" software over a local network to carry on a conversation. The judge knows there is a human in one room, and a computer in the other, but does not know which room holds the computer. The object is for the judge to ask questions of both participants, and then decide which one is the computer.*

be in two different rooms, behind closed doors, labeled *A* and *B*. In one room there is a human, and in the other there is a computer, but the judge will not know which participant is in which room.

The judge is allowed to direct questions at either of the other participants. For example, the judge might type "*A*, what color is a fir tree?" or "*B*, should the NCAA establish a playoff system for all levels of college football?" *A* and *B* each try to convince the judge they are human. If, at the end of the conversation, the judge can't decide which room holds the computer, we can conclude the machine possesses a high degree of human intelligence.

It is interesting to note that 60 years after Turing's paper was published his test is being applied in a real way. If you participate in online chats at any of the popular social networking websites, you might have seen a posting by a "chatbot," a computer program that generates posts. Malicious chatbots generate spam messages that fill chat rooms with advertising, and some chatbots have reportedly tried to pose as real users to fool people into revealing credit card numbers and other financial information. Chatbots don't need to be able to converse about subjects in general, and they don't even need to fool a high percentage of people. If just a few users are tricked, even temporarily, the chatbot will have served its purpose.

Issues related to writing computer programs that can converse with humans are part of an active area of research, with many unsolved problems and open questions. The goal is to develop methods for representing everyday knowledge, concepts we take for granted, and to design algorithms that use this knowledge to carry on a conversation. The name of this research area is **natural language processing** to distinguish it from another area of computer science, which is concerned with computer programming languages.

One of the first programs to attempt to converse in English was named ELIZA . It was written by Joseph Weizenbaum (1923–2008), a computer scientist at MIT, in 1966. ELIZA was a remarkable program, not only because it was able to generate realistic English sentences, but also because of how people reacted to it. Weizenbaum was amazed by how often people were willing to open up and converse at length with ELIZA, even when they knew they were getting responses from a computer and not another human being. In *Computer Power and*

Human Reason, a book he published in 1976, Weizenbaum compared this "ELIZA effect" to going to the theater: we know the people on stage are actors, but we suspend our disbelief for a few hours and think of the characters as real people with real lives. When users were typing sentences into ELIZA, they seemed perfectly happy to carry on a conversation, even when they knew it was a computer on the other end.

This chapter is an introduction to natural language processing. We will use a version of ELIZA, written in Python and included as part of the PythonLabs software package, to explore some of the issues faced by programs that attempt to use natural language. Unlike the other chapters, where the goal is to show how a problem can be solved by computation, the project in this chapter will raise new questions instead of illustrating various solutions. As we look at how our version of the program works, we will start to see how difficult it is to even define the problem, much less figure out how to solve it computationally. Natural language processing is a huge challenge, and there is still a long way to go before computers routinely converse in English or any other human language.

10.1 Overview of ELIZA

Weizenbaum's idea for ELIZA was to make a program that would respond by simply rearranging the words in an input sentence. The rearrangements were defined by a set of **rules**. For example, if a user types "I don't like rain," the computer could simply pick out important words, apply a rule that reorders them, and respond with "Why don't you like rain?" To test this idea, Weizenbaum developed rules that would have his system mimic a Rogerian psychologist, so the rules were designed to look for words or phrases a person might use during a therapy session.

The simplest type of rule is to have the program recognize key words and reply with one of several predefined responses for that word. For example, if the input contains "mother" or "father" or other family relationships, the response could be one of several generic sentences about families. Here is a part of a transcript from a session with Eliza, the PythonLabs implementation of ELIZA:

```
H: My father wouldn't buy me a puppy.
C: Tell me more about your family.
```

Whenever Eliza sees any sentence with a word that refers to a family member, it can respond with a generic sentence like the one above. The word "dream" is another key word, and the program will respond to any sentence containing the word "dream" with something like "What does that dream suggest to you?" or "Do you dream often?"

More complicated rules extract a large chunk of the input and use it as part of the response. As an example of how this works, consider statements like "I'm sad because my team lost the big game" or "I'm sad I never got a puppy." In general, sentences that start out with "I am sad..." continue with a description of a situation or a reason the speaker is sad. The rule Eliza uses to respond to this type of sentence requires the program to find the words following "sad" and use them as part of the response:

```
H: I'm sad I never got a puppy.
C: Can you explain what made you sad you never got a puppy?
```

Conversations with Eliza

To help distinguish between the original program defined by Weizenbaum and the Python module you will use for the tutorial project in this chapter, the name of the original program will be written as it was spelled by Weizenbaum, in all capital letters: ELIZA. The Python module will be referred to as Eliza, using the Python convention that module names start with an uppercase letter.

There are several examples throughout the chapter that show a transcript from a conversation with Eliza. In these examples, lines that begin with H indicate sentences typed by a human, and lines that begin with C are responses printed by the computer:

```
H: I am afraid of cows.
C: Is it because you are afraid of cows that you came to me?
```

When you run Eliza in the tutorial projects, Eliza will print an H as its prompt character, to indicate that it is ready for you to type a sentence.

In this second example, the key word is "sad." The program took the words following "sad," in this case "I never got a puppy," made a simple transformation that changed "I" to "you," and used the result with its predefined response of "Can you explain what made you..."

Of course, not all sentences that start out "I am sad..." fit this general pattern, and this rule can easily construct a response that looks pretty strange:

```
H: I'm sad we never had a chance to play fetch.
C: I am sorry to hear you are sad we never had a chance to play fetch.
```

The problem, of course, is that the word "we" can be interpreted two different ways. A psychiatrist hearing that sentence from a patient would assume "we" is another way of saying "my puppy and I." But the response shown above seems to be from someone who misunderstood and interpreted "we" as "you and I."

The idea of applying rules to input sentences is the central concept in ELIZA. In the sections below we will explore the details of what defines a rule and how to create new rules, but for now the main idea is to realize that ELIZA's ability to participate in a conversation depends entirely on the set of rules it uses. With only a few rules, where ELIZA always responds with set phrases associated with key words, a person interacting with ELIZA would soon realize they are conversing with a machine. But with more complex rules that analyze several parts of the input, it's possible to have ELIZA generate responses that give more of an appearance that it understands the person typing the sentences.

The set of rules used by ELIZA to transform an input sentence is called a **script**. Weizenbaum chose the name "Eliza" for his program to emphasize the fact that the rules in a script determine how the program will respond. Just as Eliza Doolittle, the fictional character from George Bernard Shaw's *Pygmalion*, became more refined by learning more rules of language and etiquette, ELIZA the program should become better at conversing as more rules are added to its script.

Weizenbaum envisioned a situation where different scripts could be written for different applications; for example, one might make a script with rules based on key words used in

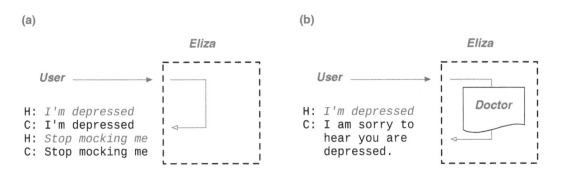

Figure 10.2: *(a)* When the `Eliza` module is first loaded, it does not have any transformation rules, so any sentences typed by a user are simply echoed. *(b)* The rules in the doctor script extract key words and phrases and transform an input sentence into a response.

cooking and baking to make a system that would give the illusion of conversing with a chef. The original script that Weizenbaum wrote, and the one that made the program famous, was named DOCTOR (Figure 10.2). The DOCTOR script that played the role of Rogerian psychotherapist contained rules based on words like "sad," "depressed," "dream," "wish," and "family."

To begin the project for this chapter we will just run Eliza using rules based on those from Weizenbaum's DOCTOR script. When you read the responses the "doctor" makes to your statements, see if you can get a sense for what sort of rules the program is applying. In the next section we will start learning how to write our own sentence transformation rules.

Tutorial Project

T1. Start an interactive session with your IDE and load the module that will be used for this project:
```
>>> from PythonLabs.ElizaLab import *
```

T2. Call the method named `Eliza.run` to start a conversation. When Eliza is running, the prompt changes to H: to let you (the human) know Eliza is waiting for you to type a sentence:
```
>>> Eliza.run()
H:
```

T3. Because Eliza does not have a script yet, it can only apply a default rule that does not change the input. Type a few sentences. Everything you type will simply be echoed right back:
```
>>> Eliza.run()
H: Hello.
C: Hello
H: How are you?
C: How are you
H: Stop repeating what I say!
C: Stop repeating what I say
```

T4. Unlike an annoying sibling, Eliza will quit when you tell it to. Type the single word bye, or quit, or hit ^D (hold down the control key while typing D) to end a conversation:
```
H:! quit
>>>
```

Note the prompt has changed back to the normal prompt for an interactive Python session.

T5. Call the Eliza module's load method to read the DOCTOR script that is included as part of the PythonLabs distribution:

```
>>> Eliza.load(path_to_data("doctor.txt"))
```

T6. Call Eliza.run again. Now the rules defined in DOCTOR will be used to transform every sentence you type into a response:

```
>>> Eliza.run()
How do you do. Please tell me your problem.
    H: I need a puppy.
    C: What would it mean to you if you got a puppy?
    H: I remember my cousin had a puppy.
    C: Do you often think of your cousin had a puppy?
```

T7. Try a few more sentences on your own. Don't forget to type quit when you are ready to continue with the next problem in the tutorial.

Eliza uses a set of key words to generate responses. In the example described earlier in this section, a sentence containing the word "mother" or "father" is transformed by a rule defined for family members to produce "Tell me more about your family." If a sentence does not contain a key word, Eliza just responds with something noncommittal, like "Go on."

T8. To see a complete list of key words in the current script call Eliza.info:

```
>>> Eliza.info()
Script:  PythonLabs/data/eliza/doctor.txt
    28 rules with 51 sentence patterns
    79 key words: ['alike', 'always', ... 'wish', ... 'your']
```

T9. Start a new conversation by typing Eliza.run again, and make up some input sentences on your own. You can type anything you would like, but you are more likely to keep up the illusion of chatting with a psychologist if your statements include key words from the list printed by the info method.

I'd Like to Have an Argument, Please

The title of this chapter comes from (what else?) a Monty Python skit. In *The Argument Clinic* two characters carry on a "conversation" that often degenerates into one side automatically contradicting the other (P = patient, C = clinician):

```
C: Look, if I argue with you, I must take up a contradictory position.
P: Yes, but it isn't just saying "no it isn't"!
C: Yes it is!
P: No it isn't!
C: It is!
P: Not at all!
```

An exercise at the end of this chapter is to write a script that mimics the "clinician" in the Argument Clinic. As you read the following sections, see if you can figure out how to transform sentences by replacing "is" with "isn't" and vice versa. Can you create an Eliza script that argues with the user?

10.2 Sentence Patterns

To implement a program like Eliza in Python, one of the first decisions to make is how to look for key words in sentences. The program needs to scan sentences like "My father wouldn't buy me a puppy" to look for words like "father" and to scan sentences like "I'm sad I don't have a dog" to break it into the parts before and after the word "sad."

The easiest way to look for words in a sentence is to do a **string search**. For example, suppose we have a sentence named s1 defined as

```
>>> s1 = "I was afraid of the cow"
```

A script with rules about farm animals would probably want to search through the string to see if words like "cow," "horse," and "pig" are in the sentence.

One way to search for words in a string is to use a string method named find. A call of the form s.find(w) does a linear search from the beginning of the string s to see if another string w appears as a substring. The method returns the starting location of the substring if the letters in w can be found anywhere in s, or −1 if w is not in s.

To see if a sentence contains a word, just pass the word as an argument in a call to find. This example asks if "cow" is in the sentence s1:

```
>>> s1.find('cow')
20
```

The return value of 20 means the letters in "cow" are found starting at location 20 in s1. Here is an example of an unsuccessful search:

```
>>> s1.find('horse')
-1
```

There are two problems with this simple plan, however. We don't want Eliza to respond to "My cat is scowling at me" with something like "Tell me more about your farm," which it might do with this sentence because it will find "cow" in the middle of the input string:

```
>>> s2 = "My cat is scowling at me"
>>> s2.find('cow')
11
```

What we want is for Eliza to look for "cow" as a complete word, but to not match the letters "cow" in the middle of "scowl."

The second problem is that we are going to end up with a lot of patterns if we have to create a new pattern for each word. The farm script should respond to sentences with "cow," "pig," "horse," and so on, and it would be nice if we didn't have to write a separate pattern for each word.

Both of these problems are solved by using an advanced form of string searching implemented in a Python library named re. The letters "re" are an acronym for **regular expression**. A regular expression is a pattern. When we do a regular expression search we are asking Python to find places in the string that match the pattern. We can define simple patterns that require an exact match with a particular word (like "cow"), but we can also define more complicated patterns, like "any substring that starts with a 'c' and is at least five letters long" (see the sidebar on Regular Expressions on page 322).

The ElizaLab objects we will use in our experiments are called Patterns (short for "sentence patterns"). The simplest type of sentence pattern is one that specifies a word that should appear somewhere in the sentence. Suppose we want a pattern that tells Eliza to reply to any sentence that contains the word "cow" with either "Tell me more about your farm" or "Go on." The first step is to create the Pattern object:

```
>>> p = Pattern("cow")
```

The argument is the string we want to look for in an input sentence; in this case, it's a single word, and it means the object can be applied to any sentence that contains the word "cow."

The next step is to add some response strings. We simply have to call a method named add_response, passing it a string to use as a reply. All the responses we add will be saved and used later during a conversation. Here is how to add the two responses we want for this example:

```
>>> p.add_response("Tell me more about your farm")
>>> p.add_response("Go on")
```

As a convenience, we can also specify a list of responses when we create the object. This pattern will respond to input sentences containing the word "duck" with one of the two sentences shown:

```
>>> p2 = Pattern("duck", ["I love ducks", "How cute"] )
```

◆ Regular Expressions in Python

In computer science, a **regular expression** is a type of string used in pattern matching operations. Python has an extensive library of functions for defining regular expressions and using them in string processing applications. This example shows how one might find all the numbers (strings of one or more consecutive digits) in the middle of a string:

```
>>> s = 'The farm had 31 cows and 2 ducks'
>>> re.findall(r'\d+', s)
['31', '2']
```

The first argument passed to **findall** is written in a special syntax used to describe patterns. Learning to write patterns is basically the same as learning to write programs in a special-purpose language. You will not need to learn how to use this library for the projects in this chapter because ElizaLab will automatically create regular expressions for you.

Regular expressions are often used by websites that ask you to enter information in a form. A page that expects a user to enter a phone number, date, social security number, or other information with a specific format will often check to see if the text typed by a user matches the general pattern for that type of data.

A web interface with a search box is another place regular expressions are used. For example, you might want to search an online crossword puzzle dictionary to find a 6-letter word that starts with *s* and ends with *t*. If you enter the pattern 's....t' (the period means "any character") the search will turn up *secret*, *sunset*, *spirit*, and dozens of similar words.

```
H: We had a cow.
C: Tell me more about
your farm.
```

```
'cow':
    "Tell me more about your farm."
    "Go on."
```

```
H: I never liked that cow.
C: Go on.
```

```
H: The cow jumped at me.
C: Tell me more about
your farm.
```

Figure 10.3: *A Pattern object (shown in the gray box) has a sentence pattern and a list of response strings. The* apply *method will see if an input sentence matches the pattern. If so,* apply *returns one of the responses. The object cycles through the list of responses so that each match gets a different response.*

To see if an input sentence matches the pattern we can call a method named apply. If p is a pattern object, a call to p.apply(s) will check to see if the sentence s contains the string used to define the pattern, and if the match is successful, return one of the response strings. Here is an example of our new pattern being applied to a sentence that contains the word "cow":

```
>>> p.apply("I milked the cow")
'Tell me more about your farm'
```

When there is more than one response, the object will cycle through them (Figure 10.3), so the next call to p.apply will use the second response in the list:

```
>>> p.apply("That cow was scary")
'Go on'
```

If a sentence does not match the pattern, the call to apply will return None. This sentence does not contain the word "cow" so a call to apply won't do anything:

```
>>> p.apply("There were pigs, too")
```

Recall that when the value of an expression is None Python does not print anything in the shell window.

One of the advanced features of regular expression pattern matching is the ability to restrict matches to complete words. The objects defined in ElizaLab take advantage of this feature to make sure patterns do not simple match letters that appear as substrings of other words. In this example we do not want the Pattern object to respond with one of its strings because the sentence is not about cows:

```
>>> p.apply("The cat was scowling at me")
```

This sentence has the letters "cow" but they are inside another word so the regular expression pattern search does not consider the sentence to be a match.

Another advanced feature of regular expressions is the ability to make a pattern that will match several different words. Instead of having to make separate patterns for different farm animals, we can make one pattern, using a regular expression that effectively says "match an input sentence that contains any of the following words." All we need to do is list all the words, separated by a | (vertical bar) character.

Here is an example of a pattern that can be applied to any sentence that contains "cow," "pig," or "horse":

```
>>> p = Pattern("cow|pig|horse", ["Really?", "Go on"] )
```

Now a call to apply will generate a response if any of the three words are in the input sentence:

```
>>> p.apply("The cow slept in the barn")
'Really?'
>>> p.apply("The horse jumped the fence")
'Go on'
```

A set of words separated by vertical bars is known as a **group.** As seen in this example, groups make it easy to define rules that can be applied to any one of a set of alternative words. Later in the chapter we will see how groups also let us extract part of the input sentence so words can be echoed to the user when Eliza constructs a response.

Tutorial Project

T10. Make a pattern that will apply to sentences that contain the word "cow":

```
>>> p = Pattern("cow", ["Tell me more about your farm."] )
```

Note there is only one response in this example, but it is still enclosed in brackets: it's a list with one string.

T11. Try applying the pattern to see how this Pattern object will respond to sentences containing the word "cow":

```
>>> p.apply("I milked the cow")
'Tell me more about your farm.'

>>> p.apply("That cow sleeps standing up")
'Tell me more about your farm.'
```

T12. Add more sentences to the list of responses by calling add_response:

```
>>> p.add_response("What made you think of cows just now?")
>>> p.add_response("Do cows worry you?")
```

T13. Now when you enter sentences that match the pattern the object will cycle through the responses. Type a few sentences with the word "cow":

```
>>> p.apply("The cow slept in the barn")
'Tell me more about your farm.'

>>> p.apply("She had a cow")
'What made you think of cows just now?'

>>> p.apply("No, not a real cow, she had a fit")
'Do cows worry you?'

>>> p.apply("Do you know what a cow is?")
'Tell me more about your farm.'
```

T14. Make a new pattern that will apply to more than one word:

```
>>> p2 = Pattern("Ruby|Perl|Python", ["The programming language?"])
```

T15. This new pattern should match any sentence that has one of the three words:

```
>>> p2.apply("Python is his favorite language")
'The programming language?'

>>> p2.apply("I prefer Ruby myself")
'The programming language?'
```

T16. This sentence does not match the new pattern:

```
>>> p2.apply('We used to use Java')
```

Java is the name of a programming language (and an island, among other things) but the Pattern object can only respond to words specified when the object is created.

Try some more tests on your own to explore the limits of these sentence patterns. Does the pattern care about capitalization? For example, will it respond to "how now, Brown Cow"? What about plurals? Does it match "those cows produce milk"?

10.3 Building Responses from Parts of Sentences

To add to the illusion that the computer is actually carrying on a conversation, Eliza should be able to create responses that use parts of the sentence typed by the user. For example, suppose the input is "I'm afraid of cows." Using the technique we saw in the previous section, we could make a pattern that looks for the word "afraid" and responds with a generic question like "Why are you scared?" But it would be more realistic if the reply was something like "Why are you afraid of cows?" because the response contains words from the input sentence. In order to create these kinds of responses, Eliza needs to be able to extract substrings from the input and then use those substrings as part of the response.

Once again we can turn to an advanced feature of regular expression pattern matching. If a pattern contains a group—defined in the previous section as a set of words separated by vertical bars—the part of the input sentence that matches the group is saved so it can be used later to make the response.

The diagram in Figure 10.4 shows how this process works. First note that the Pattern object has a group of three words, one of which is "cow." When the apply method finds the word "cow" in the input sentence, Python will save the word.

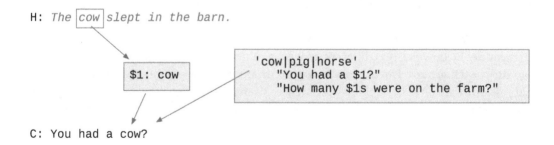

Figure 10.4: *The animal names in this pattern define a group. When an input sentence contains one of these words, the apply method saves the word in a variable so the word can become part of the response.*

Words are saved in special variables that have names of the form $n, where the *n* is the group number (because in general a pattern can have more than one group). In this example, the word "cow" is saved in $1.

Notice that the response string also has a variable name. When a response has a variable name, the apply method "plugs in" the variable's value at that location. In this example the letters "cow" were saved in $1 and the response string is "You had a $1?" so the response becomes "You had a cow?" A variable name in the response string is basically a "placeholder" that is filled in by the apply method when it generates a response.

Here are the lines from the interactive session that created the pattern shown in Figure 10.4. The first step is to make the Pattern object, supplying it with a string that has a group of words:

```
>>> p = Pattern("cow|pig|horse")
```

Next, add some response strings that have placeholders in them:

```
>>> p.add_response("You had a $1?")
>>> p.add_response("How many $1s were on the farm?")
```

Now when the apply method creates a response, it will insert the word it found in the input sentence:

```
>>> p.apply("The horse jumped the fence")
'You had a horse?'
>>> p.apply("The cow slept in the barn")
'How many cows were on the farm?'
>>> p.apply("The pig wallowed in the sty")
'You had a pig?'
```

In this example there was only one group, but in general there can be any number of groups. When a pattern has more than one group, however, there is some ambiguity in how the groups are defined. Here is an example:

```
>>> p = Pattern("hamster|guinea pig|gerbil")
```

Should Python treat this as two groups, one for the words "hamster" and "guinea" and the other for the "pig" and "gerbil"? Or is it one group, with "guinea pig" considered to be an alternative to "hamster" and "gerbil"? The answer is that Python considers this pattern to have a single group with three alternatives.

If you really do want two different groups, you need to use parentheses to surround the words in each group. Parentheses in a pattern are just like parentheses in an arithmetic expression: they alter the default precedence, so that Python evaluates the expression the way you intend and not by using its default. Here is a pattern that has two groups of words:

```
>>> p = Pattern("I (like|love|adore) my (dog|cat|ducks)")
```

When we make response strings for this pattern we can use two placeholders, because there are two groups in the pattern:

```
>>> p.add_response("Why do you $1 your $2?")
>>> p.add_response("What about your $2 do you $1?")
```

Notice how the words from each part of the input are inserted into the right place in the response:

```
>>> p.apply("I adore my cat")
'Why do you adore your cat?'
```

```
>>> p.apply("I love my dog")
'What about your dog do you love?'
```

Let's go back to the goal mentioned at the beginning of the section, where we want to define a pattern that responds to a sentence like "I'm afraid of ..." with "What is it about ... that worries you?" Now that we know about word groups, we can make a pattern that plugs in the word from the input sentence:

```
>>> p = Pattern("I'm afraid of (cows|dogs|ghosts)")
```

This will work, but unfortunately this pattern only matches the words specified in the group. If we are writing a script to play a psychologist, we have to anticipate everything a person could be afraid of.

Another very useful feature of regular expressions allows us to write patterns that save more than just a single word in a placeholder. A pattern can contain **wild cards** that match any piece of text. A wild card is written with two characters: a period followed by an asterisk. Here is how we would write the same pattern, using a wild card:

```
>>> p = Pattern("I'm afraid of .*")
```

All the words following "of" in an input sentence are saved in a variable, and they can be inserted into the response:

```
>>> p.add_response("Why are you afraid of $1?")
```

Here's what happens when we apply this pattern to some sentences:

```
>>> p.apply("I'm afraid of the dark")
'Why are you afraid of the dark?'
```

```
>>> p.apply("I'm afraid of little green men")
'Why are you afraid of little green men?'
```

Note that the wild card doesn't just match a single word; the pattern saves an arbitrarily large part of the input sentence in the placeholder variable.

To recap what was introduced in this section:

- If a sentence pattern has a group (a set of words separated by vertical bars) or a wild card (a period followed by an asterisk) the apply method saves the part of the input sentence that matches the group.

- The pieces of the input that match a group are saved in variables named $1, $2, etc.

- A response can include variable names, and as a result, the output string will contain parts of the input string.

Tutorial Project

T17. Create a pattern that has a group of words:
```
>>> p1 = Pattern("green|yellow|red|blue")
```

T18. Add a response string with a variable, so that the matching word is included in the response:
```
>>> p1.add_response("That's my favorite color, $1")
```

T19. Now call apply to see what sentence is constructed in response to the input:
```
>>> p1.apply("The sky is red")
'That's my favorite color, red'
```

T20. If the pattern is given a sentence with any one of the four colors listed, you should see a response that contains the color name:
```
>>> p1.apply("The green bird flew away")
'That's my favorite color, green'
```

T21. Make a pattern that has two groups and add a response that uses both saved substrings:
```
>>> p2 = Pattern("The (dog|cat|frog) (ran|jumped)")
>>> p2.add_response("Are you sure the $1 really $2?")
```

T22. Try applying the pattern to some sentences that contain words from both groups:
```
>>> p2.apply("The cat ran away")
'Are you sure the cat really ran?'

>>> p2.apply("The dog jumped over the fence")
'Are you sure the dog really jumped?'
```

T23. Next make a pattern that has a wild card and add a response that uses the saved text:
```
>>> p3 = Pattern("We should .*")
>>> p3.add_response("I don't want to $1")
```

T24. Try some sentences that match the pattern:
```
>>> p3.apply("We should go on a diet")
"I don't want to go on a diet"

>>> p3.apply("We should go to the beach tomorrow")
"I don't want to go to the beach tomorrow"
```

T25. Make a pattern that has three wild cards:
```
>>> p4 = Pattern("I'm .* my .* was .*")
```

T26. The response can use the matching pieces of the sentence in any order:
```
>>> p4.add_response("Really, your $2 was $3 made you $1?")
```

T27. Try out this new pattern:
```
>>> p4.apply("I'm totally disappointed my midterm exam was rescheduled")
'Really, your midterm exam was rescheduled made you totally disappointed?'
```

T28. Make a new pattern with a single wild card:
```
>>> p5 = Pattern("I like .*", ["Why do you like $1?"] )
```

T29. Try out the new pattern on some sentences:
```
>>> p5.apply("I like to solve crossword puzzles")
'Why do you like to solve crossword puzzles?'

>>> p5.apply("I like your hat")
'Why do you like your hat?'

>>> p5.apply("I like my new computer")
'Why do you like my new computer?'
```

If you look closely at the replies for Exercise T29 you'll see some of them aren't quite right. Can you tell what the odd ones have in common, and how they differ from the realistic ones?

10.4 Substitutions

The exercise at the end of the previous section showed how the simple "copy-and-paste" strategy that transfers entire fragments from an input sentence to the response string doesn't always work. The pattern object from Exercise T29 responded to "I like my new computer" with "Why do you like my new computer." If we want the "doctor" to respond to statements of the form "I like ..." with "Why do you like ...?", the program cannot simply echo the parts that match the wild card.

As an example of where the simple copy-and-paste algorithm goes wrong, suppose we have the following rule in our script:

```
>>> p = Pattern("I am (.*)", ["Are you really $1?"])
```

Here are some sentences that lead to silly responses:

```
>>> p.apply("I am happy to see you")
'Are you really happy to see you?'
>>> p.apply("I am out of my mind")
'Are you really out of my mind?'
>>> p.apply("I am sorry I dropped your computer")
'Are you really sorry I dropped your computer?'
```

Python is doing exactly what we tell it to do by copying every character after "I am" into the response string. But these responses are clearly not what we expect from a computer that converses in English.

The problem with the strange-looking replies is that they have personal pronouns. In a normal conversation, words like "I" or "you" are typically replaced by their opposite. If a patient says, "I am happy to see *you*," a real doctor would reply, "Why are you happy to see *me*?" The "you" in this sentence was replaced by "me" in the response, but the other words ("to see") are unchanged.

In order to handle situations like this, Pattern objects use an operation called **postprocessing**. After breaking the input into pieces with a regular expression, but before reassembling the pieces into a response, Eliza does an additional pattern matching operation on each of the placeholder variables. This second pattern matching step does single-word replacements. Every "I" is replaced with "you," every "my" with "your," and so on. The result isn't perfect, but with a large enough set of replacement strings Eliza can maintain the illusion of carrying on a conversation.

In order to activate postprocessing we need to pass an optional argument to apply. This argument should be a dictionary where the keys are words to replace and their values are the words that will replace them. During the postprocessing phase, the apply method looks in the dictionary for each word in $1, $2, etc., and if a word is found it is replaced by the corresponding string.

Here is a simple example of a Python dictionary with substitutions we want to make in our example pattern. It tells the apply method to replace "I" with "you" and "your" with "my" when it is assembling the response:

```
>>> pronouns = { "I" : "you", "your" : "my" }
```

If we pass this dictionary to apply, the results are much more realistic:

```
>>> p.apply("I am sorry I dropped your computer", post = pronouns)
'Are you really sorry you dropped my computer?'
```

Here is another example, one without postprocessing and one that uses the pronouns dictionary:

```
>>> p.apply("I am happy I lost")
'Are you really happy I lost?'
>>> p.apply("I am happy I lost", post = pronouns)
'Are you really happy you lost?'
```

If we try typing the input sentence in a more colloquial form, however, we'll see we have a new problem:

```
>>> p.apply("I'm happy I lost", post = pronouns)
```

Nothing was printed because the method returned None. The reason: this input starts with the contraction "I'm" but the rule only matches sentences that begin with "I am."

To fix this problem, the apply method lets us pass a second dictionary that will be used to expand contractions. Here is a simple dictionary that will expand "I'm" into "I am" so the sentence matches the pattern in the rule:

```
>>> contractions = { "I'm" : "I am" }
```

If we pass this dictionary in the call to apply we will get the response we expect:

```
>>> p.apply("I'm happy I lost", pre = contractions, post = pronouns)
'Are you really happy you lost?'
```

Because these substitutions are made before seeing if the rule can be applied this operation is called **preprocessing**. This example was motivated by the need to expand contractions into more formal English, but in fact any type of substitution can be specified, for example, expanding acronyms or replacing a person's nickname with their full name.

A script can include the definition of preprocessing and postprocessing dictionaries. When the script is loaded, the dictionaries are saved as Eliza.pre and Eliza.post. If you want to use the dictionaries defined for the DOCTOR script in your experiments, simply load the script and then pass one or both of the dictionaries in a call to apply:

```
>>> Eliza.load(path_to_data("doctor.txt"))
>>> p = Pattern("You are (.*)", ["I am not $1"])
>>> p.apply("You're crazy", pre = Eliza.pre)
'I am not crazy'
```

The complete process for transforming a sentence is illustrated in Figure 10.5. The first step is preprocessing, where the apply method sees the string "you're" in the input and replaces it with "you are." Now the sentence matches the pattern, and the words following "are" are saved in $1. The postprocessing step replaces the word "you" with "me." When the pieces are put back together, the contents of $1 are substituted into the response string to create the output.

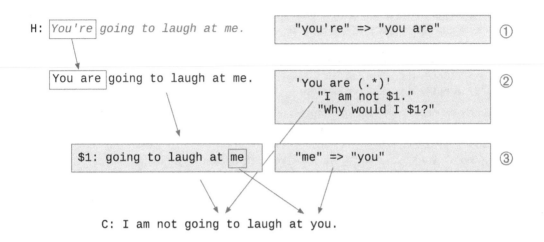

Figure 10.5: *Preprocessing and postprocessing help transform an input sentence into a response. The first step in responding to an input sentence is to expand contractions in the input, using rules in the preprocessing dictionary. Next, see if the input matches the regular expression in a sentence pattern, and if so, save the parts of the input that match wild cards or word groups. Finally, postprocessing replaces pronouns according to associations in the postprocessing dictionary.*

Tutorial Project

T30. Make a pattern to use for testing preprocessing and postprocessing:

```
>>> p = Pattern("You are (.*)", ["I am not $1"])
```

T31. Because the pattern starts with "you are" it will not match a sentence that uses "you're":

```
>>> p.apply("You're kidding")
```

Nothing is printed, meaning the method returned None.

T32. Load the DOCTOR script so you can use its preprocessing and postprocessing dictionaries:

```
>>> Eliza.load(path_to_data("doctor.txt"))
```

T33. Print the contents of the preprocessing dictionary:

```
>>> Eliza.pre
{"you're": 'you are', "i'm": 'I am', ...}
```

T34. Look up one of the contractions to see how it will be expanded:

```
>>> Eliza.pre["you're"]
'you are'
```

T35. Because this is a Lexicon (see the sidebar on "Lexicon Objects") we can also look up the same item using uppercase letters:

```
>>> Eliza.pre["You're"]
'you are'
```

T36. Apply the pattern to some sentences:

```
>>> p.apply("You're angry", pre = Eliza.pre)
'I am not angry'

>>> p.apply("I hope you're wrong", pre = Eliza.pre)
'I am not wrong'

>>> p.apply("She said you are boring", pre = Eliza.pre)
'I am not boring'
```

Do you see why the preprocessing dictionary is necessary to make the first two examples work? And how the pattern now works equally well for sentences with "you are" or "you're"? Note also that it didn't matter whether or not "you" was capitalized because Eliza.pre is a Lexicon object.

T37. Using the same Pattern object, notice how this sentence is transformed:
```
>>> p.apply("You're going to laugh at me", Eliza.pre)
'I am not going to laugh at me'
```

T38. Clearly we need to do some postprocessing to change the "me" to a "you." Ask Python to print the contents of the postprocessing dictionary:
```
>>> Eliza.post
{'me': 'you', ... 'my': 'your', 'your': 'my'}
```

T39. Call apply again, this time passing the postprocessing dictionary:
```
>>> p.apply("You're laughing at me", pre = Eliza.pre, post = Eliza.post)
'I am not laughing at you'
>>> p.apply("You are making my point for me", pre = Eliza.pre, post = Eliza.post)
'I am not making your point for you'
>>> p.apply("You are contradicting yourself", pre = Eliza.pre, post = Eliza.post)
'I am not contradicting myself'
```

Try some more sentences on your own. See if you can come up with examples of sentences that have the phrase "you're" or "you are" in them and see how the pattern object handles them, both with and without using the postprocessing dictionary.

As you test your examples, you will probably find cases where a new pronoun needs to be added to the dictionary. Can you figure out how to add new key-value pairs to Eliza.post to fix your sentences?

Lexicon Objects

As we saw in Chapter 6 one of the difficulties of writing programs that do text processing is figuring out how to handle input that might contain either upper or lowercase letters. The strategy we took there was to convert all strings to lowercase before we saved them in a dictionary.

The functions in ElizaLab take a similar approach. A special type of dictionary called a Lexicon behaves exactly like a regular Python dictionary except it converts words to lowercase, both when storing words in the dictionary and when looking up words.

| **Standard Python (class = dict)** | **ElizaLab (class = Lexicon)** |

```
>>> d1 = { }
>>> d1["you're"] = "you are"
>>> d1["You're"]
KeyError: "You're"
```

```
>>> from PythonLabs.ElizaLab import Lexicon
>>> d2 = Lexicon()
>>> d2["you're"] = "you are"
>>> d2["You're"]
'you are'
```

A regular dictionary (left) is case sensitive, but Lexicon objects (right) allow lookups with upper or lowercase

10.5 An Algorithm for Having a Conversation

To summarize what we've seen so far:

- A Pattern object is defined by a string that has a key word, groups of words, or wild-cards

- We can specify any number of response strings, and the object's `apply` method will return one of these strings if a sentence matches its pattern

- Pattern matching variables (`$1`, `$2`, *etc.*) along with preprocessing and postprocessing help the `apply` method extract parts of the input sentence and reuse them as part of the response.

As you might imagine, there are many more enhancements we might make to give the `apply` method more flexibility in the types of sentences it can transform. At this point, however, we will turn our attention to the problem of how to use patterns as part of an algorithm that will carry on a conversation.

A straightforward algorithm would be to simply try patterns until we find one that matches an input sentence. We could make a list of Pattern objects, and then write a program that iterates over the list to try the patterns in order. The first time we find a pattern that applies to the input sentence, we would use the response generated by that object. The problem with this approach is that patterns will always be applied in the same order. Patterns at the front of the list will be used more often, and patterns near the end of the list might never be used at all.

Weizenbaum's solution to this problem was to assign a **priority** to each word. He expected people would ask the program questions like "Are you a computer?" or "Am I talking to a machine?" so the original DOCTOR script had patterns to respond to inputs with the words "computer" or "machine." To make sure ELIZA responded directly to these questions, no matter where they occurred in the list of rules, Weizenbaum developed an algorithm that tried to match inputs to high-priority words like "computer" before trying any patterns for sentences containing more common words like "are" or "you."

Another issue that needs to be addressed is that some common words are going to appear in several different patterns. For example, there might be different sentence patterns containing the word "I" to respond to inputs like "I am worried ...", "I remember ...", or "Why can't I ...". We want to make sure Eliza tries all the patterns for a word before moving on to a lower-priority word.

A new type of object, called a Rule, is simply a list of pattern objects. When we say Eliza has a "rule for *x*" we mean there is a Rule object that has a set of patterns that all pertain to sentences containing the word *x*. When *x* is found in an input sentence, Eliza will try to match the sentence with each of these patterns.

After a script has been loaded, we can call a method named `rule_for` to see whether there is a rule for a particular word. For example, to see if the script has a rule for the word "remember" you would type

```
>>> Eliza.rule_for("remember")
PythonLabs.ElizaLab.Rule [5]
    'I remember (.*)'
    'do you remember (.*)'
```

The first line printed by this method shows the class name and the rule's priority (the higher the number, the higher the priority). This rule has a priority of 5, which will put it near the front of the queue most of the time. The remaining lines show the regular expression patterns for each of the Pattern objects that are part of this rule.

The algorithm for transforming an input sentence into a response uses a function and a class that were developed for projects. A function named tokenize that was written for the spam filtering project breaks a string into words, stripping away all the spaces and removing punctuation marks. A priority queue (used both in spam filtering and in the Huffman tree algorithm) is a type of list where items are always sorted.

Given these pieces, writing a function that takes a string containing a sentence as input and returns a response string is straightforward. The Python code for our function is shown in Figure 10.6. The algorithm consists of three phases: initialization (lines 6–7), the initial scan over the tokens (lines 9–12), and rule processing (lines 14–18).

The initialization phase creates the priority queue and calls a function defined in ElizaLab that performs the preprocessing step. The assignment statement on line 7 is a common Python idiom. Because we don't need the original input sentence after the contractions are expanded we can simply save the new sentence in the same variable.

```
1   from PythonLabs.Tools import tokenize
2   from PythonLabs.ElizaLab import *
3
4   def transform(sentence):
5       "Generate a response to an input sentence using rules from a script"
6       queue = PriorityQueue()
7       sentence = Eliza.preprocess(sentence, Eliza.pre)
8
9       for word in tokenize(sentence):
10          rule = Eliza.rule_for(word)
11          if rule is not None:
12              queue.insert(rule)
13
14      while len(queue) > 0:
15          rule = queue.pop()
16          response = rule.apply(sentence, post = Eliza.post)
17          if response is not None:
18              return response
19
20      return None
```

Download: eliza/transform.py

Figure 10.6: *Eliza's algorithm for transforming an input sentence into a response string has three steps. The preprocessor expands contractions and does other simple substitutions defined by a dictionary. Then each word is checked, and if there is a rule for a word, the rule is added to a priority queue. Finally, the rules are applied in order of decreasing priority.*

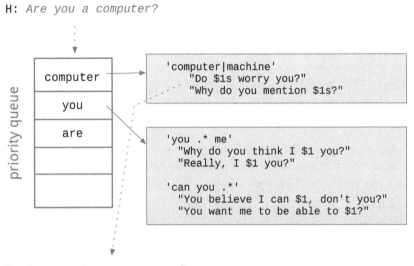

H: *Are you a computer?*

C: Do computers worry you?

Figure 10.7: *The rules for each word in an input sentence are saved in a priority queue. Eliza tries the rules in order, starting with the highest-priority word at the front of the queue. In this example, the word "computer" has the highest priority of all the words in the input sentence, and the response comes from a pattern associated with the word "computer."*

The loop that iterates over the tokens in the input sentence calls rule_for(w) to see if the script contains a rule for handling a word w. If so, the function returns a reference to the Rule object, and the rule is added to the queue. Note that rules are automatically sorted in the queue, with the highest-priority rules at the front.

The final step is to remove the rules one by one from the queue. For each rule the program calls the rule's apply method to see if it matches the input sentence. The first time apply succeeds, the string it returns is passed back as the result of the call to transform. If the loop gets through all rules without finding any that apply to the input sentence then the result is None to indicate the script has no rules for transforming this sentence.

An example of how the algorithm processes a sentence is shown in Figure 10.7. The input is "Are you a computer?" The DOCTOR script has patterns for sentences with the words "computer," "you," and "are," so rules for these words are saved in the priority queue. Because "computer" has the highest priority, it will be at the front of the queue, and on the first iteration the transform function will try to match the input to the regular expression for this rule (which matches any string that contains "computer" or "machine"). The regular expression matches the input, so the response generated by this rule ("Do computers worry you?") is the sentence that will be printed as the output.

The project in this section will give you a chance to experiment with the version of transform that is implemented as part of ElizaLab. Exercises at the end of the section have suggestions for how you can implement and test your own version based on the code in Figure 10.6.

```
>>> Eliza.transform("I remember you laughed at me")
Preprocessing...
  before: 'I remember you laughed at me'
  after:  'I remember you laughed at me'
Scanning... I*, remember*, you*, laughed, at, me
Queue: ['remember', 'you', 'i', '$noncommittal']
Rule for 'remember'
  /\bI remember (.*)/
Reassembling 'Why do you remember $1 just now?'
  postprocess 'you laughed at me' => 'I laughed at you'
'Why do you remember I laughed at you just now?'
```

Figure 10.8: *This transcript, copied from an interactive session, shows how Eliza transforms an input sentence. This sort of detailed trace is printed when* Eliza.verbose *is set to* True.

An advantage of using the ElizaLab version of transform is that we can enable a "debugging mode" that will have the function and its helper function print a detailed trace as it processes the input sentence. If you set a variable named Eliza.verbose to True you will see the contents of the priority queue as new rules are added, and then the pattern strings that are tried as the function works its way through the queue.

An example of how Eliza processes a sentence with debugging mode turned on is shown in Figure 10.8. The first three lines show the preprocessing examples; in this case there are no contractions in the sentence so no changes were made to the input sentence.

The line that starts with Scanning... shows the tokens from the sentence. An asterisk next to a word means the DOCTOR script has a rule for that word. The next line shows the queue after adding all the rules. Because "remember" has a high priority it is at the front of the queue.

The next set of lines shows how rules are processed. In this case the first pattern for the first rule matches the input sentence so there is not much to see, but in general we will see several different rules, and within each rule, each different pattern as it is applied.

Finally, the last part of the output shows how postprocessing is applied to the response string. This response has a single variable in it, and the variable contains the four words following "remember" in the input sentence. One of those words ("me") is in the postprocessing dictionary, so it is replaced with "you" before the final output string is produced.

There is one last detail to explain about the transcript in Figure 10.8. The last rule in the queue has the name $noncommittal. This is a special rule that the ElizaLabs version of transform always adds to the queue. It has a lower priority than any rule from a script, so it always goes at the end of the queue. It is the rule that generates non-committal responses, things like "go on" or "tell me more." You will see one of these generic response strings if the script does not have any patterns that match the input sentence.

Tutorial Project

If you started a new terminal session since doing the exercises in the previous section, import ElizaLab and enter the command that loads the DOCTOR script.

T40. Look up the rule for the word "if" and save it in a variable named r:
```
>>> r = Eliza.rule_for("if")
```

T41. Ask Python to print the value of r:
```
>>> r
PythonLabs.ElizaLab.Rule [3]
  'if (.*)'
```
This rule has priority 3, and it has only one pattern.

T42. A method named `patterns` will return the list of Pattern objects defined for this rule:
```
>>> r.patterns()
[<PythonLabs.ElizaLab.Pattern 'if (.*)'>]
```

T43. Save a reference to the first pattern in the list:
```
>>> p = r.patterns()[0]
```

T44. Print a list of the responses associated with this pattern:
```
>>> p.responses()
["Do you think it's likely that $1?", 'Do you wish that $1?', ...]
```

T45. Turn on debugging mode:
```
>>> Eliza.verbose = True
```

T46. Call the `transform` method to see how Eliza responds to a sentence that contains "if":
```
>>> Eliza.transform("If I could borrow your car")
Preprocessing...
  before: 'If I could borrow your car'
  after:  'If I could borrow your car'
Scanning... If*, I*, could, borrow, your*, car
Queue: ['if', 'your', 'i', '$noncommittal']
Rule for 'if'
  /\bif (.*)/
Reassembling 'Do you think it's likely that $1?'
  postprocess 'I could borrow your car' => 'you could borrow my car'
  "Do you think it's likely that you could borrow my car?"
```
Make sure you understand which rules were added to the queue, how the pattern match succeeded, and how the postprocessing step modified the string in $1 to create the final answer.

T47. Modify the sentence so it includes both "if" and "remember":
```
>>> Eliza.transform("If I remember how to drive")
...
'Do you often think of how to drive?'
```
This output has been edited, but you should see a full trace in your interactive session. Do you see how "remember" has a higher priority, so the rule for "remember" is applied before the rule for "if"?

Try calling `transform` with a few more sentences of your own. In each case, you should see one or more words being added to the priority queue, and then see how Eliza tries the rules for those words in order, from highest priority to lowest. If you want to see what the rule is for any of the words in the queue, just call `Eliza.rule_for` (or you can use your IDE to open the text file for the DOCTOR script).

T48. To turn off debugging mode set `Eliza.verbose` to `False`:
```
>>> Eliza.verbose = False
```

10.6 ♦ Writing Scripts for ELIZA

Making patterns and experimenting with them by creating Pattern objects in an interactive session is a good way to learn how Eliza uses regular expressions to break an input sentence into smaller parts. It also illustrates how Eliza creates an output sentence by transforming and reassembling the pieces of a sentence. But to get Eliza to carry on a conversation, there will be too many patterns to type in interactively. What we need to do is write down the patterns in the form of a script. Then, each time we want to try out the transformation rules, we just have to load the script. All the patterns defined in the script will be available for Eliza to use.

One way to get started on your own script is to make a copy of the DOCTOR script that comes with Eliza. A portion of this script is shown in Figure 10.9. Script files are plain text files that can be edited using your IDE. You can use the path_to_data method to find out where this script is installed on your system:

```
>>> path_to_data("doctor.txt")
'PythonLabs/data/eliza/doctor.txt'
```

Open the script with your IDE and then save it with a new name in your project directory. For example, if you save the script in farmer.txt, you can then load it just by passing that name to Eliza.load:

```
>>> Eliza.load("farmer.txt")
```

As you develop your script you can add new rules or modify existing rules.

The main part of the file will be a set of rule definitions, but there can also be other information, for example, a specification of which words are expanded during the preprocessing phase and the substitutions to perform during postprocessing.

Scripts have a very simple syntax. There are only two kinds of things in a script: **directives**, which are commands for Eliza, and **rules**, which are specifications for how to transform an input sentence.

Directives are words that start with a colon. The two directives you should know about are :pre, which tells Eliza to add a new word to the set of words it expands during the preprocessing step, and :post, which tells Eliza to add a new item to the association list it uses during postprocessing (described in Section 10.4). Some examples from the DOCTOR script are

```
:pre don't "do not"
:post your "my"
```

The :pre command tells Eliza that whenever the letters don't are seen in an input sentence they should be replaced by do not. The :post command says that "your" should be replaced by "my" in the parts of the input that are echoed back to the user, for example, when converting "I like your hat" into "Why do you say you like my hat?"

Directives like those shown above are specified on a single line, but rules need several lines. The first line has the word that triggers the rule. For example, if you are writing a rule that will process sentences related to cows, the first line in the rule will just have the word "cow." The default priority for a rule is 1. If you want the rule to have a higher priority, put the priority number right after the word.

```
# A portion of the "Doctor" script for the PythonLabs version of Eliza

 :pre you're "you are"

 :post me "you"
 :post myself "yourself"
 :post my "your"

 was 2
   /was i (.*)/
     "What if you were $1?"
     "Do you think you were $1?"
   /i was (.*)/
     "Were you really?"
     "Why do you tell me you were $1 now?"

 remember 5
   /I remember (.*)/
     "Do you often think of $1?"
     "Does thinking of $1 bring anything else to mind?"
     "What else do you remember?"
   /do you remember (.*)/
     "Did you think I would forget $1?"
     "Why do you think I should recall $1 now?"
```

Figure 10.9: *An excerpt from the "doctor" script. Lines beginning with # are comments, lines beginning with a word that starts with a colon are directives, and the remaining lines are all parts of rules. A rule starts with a line that has a single word. Sentence patterns begin and end with a slash character. Response strings are surrounded by double quotes. Although it's not strictly necessary, if you indent regular expressions by two spaces, and response strings by four spaces, as shown here, it is easier to tell which strings go with which patterns, and which patterns go with which words.*

Following the line with the key word there can be one or more sentence patterns. The first line in a sentence pattern is the regular expression that defines the pattern, and following that are the response strings Eliza will use when an input sentence matches the regular expression. There can be any number of response strings for each regular expression and any number of patterns for any word.

A complete description of how script files are organized and a few more examples can be found in the Eliza section of your Lab Manual.

10.7 ELIZA and the Turing Test

From the projects in the previous sections it may seem that all Eliza needs to pass the Turing Test is a script with rules to respond to a wider variety of questions, and perhaps a few more features similar to the preprocessing and postprocessing stages to deal with new special cases. Human languages are very complex, however, and there are limits to the sorts of things we can "teach" Eliza to respond to.

The first barrier for Eliza is that it does not take into account the structure of a sentence. It simply breaks each input into individual words, without trying to figure out whether a word is an important part of a sentence. When faced with any sentence consisting of more than a few words, Eliza is not likely to make a good choice:

> *You said I wouldn't be afraid of cows if I visited a dairy, but when I was there I was chased by a duck, and now I'm having nightmares about birds.*

It's unlikely that any ordering of the words in this sentence will lead to a realistic response. A human would recognize there are four separate topics (a suggestion for overcoming a fear, a trip to a dairy, an episode with a duck, nightmares) and try to make an appropriate response based on the topics. Eliza just dumps all the words in a queue, and the highest priority rule is applied, regardless of the context of the rest of the sentence. If you typed any long sentences when you were first experimenting with Eliza, you no doubt realized it did not do a very good job with complex sentences.

There are algorithms, called **parsers**, that will use rules of English grammar to break sentences into their constituent parts. But even if we replace Eliza's word-oriented strategy with a more complicated algorithm, one that uses a parser to break a sentence into noun and verb phrases, we would soon run into another problem. Consider this short conversation:

> *The cows were lying down in the field.*
> Go on.
>
> *They were quiet.*
> Why do you say the cows were quiet?

In order for the computer to make this second response, it would need to have some sort of memory to keep track of the fact that the conversation was about cows. Eliza has no way to connect the word "they" in the second sentence with "cows" in the first sentence. Every input is the start of a new conversation for Eliza, and all it can do is make a response from the words in the current sentence.

Here is an example of another conversation that is not possible for a computer that uses Eliza's rule-based algorithm:

> *That cow was huge.*
> How big was it?

This response requires the computer to understand the *meanings* of words. English speakers know the word "huge" means "very large," and a response that substitutes "big" for "large" is natural. Eliza, of course, knows nothing about the meanings of words. It has no data structure to correlate words with possible meanings, and no algorithm for considering whether or not to replace a word with a similar one.

Researchers in natural language processing refer to the general problem of assigning meanings to words as **semantics**. Semantics is about much more than meanings of individual words, however. It is not simply a matter of giving Eliza a dictionary so it can look up synonyms. Meanings are based on what computer scientists refer to as "real-world knowledge" or "common-sense knowledge." Here is another simple example that illustrates the challenge:

> *The cow fell into the canal.*
> Was it able to get out?

A dictionary that simply associates "canal" with "channel used to convey water for irrigation" would not help a program make this response. The reply is based on knowledge that canals have water, the cow was in danger because it was in the water, and the cow might have had some difficulty because canals usually have steep walls. A duck is also an animal commonly found on a farm, but a human's response to "the duck fell into the canal" would be very different. All of this knowledge is something we, as humans, have built up over many years of learning and using language, and without this sort of knowledge, a computer would not be able to carry on a natural-sounding conversation.

Even if we were able to put all the pieces described above into a program, so we have an application that can parse complex sentences, can keep track of statements in order to remember the context of a conversation, knows the meanings of words, and has a substantial amount of real-world knowledge, there is yet another difficult challenge. When we see or read a sentence, we are continually making **inferences**. Inference is such a common phenomenon that we take it for granted and we don't even realize we are doing it. Consider this simple statement:

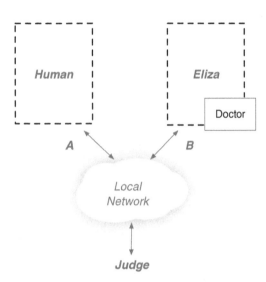

*The cows walked across the pasture
while my dad waited at the barn.*

When you read that sentence, did you immediately think that the cows were walking toward the barn? A perfectly reasonable response to that statement might be, "What happened when the cows got to the barn?" But there is nothing in the statement that said anything about where the cows were headed. We simply inferred that the cows were walking in the direction of the barn because the phrases "cows walked" and "waited at the barn" were both part of the same sentence.

If you're not convinced that you were making inferences about where the cows were walking as you read that previous sentence, consider this variation:

The cows walked across the pasture while I hid behind the barn.

The only difference is the new sentence has "I hid behind" instead of "my dad waited at." For this second sentence, did you infer the cows were walking toward the barn? Or were they just walking across the pasture?

Depending on the larger context, there could be several different interpretations. Dad could have been waiting for a taxi, or waiting for someone to bring his lunch. Even if the program knows the meanings of the words, and has built up a context from earlier statements, it will have to be able to make inferences in order to respond like a human. People are constantly "connecting the dots" when they hear or read a sentence, and a response from a human will use inferences based on the sentence and the context.

Tutorial Project

The conversation shown below is from a hypothetical program that pretends to converse in English. The responses from this program are each examples of replies that Eliza is unlikely to or unable to generate. Can you explain why Eliza would not make these responses?

If it helps, you can type a sentence into the PythonLabs version of Eliza (after loading the DOCTOR script), or use the `rule_for` method to see if DOCTOR has a rule for a particular word.

```
H:  I remember when I lived in the country and there were
      cows and chickens everywhere.
C:  Why do you mention cows just now?

H:  They were always looking at me.
C:  The cows were looking at you?

H:  Our parents got mad when we tipped the cows.
C:  Did you really push over a cow?

H:  The one with the big horns was the scariest.
C:  Are you referring to a bull?

H:  Yeah, that guy was pretty aggressive.
C:  Did the horns seem dangerous?
```

10.8 Summary

In his 1950 paper on the nature of intelligence, Alan Turing was optimistic about the prospects for natural language processing. He wrote:

> I believe that in about fifty years time it will be possible to programme computers ... to make them play the imitation game so well that an average interrogator will not have more than 70 per cent chance of making the right identification after five minutes of questioning.

Another famous "prediction" that natural language processing would be well established by the year 2000 was Arthur C. Clarke's *2001: A Space Odyssey*. The movie, which was released in 1968, featured a computer named Hal that spoke flawless English with the crew of a spaceship on a journey to the planet Jupiter.

Actual achievements in natural language processing have fallen far short of these predictions. There have been some notable successes, but only in very limited areas. Reservation systems for airlines and call centers for a variety of businesses are examples where natural language interfaces are being used effectively. None of these systems are able to converse on a wide variety of subjects, however.

One area where there has been some success is in parsing sentences. If you do an Internet search for "English parsers" you will find several websites at research labs that are working on methods for analyzing complex sentences. Many of these sites allow you to type a sentence into a form in your browser and submit it for analysis. The parser will return a page that shows the overall sentence structure, with noun phrases, verb phrases, and other parts of speech.

The Turing Test itself has become a source of controversy in recent years. One criticism is that programs can do well in a contest simply by appealing to the judges, and do not actually have to carry on a conversation. If a programmer knows judges are influenced by quirky responses, even to the point of changing topics at random, then a program that spits out

---

Concepts and Terminology Introduced in This Chapter

natural language processing	The area of computer science research concerned with algorithms for understanding human language
Turing Test	A proposal for assessing machine intelligence; a program would pass the test if it could carry on a natural language conversation
ELIZA	The first program to attempt to converse in English; it used simple rules to transform input sentences into responses
regular expression	A type of expression used for pattern matching; used by the PythonLabs implementation of Eliza to process input sentences
semantics	The meaning of a word or sentence; a difficult challenge for natural language processing

---

non sequiturs every now and then has a better chance of winning. Is this the sort of thing people do, during a conversation, or simply a strategy to win a contest? Should the test be amended, to require programs to try to emulate how people think and talk? Or should Turing's original rules still apply, so that any program that carries on a realistic conversation for five minutes will earn the distinction of being called "intelligent"?

A computer program that truly does converse with a human would be one that remembers what was said earlier in the session, knows the meanings of words, and is capable of making inferences based on what it hears. It will be a long time before such programs are available.

Exercises

The first set of questions ask, "What would Eliza do?" Try to answer these questions based on your experience with Eliza in the tutorial projects. To check your answers, you can either look at the text for the DOCTOR script, or start an interactive session with Eliza.

10.1. Here are some sentences typed by a user, and the responses generated by Eliza (after loading the DOCTOR script). Explain why Eliza responded the way it did.

```
H: I can't remember the date.
C: How do you know you can't remember the date?

H: I can't remember her name.
C: I am not interested in names.

H: There is no beer left.
C: Are you saying 'no' just to be negative?
```

10.2. Does Eliza handle plural forms of nouns? Would it generate the same response for "My brother lives in Hawaii" and "Both of my brothers are younger"?

10.3. Does Eliza work equally well with sentences that have the past, present, and future forms of a verb?

10.4. Some of the reasons a program like Eliza would have difficulty passing the Turing Test were presented in Section 10.7. Can you think of an example sentence, or a sort of conversation, that illustrates each situation?

   a) A sentence structure that is too complex, that is, it would require a parsing algorithm.

   b) A conversation that depends on remembering something from a previous sentence.

   c) A response that requires Eliza to know the meanings of words.

   d) A response that requires real-world knowledge.

   e) ♦ A response based on an inference.

10.5. Here is the definition of a sentence pattern in a session with ElizaLab:

```
>>> p = Pattern("I (hope|wish|want) (.*)")
>>> p.add_response("Do you really $1 $2?")
>>> p.add_response("Have you always $1ed $2?")
```

Show what Python will print for the following expressions (note that no postprocessing rules are specified):

```
>> p.apply("I want a new computer")
>> p.apply("I wish you would stop putting words in my mouth")
```

10.6. What postprocessing rules are necessary for Eliza to respond the way it does in the following examples (using the pattern object defined in the previous exercise)?

   H: *I want my computer to understand me.*
   C: Do you really want your computer to understand you?

   H: *I wish you would be more sympathetic.*
   C: Have you always wished I would be more sympathetic?

   H: *I hope you can help me understand why our cow used to scare us.*
   C: Do you really hope I can help you understand why your cow
      used to scare you?

10.7. Define sentence patterns that would "teach" Eliza how to respond to the following types of sentences:

   a) Reply to sentences containing the word "coffee" with "No thanks."

   b) Reply to sentences containing "coffee," "tea," or "latte" with "No thanks, I don't drink x," where x is one of the three key words.

   c) Reply to sentences of the form "I drink x" with "Tell me why you like x," where x can be any word or phrase.

   d) Reply to sentences containing both drinks and food, for example, respond to "I drink coffee with a scone in the morning" with "Do you really have a scone with your coffee?" Make up your own list of drinks and foods.

   e) Reply to sentences with the words "comics," "books," or "web site" with "Why do you mention x?" or "Do you often read x?" or "I don't like x, myself."

10.8. Discuss some of the issues that would have to be resolved before Eliza could be put to work at a call center to handle telephone calls from customers (assuming there is a program that will accurately translate speech into text). Here are some situations to consider:

   • Credit card customers inquiring about their current balance, or about specific transactions.

   • Airline customers calling to make reservations, or ask about itineraries.

   • Customers who purchased software who have questions about how the software works or are reporting errors.

10.9. The Python implementation of Eliza's algorithm for transforming a sentence was given in Figure 10.6 on page 334. Could this algorithm be considered a type of search? How does it compare to a linear search?

10.10. Extend the DOCTOR script by adding rules for farm animals, movies, books, or any other topic a patient might discuss with a psychiatrist.

10.11. Write a new script that replies to sentences by generating a *non sequitur*. The script should have patterns that look for a word in an input and a set of response strings that contain that word. For example, a pattern for the word "dream" would reply to any input sentence containing "dream" with a random sentence or quote related to dreams:

```
H: My dream date is a romantic dinner.
C: I dream of electric sheep.
H: In my dream I am sitting on a beach in Hawaii.
C: There are more things in Heaven and Earth, Horatio,
   than are dreamt of in your philosophy.
```

10.12. ◆ Write a new script for "small talk" at a party. Add rules for a few specific items, like your favorite sports team, but if the input does not match one of the script patterns have it reply with something like "strange weather we've been having" or "did you watch the game last night?"

10.13. ◆ Search the Internet for a transcript of the old Abbott and Costello routine called "Who's on First?" See if you can devise a set of rules so that when a person types something like "Tell me, what is the name of the first baseman?" your script replies "What's on second." How many other formulaic responses can you generate?

10.14. ◆ Search for a transcript of the Monty Python skit called "The Argument Clinic." In this skit, a "clinician" tries to carry on an argument by negating every sentence spoken by a "patient." If the input is "Yes it is" the response is "No it isn't" and vice versa. Can you make a script that argues with the user?

Splitting and Joining Strings

Some of the programming projects in this chapter are functions that find placeholders in a piece of text and replace them with values from a dictionary. For example, suppose you define the following string, containing two placeholders, and a dictionary that defines the substitutions:

```
>>> s = '$name lives at $address'
>>> d = { 'name' : 'Bilbo', 'address' : 'Bag End, Hobbiton' }
```

After you substitute the dictionary values for the placeholders you will have this string:

```
'Bilbo lives at Bag End, Hobbiton'
```

Here are some hints for these projects:
- Use split (described in Chapter 6) to break the input sentence into individual words.
- Iterate over the list of words to find strings that starts with $, then find its replacement in the dictionary.
- This expression will concatenate the words in a list named a into a single string separated by spaces:

```
>>> " ". join(a)
```

Programming Projects

Several of the programming projects in this chapter ask you to write functions that find placeholders in a piece of text and replace them with values from a dictionary. The sidebar on "Splitting and Joining Strings" on the previous page has some hints for how to write these programs.

10.15. Write a function named `read_patterns` that will create a list of Pattern objects given the name of a file containing pattern specifications. Lines in the file should either start with a slash, in which case they are regular expressions used to create a Pattern object, or with a quote mark, in which case they are response strings. You can test your function using a file named `patterns.txt` in the PythonLabs data directory:

```
>>> read_patterns(path_to_data('patterns.txt'))
[<Pattern 'i (eat|drink) (.*)'>, ..., <Pattern 'am i (.*)'>]
```

10.16. Write a program named `converse` that will use a set of patterns to carry on a conversation. Use the `read_patterns` function from the previous problem to read a list of pattern specifications from a file. The main loop of the program should ask the user to enter a sentence in their terminal window (see the sidebar on "Interactive Programs") and then scan the list of patterns to find one that applies to the sentence. Print the response returned by the first pattern that applies to a sentence, or the word None if the sentence doesn't match any pattern.

10.17. Write a function named `mad_lib` that plays a game called Mad Libs. The argument passed to the function should be a sentence with placeholders, for example

```
>>> s = "The $adjective cow $past-tense-verb in $location"
```

The program should ask the user to supply a string for each placeholder and then substitute the placeholders into the sentence and print the result:

```
>>> mad_lib(s)
adjective: vibrant
past-tense-verb: swam
location: the gym
'The vibrant cow swam in the gym'
```

10.18. Write a program named `play_mad_libs` that will play a series of Mad Libs games using the `mad_lib` function (Problem 10.17) as a helper function. An argument passed on the command line should be the name of a file with several sentences that each contain placeholders. To play the game, select a random sentence pass it to `mad_lib`.

Interactive Programs

Python has a built-in function named `input` that can be used to read strings typed in a terminal window. The function will print a prompt (which is passed as a parameter) and return the string typed by the user:

```
>>> s = 'enter your name: '
>>> t = input(s)
enter your name: frodo
>>> t
'frodo'
```

10.19. Write a function named `blurb` that will generate a random idea for a new movie. Every movie description will have the same format:

```
>>> formula = 'A $adjective $genre about a $hero and a $sidekick who $verb'
```

Create a dictionary like the one shown in the sidebar on splitting and joining strings, using the placeholders in the sentence above as keys, but the values should be lists of strings:

```
fillers = {
    'adjective' : ['epic', 'stirring', 'romantic', 'crazy'],
    'genre' :    ['adventure', 'romance', 'comedy'],
    ...
}
```

When constructing the output sentence, choose a random word from each category. A sample output will look like this:

```
>>> blurb(formula, fillers)
'A crazy romance about a chef and a grad student who hunt vampires'
```

Complete the definition of the dictionary shown above, adding `hero`, `sidekick`, and `verb` entries, and add more strings to each category.

10.20. Some of the sentences produced by the `mad_lib` and `blurb` functions won't be grammatically correct, for example:

```
>>> blurb(formula, fillers)
'A epic comedy about a astronaut...'
```

Write a function named `check_grammar` that will replace `'a'` by `'an'` if the next word starts with a vowel. The easiest way to implement this function is to have it take a list of words and have it return an updated list so the other functions can call `check_grammar` as a helper before they join the words into the final output. Are there any other mistakes this function needs to correct?

Chapter 11

The Music of the Spheres

Computer simulation and the N-body problem

Before the seventeenth century, comets were thought to be random, unrelated occurrences. Astronomers who studied the motions of comets thought they moved in a straight line through space, and there was no reason to suspect a strange object appearing in the sky was the same body that had last shown up dozens of years earlier. But by 1687, when Isaac Newton (1643–1727) published *Principia Mathematica*, astronomers realized comets, like planets, were celestial bodies that orbited the Sun. Newton, using his new theory of gravitational attraction, showed that comets followed highly elliptical orbits, so they were only visible from Earth for short periods when they were close to the Sun.

A few years later, in 1705, Edmond Halley (1656–1742) analyzed records from previous comet sightings, and proposed that a body most recently observed in 1682 was a comet that orbited the Sun every 76 years. If this was the case, Halley needed to account for large differences in the lengths of previous orbits. The time between appearances in 1531 and 1607 was 76.1 years, but the period between the sightings in 1607 and 1682 was only 74.9 years (Figure 11.1). Being familiar with Newton's new theory, Halley proposed that these differences were the result of the gravitational pull of Jupiter and Saturn. Comets are very small, so their orbits would be easily influenced by the two largest planets in the solar system. If Halley was right, the comet would appear again some time around 1758. It was the first time anyone had tried to predict a future appearance of a comet.

One of the landmark achievements in the history of computation, well before there were any computing machines, was the calculation of the orbit of Halley's Comet by the French astronomer Alexis-Claude Clairaut (1713–1765). In the summer of 1758, before the comet was sighted, Clairaut decided to test Newton's equations and Halley's prediction by computing the exact orbit of the comet. He began by assuming the comet traveled mainly along a smooth elliptical orbit, and then performed a series of adjustments to account for the gravitational effects of the two large planets. Clairaut spent five months, working with his colleagues Joseph Jérôme Lalande (1732–1807) and Nicole-Reine Lepaute (1723–1788), on detailed corrections to the orbit, each step depending on the result of earlier calculations.

The group began their work in June, and by November Clairaut and his colleagues had determined that Jupiter and Saturn would cause the current orbit to be 618 days longer than the previous one. On November 14, 1758, Clairaut predicted Halley's Comet would be closest to the Sun in the middle of April, with a margin of error of one month. When the comet finally did reappear, astronomers were able to record the actual date it was closest to the Sun, which was March 13, 1759.

The calculations by Clairaut are interesting not only because they are an example of a significant computation from a time well before there were electronic computers, but because they were an important scientific result, as well. Newton's equations of motion were still not universally accepted, and the computation of the gravitational effects of the outer planets was one of the first major confirmations of Newton's theory.

Before Newton, astronomers had attempted to describe the motions of planets by geometric equations. Most ancient philosophers believed the planets rotated around the Earth, following circular paths. In the Middle Ages, astronomers began to question this view, and gradually adopted the position of Nicolaus Copernicus (1474–1543), who argued that the planets revolved around the Sun. Johannes Kepler (1571–1630) made an important contribution when he proposed that the orbits were elliptic, rather than circular. This allowed him to simplify the equations for the orbits, and as a result the apparent movements of the planets were described with more accuracy.

Newton's theory, that planets and other bodies simply move according to the gravitational effects of the objects around them, was a major break from previous descriptions of the motions of the planets. According to Newton, the planets appear to move with such regular circular or elliptical orbits because the Sun is so much larger than all the planets. The mass of the Sun is about 10^{30} kilograms, which is more than 300,000 times the mass of the Earth and 1000 times more than Jupiter, the largest planet. This large mass causes the planets to circulate in orbits that are, for all practical purposes, ellipses.

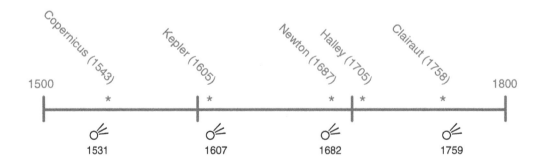

Figure 11.1: *This time line shows the dates of important results in the study of comets. In 1705 Edmond Halley predicted an object last seen in 1682 would reappear in 1758. Alexis-Claude Clairaut used Isaac Newton's law of gravitational attraction to compute the effects of Jupiter and Saturn on the orbit of Halley's comet. Before the comet was visible Clairaut successfully predicted the date it would be closest to the Sun.*

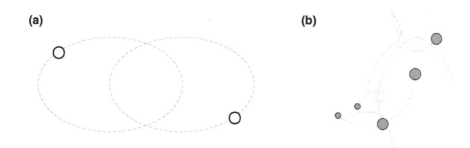

Figure 11.2: *(a) In a binary star system, with two bodies that are approximately the same size, one can solve a set of equations to predict the future location of each star. (b) The more general case is an N-body system. There is no set of equations that will describe the motion of each body, and the only way to predict the future location of each body is through computation.*

The fact that bodies do not follow a prescribed orbit, but simply move according to the force of gravity, means there is no equation that will tell us precisely where any body will be at some arbitrary point in the future. One of the first formulas students learn in a science class is $d = v \times t$, which describes the distance d an object will travel when it moves with constant velocity v for a period of time t. If we know how fast an object is moving, and we assume it keeps moving at the same constant speed, we can use this equation to predict where the object will be at any point in the future by simply multiplying the velocity by the time. But for planets, moons, and comets moving under the influence of gravitational forces there is no such formula.

For certain special situations—a binary star system where two similarly sized objects orbit each other, or a small falling object near the Earth's surface—there are equations that describe the motions, and it is possible to solve these equations to predict the future locations of the objects. But in the general case there are no equations to predict exactly where each body in a system will be at any future date. Astronomers use the term **chaotic** to describe this sort of motion. When used in this technical context, chaotic basically means "unpredictable." One of the consequences of a system being chaotic is that very small changes in the current conditions may lead to very different positions after a period of time.

As soon as Newton published his theory of gravity, he and others began an effort to develop a set of equations to describe the relative motions of the Earth, Moon, and Sun, in what became known as the "3-body problem." Eventually mathematicians were able to prove that it would not be possible to derive simple equations that would predict exactly where any one of $N \geq 3$ bodies would be at any arbitrary point in time. Physicists today refer to the problem of trying to describe the motion of a collection of bodies as the **N-body problem** (Figure 11.2).

While it may not be possible to solve an equation to determine the precise locations at any point in the future, it is possible to estimate where a set of bodies will be by doing a series of calculations. The idea is to start with the current locations and headings of each body. We can compute the effect of gravity on each of the objects and adjust their headings accordingly. We then assume the bodies will move in the direction defined by their updated heading, which will tell us where the objects will be a short time later. By repeating this process, recomputing the gravitational attractions from each new position and computing new positions a short time later, we can estimate where the bodies will be at some future point in time. Mathematicians refer to this type of operation as **numeric integration**.

Clairaut used this technique when he computed the orbit of Halley's comet in 1759. He began by assuming the comet would follow an orbit that was basically an ellipse, and when the comet was far away from any planet, he computed its trajectory using formulas for elliptical paths. When the comet was relatively close to Jupiter or Saturn he applied numeric integration to make a series of corrections based on the gravitational attraction of the planet.

Our project for this chapter will explore computations that predict the motions of the planets in the solar system. SphereLab, the Python module with the software we will use, defines a new type of object, called a Body, that can be used to represent planets and other celestial bodies. When we make a body object, it will encapsulate all the necessary information, such as the planet's mass and position. We will then be able to call methods that compute the gravitational attraction between bodies and update their positions in response to these forces. By repeatedly computing the forces acting on the planets, and moving them for a small amount of time, we will see how an algorithm can paint an accurate picture of the orbits of the solar system.

Using computation to solve problems like predicting the future positions of planets is an example of **computer simulation**. Simulation is an important area in applied computer science, used by professionals in a wide variety of areas, including science, medicine, engineering, and business. Computer-generated imagery (CGI), a growing part of the entertainment industry, is also a form of computer simulation.

Today astronomers use computers to calculate the orbits of several thousand bodies, including comets and asteroids. Simulations are used to predict the locations of satellites, to help plan launches or return trips from the International Space Station, and to track asteroids that are potential threats to Earth. The project in this chapter is a scaled-down and simplified version of the algorithms used by astronomers, but it is nonetheless a good introduction to some of the issues in computer simulation.

The Music of the Spheres

Johannes Kepler, like most astronomers before him, believed the planets traveled in regular orbits that could be described by mathematical equations. Kepler's main contribution was the realization that the orbits were elliptical and not circular. In his work *The Harmony of the Worlds* (1619), Kepler described a correspondence between harmonic ratios in musical notes and ratios of the orbital speeds of planets. But unlike vibrating strings on a musical instrument, the motions of planets cannot be described by equations that will predict a location at an arbitrary point in time. Instead, computer simulations, based on calculations of motions over very small time steps, are used to track the positions of planets, comets, and asteroids.

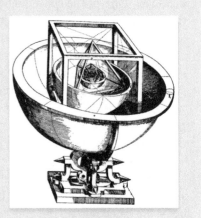

11.1 Running Around in Circles

One of the decisions Clairaut had to make when planning his computation of the orbit of Halley's Comet was how many times to adjust the comet's orbit to account for the gravitational attraction of Jupiter and Saturn. A single adjustment half-way through the orbit might have improved the estimate of the full orbit time, but the calculation would be more accurate if several small adjustments were made at points evenly spaced throughout the orbit. On the other hand, planning to do a calculation for every 24-hour period would lead to far too many steps in the computation. That level of detail might not be necessary, as the effects of Jupiter and Saturn are negligible when the comet is at the farthest reach in its orbit. Clairaut eventually settled on a combination of the two strategies, making detailed calculations when the comet was closest to Jupiter and Saturn, but using the elliptical orbit when the comet was far away.

To begin the project for this chapter, we will explore the trade-off between how often positions are calculated and the overall accuracy of the computation. For the first set of experiments, imagine a situation where a robot has landed on a distant planet, and we want to send it out on a mission to explore its surroundings. We want the robot to travel a path that is a perfect circle, and we want it to end up at the same location where it started. We can transmit instructions to the robot to tell it to move straight ahead, or to turn so it is heading in a new direction. When we tell the robot to move, it will advance straight ahead for a specified amount of time and then wait for further instructions.

Figure 11.3 shows how we can get the robot to make a trip that will move along a circular path that brings it back to its starting point. With only six corrections the path would be a hexagon. If the goal is to make a smooth circle, we need to correct the path more often. The trade-off is that the more corrections we make, the longer it will take the robot to make the circuit, because it has to pause to wait for a new heading at the end of each leg.

For the lab project, we will monitor the progress of the robot in the PythonLabs graphics window. The function that initializes the graphic window also creates an object that represents the robot, so when we call `view_robot` we will want to save the return value, as shown here:

```
>>> robot = view_robot()
```

The function initializes the window to show a map of a square piece of terrain, 400 meters on each side. The robot is initially on the west (left) edge of the map.

To get the robot to move we need to call a method associated with the object that represents the robot. Because we saved a reference to the object in a variable named `robot`, the statements that tell it what to do are all of the form `robot.x()`, where x is the command to execute.

Figure 11.3: *If we tell the robot to correct its heading six times, the path it travels will be a hexagon. With more frequent updates the path will look more like a smooth circle.*

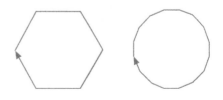

Command	Action
advance(t)	Move ahead for t hours
turn(a)	Turn clockwise by a degrees
plant_flag	Place a marker at the current location
orient	Rotate until the marker is 90° to the right
heading	Return the current direction (in degrees) the robot is headed
speed	Return the robot's velocity (in meters per hour)
track(:on)	Turn on the tracking option
track(:off)	Turn off tracking

Table 11.1: *Commands that can be transmitted to the robot explorer correspond to names of methods that are called to update the robot object.*

For example, this command tells the robot to turn clockwise 30°:

```
>>> robot.turn(30)
```

The complete list of commands is given in Table 11.1.

When we send the command that tells the robot to move forward, we need to tell it how long to move before it stops to wait for the next command. The distance an object moves in an amount of time t is defined by the simple formula $d = v \times t$, where v is the velocity at which the object moves. Our robot will move at 10 meters per hour. So if we want it to move 50 meters, we have to send the command to tell it to advance straight ahead for 5 hours:

```
>>> robot.advance(5)
```

After the robot advances for the specified amount of time it will stop. At this point, if we want it to move in a circle, we need to send it a command to turn clockwise before it advances again. The question is, how far should we tell it to turn?

One way to answer this question is to imagine there is a flag in the middle of the circle, and the robot has an arm sticking straight out from its right side, at a 90° angle from the direction it is heading. If the robot moves in a perfect circle, the arm will always be pointing at the flag. After advancing in a straight line for a while, the flag will still be off to the right of the robot, but it will be slightly behind. To get the robot back on track, we want it to turn to the right until the arm is pointing at the flag again.

Figure 11.4 illustrates how the robot moves and turns. At the start of its journey, the robot is facing north and its arm is pointing due east at the flag. After moving straight for a while, the flag has been left behind. Before continuing, the robot should turn until the arm is once again pointing at the flag.

There are two methods that will help us navigate according to this system. One, called plant_flag, will leave a flag at the robot's current location. After a flag has been set, we can call a method named orient that will have the robot turn according to the rules outlined above.

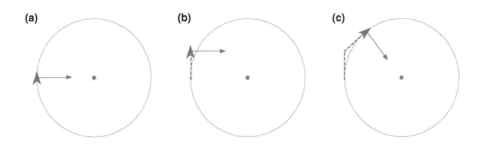

Figure 11.4: *(a) The initial position of the robot explorer. (b) The robot's position after moving straight ahead. (c) After reorienting and moving ahead again, the robot is back on the circular path and the arm is again pointing at the flag.*

The first step in the project is to plant a flag to mark the center of the circular path. When the robot first lands, it will be facing due north. We want it to move some distance to the east, plant the flag, turn around, and move back to the starting position. Then we'll send it out to the north, and periodically have it reorient as it moves clockwise around the flag. If we repeatedly tell the robot to advance and reorient, it will eventually move in a complete circle and end up back where it started.

Tutorial Project

T1. Start a shell session and load the module that will be used for this project:
```
>>> from PythonLabs.SphereLab import *
```

T2. Call `view_robot` to initialize the canvas and create an object to represent the robot:
```
>>> robot = view_robot()
```
You should see a new window, with the robot on the left side, pointing north (straight up).

T3. The robot object is an instance of a class called Turtle:[1]
```
>>> type(robot)
<class 'PythonLabs.SphereLab.Turtle'>
```

T4. Call the `location` method to get the *x* and *y* coordinates of the robot:
```
>>> robot.location()
(40.0, 200.0)
```
The coordinates represent distances from the point at the southwest corner of the map. The robot starts out at a point 40 meters in from the left and 200 meters up from the bottom.

T5. The `speed` and `heading` methods will tell you how fast the robot will move (in meters per hour) once we tell it to start and the direction it is currently headed (360° is north):
```
>>> robot.speed()
10.0
>>> robot.heading()
360.0
```

To plant the flag in the center of this territory, we need the robot to turn so it is heading east, travel 160 meters (the robot's current *x* coordinate is 40, and the center is at $x = 200$), plant the flag, turn back to the west, and travel the same amount of time to get back to the starting point.

---

[1]The name comes from a language named Logo that introduced this style of interactive graphics.

T6. Point the robot to the east:

```
>>> robot.turn(90)
```

You should see the arrow that represents the robot on the screen turn so it points toward the center of the map.

T7. The robot moves at 10 meters/hour, so we have to tell it to travel for 16 hours to get to the center of the map:

```
>>> robot.advance(16)
```

T8. Get the new location of the robot:

```
>>> robot.location()
(200.0, 200.00000000000003)
```

Can you see how it has moved 160 meters to the east? The previous x coordinate was 40, and the new one is 200.

T9. Plant the flag:

```
>>> robot.plant_flag()
```

T10. Turn the robot around, send it back to the starting point, and tell it to turn to the north again:

```
>>> robot.turn(180)
>>> robot.advance(16)
>>> robot.turn(90)
```

T11. It will be easier to follow the robot's progress on its circular trip if it leaves a track behind as it moves. Type this command to turn on the tracking option:

```
>>> robot.tracking = True
```

T12. Tell the robot to move for 3 hours in the direction it is currently heading:

```
>>> robot.advance(3)
```

You should see the robot icon move a short distance toward the top of the map, leaving a line segment on the canvas to show where it has been.

T13. Tell the robot to turn 90°:

```
>>> robot.turn(90)
```

T14. You can turn the robot counterclockwise by passing a negative number to turn:

```
>>> robot.turn(-30)
```

T15. After turning 90° clockwise, and then 30° counterclockwise, the new heading should be 60°:

```
>>> robot.heading()
59.999999999999986
```

Is the robot icon on your screen pointing to 60° (toward the northeast)?

T16. Move the robot for 10 hours along its new heading:

```
>>> robot.advance(10)
```

Your canvas should now look something like the screen snapshot in Figure 11.5a.

♦ Get the robot's new location:

```
>>> robot.location()
(126.60254037844385, 280.0)
```

Use your calculator to check to see if this location is correct (see Figure 11.5b).

Do some more experiments on your own. Can you use calls to advance_robot and turn_robot to have the robot travel a circular path around the flag in the middle of the terrain and return to its starting point?

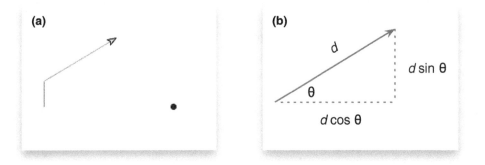

Figure 11.5: *(a) The location and orientation of the robot after moving for 3 hours, turning 60°, and moving another 10 hours. (b)* ♦ *The second part of the path traveled by the robot is the hypotenuse of a right triangle. The distance moved in the x and y dimensions corresponds to the cosine and sine of the angle adjacent to the hypotenuse.*

T17. Clear the canvas and reposition the robot at its starting location by calling `view_robot` again. Because the remaining experiments will want to track the motion of the robot around the flag in the middle of the map, we can pass options to `view_robot` so the screen is initialized with a flag and with tracking turned on:

```
>>> robot = view_robot(flag = True, track = True)
```

T18. Tell the robot to move straight ahead for 5 hours:

```
>>> robot.advance(5)
```

T19. Call `orient` to have the robot turn so its arm it pointing at the center again:

```
>>> robot.orient()
```

T20. Call the `heading` method to see how the call to `orient` changed the robot's direction:

```
>>> robot.heading()
34.70804927252268
```

T21. Enclose the calls to `advance` and `orient` in the body of a `for` statement so they are executed 19 more times:

```
>>> for i in range(19):
...     robot.advance(5); robot.orient()
...
```

Is the robot progressing on a path that will take it back to its "home base"?

T22. Clear the canvas and create a new robot. Repeat the previous exercise, but this time have the robot travel for only 2 hours instead of 5 before it reorients, and have it make a journey involving 50 segments instead of 20:

```
>>> robot = view_robot(flag = True, track = True)
>>> for i in range(50):
...     robot.advance(2); robot.orient()
...
```

In the first experiment, the path was adjusted once every 5 hours, but in the second experiment the robot turned once every 2 hours. Do you see that when the robot is told to adjust its direction more frequently the path looks more like a circle?

♦ When we call `robot.orient` the robot actually turns an amount that puts the flag slightly in front of where the arm is pointing; if you're curious about the exact calculation, refer to the SphereLab documentation in your Lab Manual.

♦ What do you suppose would happen if the time interval is increased, instead of decreased? Think about where the robot would go if it moved north for 12 hours and then reoriented and advanced again. Would the line segments that make up the path still be tangent to the circular path we want the robot to make?

♦ Check your answer to the previous question by repeating Exercises T17 and T22, passing 12 in the call to robot.advance.

♦ An even more extreme path is made if the robot travels for 24 hours or more, but there isn't room to show this path on the map. You can get the same effect if you move the robot closer to the flag before it starts, which will have it make a smaller circle. Type this after initializing the map but before you enter the for statement:

```
>>> robot.turn(90); robot.advance(8); robot.turn(-90)
```

Now send the robot off on a journey that moves for 12 hours before reorienting. What do you see? Is it a circle?

♦ Suppose the robot leaves a monitoring instrument each time it stops. Describe the pattern defined by the locations of the instruments. Are they all equally distant from the flag? Do the locations define a circle?

11.2 The Force of Gravity

To simulate the movement of the planets in the solar system we are going to use the strategy introduced in the previous section. There we told a robot to move in a straight line for a short period and then corrected its heading. The solar system simulation starts with the current location and heading of each planet. The basic equations of motion predict where planets would be a short time later if they all traveled in a straight line. From these new positions the simulator uses gravitational forces between bodies to calculate adjustments to headings, and we will simulate straight-line motions for another short period of time.

The amount of time the bodies move is called the **time step**. As was the case with the robot experiment, the size of a time step is a critical factor in the accuracy of the simulation. The more often new trajectories for the planets are computed, the more accurate their simulated movements will be.

One difference between the robot experiment and the planet simulation is that we do not have any preconceived notion of where the planets are supposed to move. The adjustments to a planet's heading will be determined only by the gravitational forces from all the other bodies in the system.

Another difference is that the robot always moves with a constant velocity. We simply assumed that once it receives an advance command the robot will immediately start off at 10 meters per hour. A real robot, like a car, will need some time to accelerate before it reaches the final velocity. Similarly, planets and other bodies in the solar system change velocity, speeding up the closer they get to the Sun.

The next step in the development of our simulation is to explore situations where velocity changes. When a body changes velocity, it is because some external force is being applied that causes the body to accelerate. In everyday usage, "acceleration" means an increase in velocity, and "deceleration" refers to a decrease in velocity. To a physicist, acceleration simply means a change in velocity, whether it increases or decreases.

The project in this section will show how we can simulate the motion of a body when it is accelerating as a result of gravitational forces. Our experimental system will have just two bodies: a watermelon and the Earth. We will create an object for each of these two bodies and then simulate the motion of the watermelon as it falls toward the Earth.

The advantage in using a simple "two-body" system for our initial experiment is that there is an equation that predicts how far the melon will fall in a given amount of time, and we can use this equation to check the accuracy of the simulation. Recall from the discussion in the introduction to this chapter that for a general situation, with $N \geq 3$ bodies, there are no formulas to explain how each body will move, but a small object falling toward the Earth is a special case that can be described precisely by a simple equation (Figure 11.6).

Assuming an object is initially stationary, and that it does not slow down because of air resistance, the equation that predicts how far it will fall in an amount of time t (measured in seconds) is $d = 1/2 \ g \times t^2$. The g in this equation stands for the **acceleration due to gravity**, which is the effect of the gravitational force exerted by the Earth.

There are slight variations in this force, depending on where one stands on the surface of the Earth, but we will ignore these small differences and treat g as a constant. We are measuring distances in meters, so the value of g is 9.8 m/sec^2. As an example of how this equation can be applied, if the watermelon is dropped from a balcony of a tall building, the distance it falls in 2.0 seconds is $d = 1/2 \times 9.8 \times 2.0^2$, or 19.6 meters.

By rearranging the terms in the equation, we can turn it into a formula that defines t as a function of d:

$$t = \sqrt{2 \times d/g}$$

In this form, the equation tells us how long it will take an object to fall a distance of d meters. For example, if we know the balcony is 35 meters high, we can predict the time it will take the watermelon to hit the ground as $\sqrt{2 \times 35/9.8} = 2.67$ seconds. This prediction will be accurate if the watermelon is perfectly stationary when it is let go (the person dropping it doesn't give it a downward push in order to make a bigger splat) and if there is no air resistance to slow the watermelon's fall.

The SphereLab module includes a class named Body where instances represent small bodies like the watermelon, as well as larger bodies like the Earth, Sun, and planets. Each body object has a mass, a position, and a velocity. The methods defined for bodies will do all the necessary calculations for us; all we need to do is create the objects and then call the methods to update their velocities and positions.

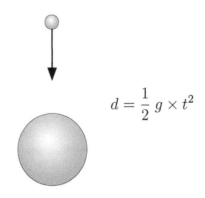

Figure 11.6: *A small object near the surface of the Earth is pulled toward the center of the Earth by the force of gravity. A simple equation predicts how far the object will move in t seconds.*

$$d = \frac{1}{2} \ g \times t^2$$

A function named make_system sets up a computational experiment for a system of bodies. The argument passed to the function is the name of a file containing the initial conditions for a set of bodies. SphereLab includes several test data sets, including one for the two-body Earth/watermelon system. Here is the call to make_system that sets up the watermelon experiment:

```
>>> b = make_system(path_to_data('melon.txt'))
```

The function creates a list containing two Body objects, one for the watermelon and one for the Earth. If we print one of the bodies this is what we'll see:

```
>>> b[1]
earth m: 5.97e+24 r: (0, 0, 0) v: (0, 0, 0)
```

The string printed for a Body object shows its name and mass, followed by the object's position and velocity. The details of how the position and velocity are represented will be explained in the next section; for now we just need to know that they are included as instance variables of each body.

We can monitor the progress of the watermelon experiment by watching the bodies move on the PythonLabs canvas. To initialize the drawing, pass the list of Body objects to a method named view_melon:

```
>>> view_melon(b)
```

The drawing will show the watermelon is initially on the Earth's surface. The first step in the experiment is to raise the melon to a specified location above the ground. To do this, we call a method named set_height. Because the melon is the first Body object in the system, the call is

```
>>> b[0].set_height(50)
```

In the view of the simulation the circle that represents the melon should now be up off the ground.

The key step in the simulation is performed by a method named step_system, which simulates the motion of the bodies in the system for a specified length of time. The argument passed to step_system is the size of the time step. The method will use the force acting on the melon to update its trajectory and compute the new position after the specified amount of time. This statement predicts the location of the melon using a time step of 0.5 seconds:

```
>>> step_system(b, 0.5)
```

If we want to know the new location of the melon we can call its height method:

```
>>> b[0].height()
47.54596951138228
```

The melon was initially 50 meters above the ground, and the result of this call shows the simulator has determined that after falling for one half second the melon will be about 47.5 meters above the ground.

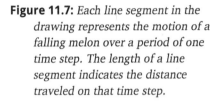

Figure 11.7: *Each line segment in the drawing represents the motion of a falling melon over a period of one time step. The length of a line segment indicates the distance traveled on that time step.*

If we repeat the call to step_system and ask for the height of the melon again we'll see something interesting:

```
>>> step_system(b, 0.5); b[0].height()
42.63790664356202
```

As expected, the new height is lower than it was at the end of the previous time step, but notice how much more the melon moved. In the first time step, the melon moved $50 - 47.54 = 2.46$ meters, but in the second time step it moved $47.54 - 42.63 = 4.91$ meters. This is exactly what we expect, since the melon is accelerating. Its velocity is continually increasing as it falls, and the longer it falls the faster it should move.

As we watch the melon's progress on the canvas, we will see a series of dashed lines. The length of a line segment corresponds to the distance the melon moved during one time step. Because the melon is falling faster and faster, it is moving farther on each time step, and we should see the dashes growing longer with each new time step (Figure 11.7).

A function named drop_melon will repeatedly call step_system and return as soon as the melon hits the ground. The value returned by drop_melon is the amount of time, according to the simulation, required for the melon to drop to the Earth's surface. For example, after initializing the experiment and positioning the melon at 50 meters, a call to drop_melon shows it takes 3.0 seconds for the melon to hit the ground:

```
>>> drop_melon(b, 0.5)
3.0
```

Note that the return value will always be a multiple of the time step size. One way to interpret this result is that drop_melon had to call step_system six times before the melon dropped all the way to the ground, because six times steps at 0.5 seconds per step adds up to 3.0 seconds.

We can check the accuracy of this result by plugging the melon's initial height into the equation that predicts the time it will take an object to fall this distance:

$$t = \sqrt{2 \times d/g} = \sqrt{2 \times 50/9.8} = 3.19$$

So the simulation was off by 0.19 seconds, which is a sizable error. The project in this section will experiment with shorter time steps to see if they lead to a more accurate simulation.

Tutorial Project

T23. Use the make_system function to create the Body objects for the "two-body" problem with a watermelon and the Earth:

```
>>> b = make_system(path_to_data('melon.txt'))
```

T24. The variable b is now a list with two objects:

```
>>> b[0]
melon m: 3 r: (0, 6.371e+06, 0) v: (0, 0, 0)
>>> b[1]
earth m: 5.97e+24 r: (0, 0, 0) v: (0, 0, 0)
```

The melon has a mass of 3 kilograms (about 6.6 pounds), and the Earth has a mass of 5.97×10^{24} kilograms (a lot bigger). The other information printed after the mass are the location and velocity vectors (these will be explained in the next section).

T25. Call view_melon to make a drawing showing the two bodies in this system:

```
>>> view_melon(b)
```

You should see the melon resting on the surface of the Earth.

T26. Type this statement to raise the melon to a position 100 meters above the Earth's surface:

```
>>> b[0].set_height(100)
```

You can specify any height between 0 and 100 meters. On the canvas, the circle representing the melon should have moved to the top of the drawing.

T27. Print the current height of the melon to verify it was raised to 100 meters:

```
>>> b[0].height()
100.0
```

T28. Call step_system to compute the melon's new position after a time step of one second and print the melon's height again:

```
>>> step_system(b, 1.0); b[0].height()
90.18403211794794
```

You should also see the melon move toward the ground in the drawing.

T29. Repeat the previous call:

```
>>> step_system(b, 1.0); b[0].height()
70.55206610728055
```

Do you see how the melon accelerated during this second time step? The first time step moved it about 10 meters, and the second time step moved it about 20 meters. On the screen, the circle moved twice as far on the second update.

If you repeat the previous statement a few times, you will see the distance traveled increases by about 10 meters on each successive call (but step_system will not move the melon any further once it reaches a value at or below the Earth's surface).

The equation for the distance traveled by a falling object predicts that in 1.0 seconds the melon should fall $1/2\, g\, t^2 = 1/2 \times 9.8 \times 1.0^2 = 4.9$ meters. But in our simulation the melon fell 9.81 meters in the first time step. The reason for this large error is because the time step was too big. If we move the melon gradually, over smaller time steps, the simulation will be more accurate.

T30. Call view_melon to clear the dashes from the screen and position the melon 100 meters above the Earth again:

```
>>> view_melon(b)
>>> b[0].set_height(100)
```

T31. Call `step_system` with a time step of 0.1 seconds:

```
>>> step_system(b, 0.1); b[0].height()
99.90184032078832
```

As expected, the melon moves a much shorter distance in 1/10 second.

T32. Call the method that updates the position nine more times and get the height of the melon:

```
>>> for i in range(9):
...     step_system(b, 0.1)
...
>>> b[0].height()
94.60121616721153
```

So in this second simulation, using 10 time steps of 0.1 seconds each, the melon moved about $100 - 94.6 = 5.4$ meters. This is much closer to the real-world value of 4.9 meters.

You're now ready to run the complete simulation using the `drop_melon` function.

T33. Reinitialize the drawing and reposition the melon at 50 meters:

```
>>> view_melon(b)
>>> b[0].set_height(50)
```

T34. Call `drop_melon` to run the simulation with a time step size of 0.5 seconds:

```
>>> drop_melon(b, 0.5)
3.0
```

The simulator predicts it will take the melon 3.0 seconds to fall from 50 meters.

T35. Repeat the simulation using a smaller time step:

```
>>> view_melon(b)
>>> b[0].set_height(50)
>>> drop_melon(b, 0.1)
3.2
```

This time, the simulator says 3.2 seconds instead of 3.0 seconds. Compare these results with the value predicted by the formula $t = \sqrt{2 \times d/g}$. Is the simulation with the smaller time step more accurate?

T36. Use the formula to figure out how long a melon will fall if it is dropped from a seventh-floor balcony that is 35 meters above the ground.

If a person gives the melon a sideways push when dropping the melon, it will continue to move horizontally as it drops. We can simulate this motion by specifying a horizontal velocity after we position the melon. Gravity will cause the melon to move with increasing speed toward the ground, but it will move with a constant velocity to the right (Figure 11.8).

T37. Reinitialize the screen, reposition the melon at 100 meters above the ground, and assign a velocity of 5 meters/second in the x (horizontal) dimension:

```
>>> view_melon(b)
>>> b[0].set_height(100)
>>> b[0].velocity.x = 5
```

T38. Run the simulation using a 0.1 second time step:

```
>>> drop_melon(b, 0.1)
4.5
```

You should see something like the drawing in Figure 11.8.

T39. How far to the right should the melon move in the amount of time it takes to fall 100 meters?

Figure 11.8: *When the melon is given a shove sideways, it moves horizontally with a constant velocity while gravity is causing it to accelerate vertically. The resulting path is a parabola that gets steeper the longer the melon falls.*

T40. If the scale of the drawing is such that the vertical distance in the path shown on your screen is 100 meters, is the horizontal distance consistent with your answer to the previous problem?

In the next section we will see that the equations of motion implemented in body objects are based on Newton's law of universal gravitation, which says the force pulling the melon toward Earth is balanced by an equal force that moves the Earth toward the melon.

♦ Make a fresh set of Body objects:

```
>>> b = make_system(':melon.txt')
```

♦ Type this expression to get the (x, y, z) coordinates of the center of the Earth:

```
>>> b[1].position
(0, 0, 0)
```

♦ Raise the melon up to 100 meters and drop it:

```
>>> b[0].set_height(100)
>>> drop_melon(b, 0.1)
4.5
```

♦ Check the Earth's position again:

```
>>> b[1].position
(0, 5.1022e-23, 0)
```

The y coordinate in this output shows the Earth moved 5.1×10^{-23} meters.

♦ The Earth was pulled toward the melon, but only by a tiny distance. Look up the diameter of a hydrogen atom in a physics text or on the Internet. How many diameters did the melon cause the Earth to move?

11.3 Force Vectors

The exercise near the end of the previous section showed that an object can be moving in two dimensions at once: an initial sideways shove caused the melon to move horizontally, and gravity caused it to move vertically. In the case of planets, we need to be concerned with motions in three dimensions.

One way to describe these more complex motions is to use **vectors**. Mathematically, a vector is a set of three numbers, corresponding to x, y, and z coordinates. For example, when we assigned a horizontal motion to the melon, the result was a set of three numbers that specified the melon's initial velocity in the x, y, and z dimensions:

```
>>> b[0].velocity.x = 5
>>> b[0].velocity
(5, 0, 0)
```

The 5 in the first position means this body is moving 5 meters per second horizontally (the x dimension), and the 0 in the other two coordinates mean the body initially had no velocity in the y and z dimensions.

The Python objects that represent bodies use vectors to keep track of positions, velocities, and accelerations. For this project, we can simply think of these vectors as arrows; for example, a velocity vector is an arrow that points in the direction the body will move in the next time step. If you are curious about how the methods carry out the arithmetic operations in terms of vectors, you can read about them in your Lab Manual.

The equation that describes the force that pulls two bodies toward each other is Newton's law of universal gravitation:

$$F \propto \frac{m_1 \times m_2}{d^2}$$

The symbols m_1 and m_2 stand for the masses of the two bodies, and d is the distance between them. This equation describes a general relationship between the gravitational force (F), the size of the bodies, and how far apart they are. When the bodies are larger, the values of m_1 or m_2 will be higher and the force is larger. On the other hand, as the distance grows larger, the force will be smaller, because the value of a fraction shrinks as the denominator grows.

Newton's law is shown pictorially in Figure 11.9. The arrows in the figure represent acceleration vectors. This figure emphasizes the fact that gravity is pushing the bodies in a particular direction, indicated by the arrows that point from each body toward the other one.

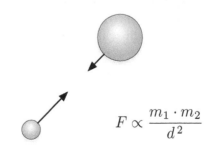

Figure 11.9: *The force pulling two bodies toward each other generates an acceleration that depends on a body's mass. The smaller body will be accelerated more, as indicated by the longer arrow.*

$$F \propto \frac{m_1 \cdot m_2}{d^2}$$

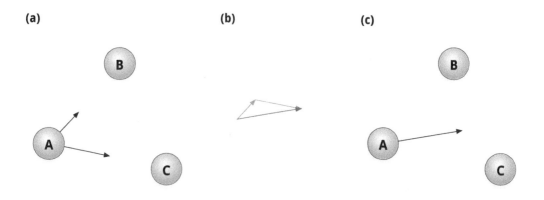

(a)   **(b)**   **(c)**

Figure 11.10: *(a) Body A is simultaneously being pulled toward bodies B and C, defined by two different vectors. (b) The two vectors can be added together to compute the cumulative force. (c) Body A will move in the direction defined by the cumulative force.*

An important point to notice is that the length of an arrow corresponds to the magnitude of the acceleration. If the two bodies were the same size, they would each experience the same acceleration. But when they have different masses, the smaller one is accelerated more. This is what we saw at the end of the previous section: the melon moves rapidly toward the Earth as a result of the large acceleration caused by gravity, while the Earth barely moves toward the melon.

We now come to one of the most important points about the physics behind the motion of the planets: the forces acting on a body are **additive.** What this means is that if we want to figure out how body A will move when it is accelerated by the force of gravity from two other bodies B and C, we can calculate the force moving A toward B, then calculate the force moving A toward C, and then add the two forces together to get the cumulative force.

One reason to use arrows to represent vectors is that it is easy to see how vectors are added together. To add two arrows, simply connect the head of one arrow to the tail of the other. You can imagine using a computer graphics application, and picking up one arrow and dragging it on your canvas until its tail lines up with the head of the other arrow (Figure 11.10). When there are more than two other bodies, the forces can be added one after the other, in any order, because vector addition is like ordinary addition: the sum $x + y + z$ can be computed by adding $x + y$ and then adding z to the result.

To get a sense of how force vectors are added together in an N-body simulation, the tutorial exercises in this section will set up an experiment with a set of random bodies. One of the bodies will have a position a little farther from the center of the system than the others. When we start the simulation, all the bodies will be stationary. We will then let the body farthest from the center "fall" toward the others. Just like the melon in the two-body experiment, this body will start slowly and gradually accelerate as it is drawn toward the other bodies.

It is important to realize that this experiment is not intended to be a simulation of a real system of bodies acting under the influence of gravity. In a real system, all the bodies would be in motion. The goal here is simply to illustrate that forces are additive, by focusing on a single body in order to see how it is pulled in a direction that is the sum of the forces acting on it by all the other bodies.

To create the initial data for this experiment, tell make_system to load a data set named fdemo (for "force demo"):

```
>>> b = make_system(path_to_data('fdemo.txt'))
```

To draw the bodies on the canvas call view_system:

```
>>> view_system(b)
```

When we want this simulation to advance by a single time step, we will call a special version of step_system named step_one. As the name implies, it will move only one Body object in the system. The parameters are the body we want to move, a list with all the other bodies, and the time step size. In this data set, the first body in the list is the "falling" body, and the remaining objects represent the stationary bodies, so this expression will simulate the motion of the falling body for a time step of 1.0 seconds:

```
>>> step_one(b[0], b[1:], 1.0)
```

To run several time steps we can pass optional parameters in the call to step_one. This statement will run the simulation for 10 time steps:

```
>>> step_one(b[0], b[1:], 1.0, nsteps = 10)
```

The main thing to watch for as we run the experiment for several time steps is that at each step the gravitational force acting on the moving body will be the sum of the forces pulling it toward all the others (Figure 11.11).

What's interesting about this experiment is that it is impossible to predict where the falling body will end up at any future time step. We will be able to predict the general direction for the first few time steps, but after that any further motion will be determined by the gravitational effects exerted at the new position, and these effects need to be recalculated at each time step.

One of the hallmarks of chaotic systems is that very small differences in the starting conditions often lead to significantly different states, even after just a few time steps. To demonstrate this effect, before we start the simulation with the falling body, we will make two copies of the object. Then we will run the simulation again, except the copies of the

Figure 11.11: *In the "falling planet" simulation, one body is allowed to move according to the sum of the gravitational forces acting on it while all the others remain stationary.*

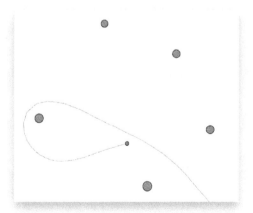

falling body will start in slightly different locations. It will be apparent after only 50 time steps that the trajectories of these bodies have started to diverge and that it is impossible to predict where they will be after the next 50 time steps.

Tutorial Project

T41. Load the data set with the six bodies used in this experiment and initialize the canvas

```
>>> b = make_system(path_to_data('fdemo.txt'))
```

You should see six circles on the canvas, similar to the snapshot shown in Figure 11.11. One circle, representing the falling body, will be red, and the other five will be blue.

T42. Save the first body in the list in a variable named f1, and make two copies of the object, calling them f2 and f3 (the argument passed to copy specifies the color of the circle representing the body):

```
>>> f1 = b[0]
>>> f2 = f1.copy('green')
>>> f3 = f1.copy('yellow')
```

T43. This expression asks Python to print the current location of the first body:

```
>>> f1.position
(81.472, 145.85, 0)
```

The three numbers are the x, y, and z coordinates of the body.

T44. Move the two copies slightly to the right by adding a small amount to their x coordinates:

```
>>> f2.position.x += 1
>>> f3.position.x += 2
```

T45. Run the simulation with the first falling body for 10 time steps. The second optional argument in this statement sets the animation rate so the canvas pauses for $1/10$ second after each new position is plotted:

```
>>> step_one(f1, b[1:], 1.0, nsteps = 10, delay = 0.1)
```

You can control the speed up the animation by passing a smaller value for delay to speed it up or a higher value to slow it down.

You should see the red circle move slightly on the screen. As was the case with the watermelon simulation, the falling body will move very slowly at first and then pick up speed. Initially the body will be pulled toward the center of the system, and soon it will pass close to one of the stationary bodies. When that happens, the moving body will speed up the closer it gets to the stationary body.

T46. Repeat the previous expression to move the falling body for 40 more time steps:

```
>>> step_one(f1, b[1:], 1.0, nsteps = 40, delay = 0.1)
```

T47. Were you able to anticipate how the body moved as it got closer to the others?

T48. Can you guess where the body will be after the next 10 time steps? Type a command to move the body for one time step:

```
>>> step_one(b[0], b[1:6], 1.0)
```

T49. Keep repeating the previous expression and watch the progress of the falling body.

When the falling body comes very close to a stationary body, it will speed up because the gravitational force from that body is stronger at shorter distances. After 20 more time steps you will see the falling body get so close to a stationary body that it accelerates fast enough to be flung out of the system in a "slingshot" effect (at which point its trajectory is very predictable).

T50. Run the experiment again, but this time use f2, the green copy of the falling body, and run it for 50 steps:

```
>>> step_one(f2, b[1:], 1.0, nsteps = 50, delay = 0.1)
```

T51. Because the tracks left by f1 are still on the screen it's easy to compare the trajectories of the two bodies. Can you see how a small change in the starting position of the body led to a pretty big difference after 50 time steps?

T52. Repeat the command, but move the third (yellow) body:

```
>>> step_one(f3, b[1:], 1.0, nsteps = 50, delay = 0.1)
```

Again the trajectory is very similar for the first few time steps, and then the path starts to diverge. The three bodies are identical, except each copy started with a slightly different x coordinate, and this small difference in initial conditions leads to large differences in locations after only 50 time steps.

As long as f2 or f3 is on the screen you can keep repeating the calls to step_one. When they are ejected from the system do they head out in the same general direction as f1? How many time steps does it take before f2 and f3 are ejected?

11.4 *N*-Body Simulation of the Solar System

We're now ready to put all the pieces together to carry out a simulation of the movement of the planets in the solar system.

What we did in the last section for one body—calculate all the force vectors influencing the body, add the vectors together to determine the cumulative force, and then move the body in the direction determined by that force—needs to be done for all the bodies in the system at each time step.

For the solar system simulation, we need a list of Body objects for the Sun and each of the nine planets. A file named solarsystem.txt has the necessary information:

```
>>> b = make_system(path_to_data('solarsystem.txt'))
```

The parameters for each body are from a solar system database and represent the positions and headings of each body on January 1, 1970. The positions have all been adjusted so the Sun is at the center of the system, that is, its x, y, and z coordinates are (0,0,0).

To monitor the simulation, we can simply pass the list to view_system. But if we do, we'll discover a slight problem: Pluto and the outer planets are so far from the Sun that the inner planets are all crowded together in the middle of the display. A better strategy for viewing the simulation is to pass only the first five bodies in the call to view_system. The simulation will still use all ten bodies, but the drawing will show only the Sun and the planets Mercury, Venus, Earth, and Mars.

To run the simulation we need to choose a time step size. To start with, we can use the number of seconds in one Earth day. The fastest-moving planet, Mercury, makes a complete orbit in 88 days, so a time step of one day should lead to an ellipse made from 88 line segments, which will be reasonably accurate for our purposes. There are $24 \times 60 \times 60 = 86,400$ seconds in one 24-hour period. If we want to be a little more precise, we should use 86,459 seconds as our time step size, since there are 365.25 days in a year.

The first set of experiments in this section will use the version of step_system that is implemented in SphereLab. A simplified version of the Python code for the function is

```
1   from PythonLabs.SphereLab import Body
2   from PythonLabs.Canvas import Canvas
3
4   def step_system(bodies, dt):
5       """
6       Carry out one time step of an n-body simulation.  The two arguments
7       passed to the function are a list of Body objects and the time step size.
8       """
9       nb = len(bodies)            # nb = "number of bodies"
10
11      for i in range(0, nb):    # compute pairwise interactions
12          for j in range(i+1, nb):
13              Body.interaction( bodies[i], bodies[j] )
14
15      for b in bodies:
16          b.move(dt)              # apply the accumulated forces
17          b.clear_force()         # reset force for next round
18
19      if Canvas.view:
20          Canvas.update()
```

Download: spheres/nbody.py

Figure 11.12: *The Python implementation of the N-body simulation algorithm. The method uses nested loops to iterate over all pairs of bodies to do $\mathcal{O}(n^2)$ force calculations on each time step.*

shown in Figure 11.12. The only difference between this version and the one in SphereLab is that it does not have options to run multiple time steps or slow down the animation.

The calculation of the gravitational forces between two bodies is carried out by a function named Body.interaction, which is called from line 10. An important detail is that this function computes two sets of forces at the same time. The force pulling body j toward body i is the same as the force pulling i to j. The function does all the calculations needed to compute the force vector acting on one of the bodies, and then the vector pointing in the opposite direction is used to update the forces acting on the other body (Figure 11.9).

If the loops in this algorithm looks familiar, it's because they have same overall structure as the nested loops in the insertion sort algorithm from Chapter 4. The step_system function will make $n \times (n-1)/2$ calls to Body.interaction as it computes the forces on a list of n bodies. As a result, the algorithm does $\mathcal{O}(n^2)$ force calculations on each time step.

Tutorial Project

T53. Call make_system to create a list of body objects for the Sun and planets:
        >>> b = make_system(path_to_data('solarsystem.txt'))

T54. There should be ten objects in the list:
        >>> len(b)
        10

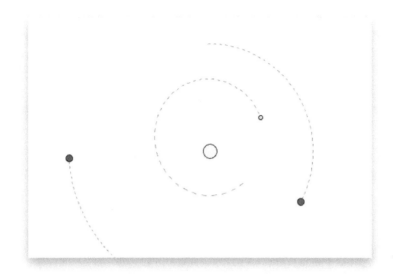

Figure 11.13: *A snapshot of the PythonLabs canvas during the solar system simulation. The simulation uses all ten bodies, but only the Sun and inner planets are shown on the canvas.*

T55. The first object represents the Sun:
```
>>> b[0]
sun m: 1.99e+30 r: (0, 0, 0) v: (0, 0, 0)
```
The values printed on this line are the mass (m), position vector (r), and velocity vector (v). The Sun starts out at the center of the system, and it is not moving.

T56. Call `view_system` to plot the positions of all the bodies:
```
>>> view_system(b)
```
The distances are scaled so all the bodies will fit on the canvas. As a result, the inner planets are all scrunched together in the center of the canvas.

T57. It will be easier to watch the orbits if you tell `view_system` to track only the first five bodies (Sun, Mercury, Venus, Earth, Mars):
```
>>> view_system(b[0:5])
```
You should see something similar to the snapshot in Figure 11.13.

T58. Run the simulation for 10 time steps, using a step size of 1 day:
```
>>> step_system(b, 86459, nsteps = 10, delay = 0.1)
```
You should see the planets move on the canvas. The length of a dash corresponds to the amount of movement in one time step. Mercury is moving much faster than Earth and Mars, as indicated by the longer dashes.

T59. Continue the simulation for another 355 days, so the total simulation runs for 365 time steps:
```
>>> step_system(b, 86459, nsteps = 355, delay = 0.1)
```

As you run the simulation, watch to see if the planets travel in realistic orbits. Are the paths ellipses? If you run the simulation for 365 time steps, does Earth end up more or less where it started, after making one complete orbit? Do the planets move faster when they are closer to the Sun?

A claim made at the beginning of the chapter was the reason the planets appear to move in nice, orderly elliptical orbits is that the Sun is much more massive than the other bodies. For the next part of the project, we'll make one of the planets much larger and see what effect that has on the motion of the other planets.

T60. Ask Python to print the mass of the Sun and the mass of Mars:

```
>>> b[0].mass
1.9891e+30
>>> b[4].mass
6.4185e+23
```

These values are displayed using Python's convention for printing scientific notation. The mass of the sun is 1.99×10^{30} kilograms and the mass of Mars is 6.42×10^{23} kilograms. Mars is roughly 3×10^6 times smaller than the Sun.

T61. Type this expression to multiply the mass of Mars by 1×10^6:

```
>>> b[4].mass *= 1e6
```

The circle representing Mars on the canvas does not change, but the instance variable representing the planet's mass is now set to $6.4185e+29$ (6.42×10^{29}).

T62. Continue the simulation for another 365 time steps:

```
>>> step_system(b, 86459, nsteps = 365, delay = 0.01)
```

Can you explain why each of the planets moved the way it did?

♦ Try some more experiments on your own. Start each experiment by making a fresh list of bodies and reinitializing the display. Some ideas:

  • Vary the step size to see what effect that has on the accuracy of the simulation. What happens if you increase the time step to the number of seconds in 3 days? In 30 days?

  • View different planets. For example, passing b[0:6] to view_system will include Jupiter in the display.

  • Can you figure out how to view only the Sun and the outer planets (Jupiter, Saturn, Uranus, Neptune, and Pluto) during a simulation?

  • How many steps (with a time step size of 86,459) would you have to run to have Jupiter make a complete orbit?

  • Multiply the mass of Mars by different values; instead of 10^6 try 10^5 or other values.

  • Increase the mass of different planets, including Jupiter or Saturn. The goal is to have two or more bodies that have roughly the same mass as the Sun.

For the remaining experiments in this section, you should either download the file nbody.py from the website, or use your IDE to create a new file and type in the code for step_system shown in Figure 11.12.

T63. Start a new shell session and load your definition of step_system.

T64. Verify the function works by reading the solar system data file and simulating a few steps:

```
>>> from PythonLabs.SphereLab import make_system, view_system

>>> b = make_system(':solarsystem.txt')

>>> for i in range(50):
...     step_system(b[0:5], 86459)

...
```

Note: The code that updates the canvas is called automatically by the move method in the Body class, so as long as the list of bodies has been displayed by view_system your code should also generate an animation.

T65. Add this line to your function so it is called from the inner `for` loop, right before the call to `Body.interaction`:

```
print(bodies[i].name, '<=>', bodies[j].name)
```

T66. Reload your function definition and make a single call to `step_system`:

```
>>> step_system(b[0:5], 86459)
sun <=> mercury
sun <=> venus
    ...
earth <=> mars
```

Do you see how the nested loops work? When i is 0, j takes on all values from 1 to 4, which accounts for the first four lines of output. Then when i is 1 j varies from 2 to 4, producing the next three lines, and so on.

T67. Is every body being compared with every other body? When five bodies are passed to `step_system`, each one should be compared with all four of the others. Does each name show up exactly four times?

T68. How many lines are printed? Since we expect $n \times (n-1)/2$ interactions, there should be ten lines for this set of five bodies.

T69. Another way to count the number of times `Body.interaction` is called is to initialize a variable named `count` to 0 before the outer `for` loop, update the variable inside the inner `for` loop, and print the value after the loop terminates. Add these statements to your code.

T70. Reload your function and run it again. Does it print 10 as the value of `count` when you pass it the five bodies in `b[0:5]`?

T71. What is the value of `count` if you pass the entire solar system to `step_system`?

```
>>> step_system(b, 86459)
sun <=> mercury
sun <=> venus
    ...
neptune <=> pluto
```

Did you get $10 \times 9/2 = 45$? Do all ten names appear exactly nine times?

11.5 Summary

The important concept from the field of computer science introduced in this chapter is the idea of computer simulation. We set up computational experiments in which a set of Python objects served as models of various real-world objects, and then we used equations of motion to move the objects around in their simulated world.

The main goal for these experiments was to learn about issues involved with computer simulation. One issue we looked at was the trade-off between the size of a time step and the accuracy of the simulation. Using a smaller time step, so positions are updated more frequently, leads to a more accurate simulation, but it comes at the cost of a higher number of calculations. The idea was introduced by simulating the movements of a hypothetical robot explorer, and we also saw how the size of a time step affected simulations of a falling object and the motions of the planets.

The key ideas used to compute the motions of bodies in the solar system simulation are the fact that bodies move according to the force of gravity and that forces are additive. To determine which direction a body is being pulled, we can compute the force that is pulling it toward each of the other bodies and then compute the sum of the individual forces.

Concepts and Terminology Introduced in This Chapter

N-body problem	The problem of trying to predict the motions of a group of three or more celestial bodies; no exact solutions are possible, so positions must be computed
acceleration	A change in velocity; for the N-body problem, acceleration is caused by the force of gravity
computer simulation	A method for solving problems like the N-body problem where computational objects represent real-world objects
time step	An amount of time simulated objects move before adjusting their trajectories
vector	In an N-body simulation, a vector is a set of three points, used to represent the (x, y, z) coordinates of a body's location or its heading
accuracy	In a computer simulation, accuracy is determined by how closely the computed results match real-world measurements

---

Forces are represented in a program in the form of vectors, which can be visualized as arrows pointing in the direction of the force. Computing the sum of forces is equivalent to connecting the arrows, drawing the tail of one arrow next to the head of another.

Several factors besides the size of a time step can have an impact on the accuracy and reliability of a computer simulation. The first, of course, is making sure the software is implemented correctly. But even assuming there are no bugs in the application, there are many other ways in which a simulation can give a misleading result. In our solar system simulation, we assumed the mass of each body remains constant, which is a reasonable assumption for planets. But comets lose a small amount of their mass each time they approach the Sun, so an accurate simulation that includes comets will need some way to update their mass. Asteroids have a very irregular shape, and a typical asteroid is very "lumpy," so its mass is not evenly distributed. That means equations that treat the asteroid as a single uniform object may not give the best results.

Accuracy is the central issue in simulation. For the falling watermelon or solar system, it's easy to check accuracy by comparing the program's results with what happens in the real world. For example, our solar system data was taken from observations of the actual positions of the planets on January 1, 1970. To see how accurate our simulation was, we could go to the same database to retrieve the recorded locations of the planets on January 1, 1971, and compare them to our simulated positions. Once we are convinced the program is working correctly, we can start adding other simulated objects, or use the model to predict future locations.

For other applications, it may be much more difficult to determine the accuracy of a simulation. One important factor is deciding which attributes of a real system need to be included in a computer model. For example, to simulate fuel consumption for a car, attributes like the color of the car can obviously be omitted, but other factors, like the size of the tires, and whether they are inflated properly, will have an effect. But adding additional attributes makes the software much more complicated, and with added complexity there is also a greater chance for programming mistakes.

In spite of the difficulty in designing and testing simulations to make sure they are as accurate as possible, modeling and simulation is an important area in applied computer science. One of the main reasons scientists and people in other fields use computer simulations is that they provide an effective way to understand the workings of a complex system.

Exercises

11.1. Exercise 1.2 on page 15 listed several fields where computers play an important role. Choose one of these areas and do some research to find out how computer simulation is used in this area.

11.2. What would you type in a shell session to tell the robot of Section 11.1 to travel in a path that would draw a square?

11.3. Can you figure out how to get the robot to draw a five-pointed star? Hint: Each leg of the star will be the same length, and the angle the robot turns will be the same at the end of each leg.

11.4. Suppose the watermelon simulation is set up so the melon is positioned 75 meters above the ground.

   a) Use the equation $t = \sqrt{2d/g}$ to calculate how many seconds the melon will fall before it hits the ground.

   b) Here is the result of a call to `drop_melon` that simulates the fall with a ¼ second time step:
   ```
   >>> drop_melon(b, 0.25)
   4.0
   ```
   How far off is the computer simulation?

11.5. Copy the picture of five bodies in Figure 11.14 to a piece of paper, or, if you have a drawing application, make a new document and add five circles of the same size and location as the ones in the figure. Sketch vectors that represent the forces acting on each body, similar to those shown in Figure 11.9 on page 365.

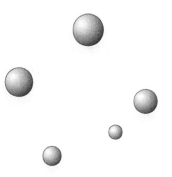

Figure 11.14: *In this drawing, the mass of a body is proportional to its size.*

11.6. Modify the drawing you made for the previous exercise to show the cumulative force acting on each body. For each body draw a new vector that is the sum of the forces pulling that body toward the others.

11.7. Suppose the solar system data is updated to include the locations and headings of 15 moons, so there are now a total of 25 bodies in the dataset. How many force calculations would be made on each time step?

11.8. In the solar system simulation with 25 bodies, how many force calculations would be made if the program runs for 350 time steps?

11.9. Do you think the Sun moves during the solar system simulation? How could you find out?

11.10. The experiments at the end of Section 11.3 were based on bodies named f1, f2, and f3. The three were identical, except f2 and f3 started out in a slightly different position:
```
>>> f2.position.x += 1
>>> f3.position.x += 2
```
What do you suppose would happen if f3 started out halfway between f1 and f2? In other words, what if the x coordinate for f3 was defined by this statement?
```
>>> f3.position.x += 0.5
```
Would its final position be somewhere in the middle between the final positions of f1 and f2? Make a prediction about where the body will go and then run the simulator to check your prediction.

11.11. If you want to use simulation to explore your answer to the previous question, set up some more experiments, using even smaller differences. For example, start f3 at 0.5, 0.25, or 0.125 units to the right of f1. Are the results still chaotic?

Programming Projects

11.12. Write a function named distance_traveled that will compute the total distance for a journey. The argument should be a list of (v, t) pairs, where v is a velocity, in kilometers per hour, and t is the time, in minutes, spent moving at that velocity. For example, this trip starts by traveling 30 km/h for 15 minutes (0.25 hours), then 120 km/h for 30 minutes, and finally 20 km/h for 6 minutes:
```
>>> distance_traveled( [(30, 0.25), (120, 0.5), (20, 0.1)] )
69.5
```
The result shows the total distance was 69.5 kilometers.

11.13. Write a function named drop_time that will compute how long it will take an object to fall a specified distance, using the equation $t = \sqrt{2d/g}$. Test the function by verifying an object will fall 35 meters in 2.67 seconds:
```
>>> drop_time(35)
2.6726124191242437
```

11.14. Write a function that will automatically run the "falling body" experiment. The arguments should be the number of experiments to run, the value to add to the initial x coordinate of the moving body, and the number of time steps to simulate. For example, this command will replicate the experiments at the end of Section 11.3, which ran three experiments, each starting the body 1 unit further to the right, and each running for 50 time steps:
```
>>> run_falling(3, 1, 50)
```

11.15. If you use the `run_falling` function from the previous exercise to run experiments for 100 time steps, it might look like the program stopped because the first body is flung from the system after 69 time steps. The simulation is still running, but since the body is not on the screen it looks like nothing is happening. Modify the function so it stops an experiment as soon as the absolute value of either the x or y coordinate of the falling body is greater than 200. Use the new version of the function to run 10 experiments for 300 time steps each:

>>> *run_falling(10, 1, 300)*

11.16. Write a function named `polygon` that will use the robot to draw a regular polygon on the PythonLabs canvas. The arguments passed to the function should be the number of sides and the length of the perimeter. Here are some example calls and the resulting figures:

`polygon(6,200)`   `polygon(8,200)`

11.17. ◆ Suppose we want to constrain the robot explorer so it stays within the 400×400 meter space where it lands. When it is on a path that takes it to a boundary, it should "bounce off" and continue as if it had hit a wall. Write a function named bounce that will compute two values: the length of time before the robot will reach a boundary if it continues on its current path and the new heading to follow once it gets there. This drawing explains the necessary calculations:

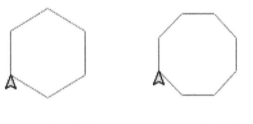

compute the angle to the wall: θ = `radians(90-heading)`

let dx *be the distance to the left or right edge and* dy *the distance to the top or bottom edge*

set tx *and* ty *to the time left before hitting an edge (equations at right)*

$$t_x = |\,d_x/(v \cdot \cos\theta)\,|$$
$$t_y = |\,d_y/(v \cdot \sin\theta)\,|$$

if tx > ty *return* (ty, 180-heading), *otherwise return* (tx, 360-heading)

Here are two examples to test your code. Each test creates a Turtle object (the robot explorer is an instance of the Turtle class) at a specified location and heading and then calls bounce. This test computes how long the robot will move on the first leg of the journey shown above:

>>> *r = Turtle(x = 200, y = 350, heading = 60)*
>>> *bounce(r)*
(10.000000000000004, 120.0)

So after 10 hours the robot will be at the northern boundary, when it should start traveling southeast at a heading of 120°. This test is for the second leg:

>>> *r = Turtle(x = 286, y = 400, heading = 120)*
>>> *bounce(r)*
(13.163586137523469, 240.0)

11.18. ◆ Write a function named pong that will send the robot on a journey that bounces off the edges of the region. The function should take a single parameter, which will define the number of legs in the journey. Start by turning the robot a random number of degrees (anywhere from 1° to 359°). Then for each leg of the journey, use the bounce function you wrote for the previous problem (or download the code from the book website) to figure out how long to move the robot until it hits the edge it is pointed at and which direction to turn it. The drawing below shows a path that was created by a call to pong(21).

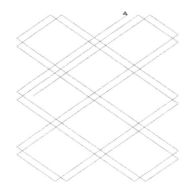

Hint: If bounce returns a value α as the new heading, you want to pass $\alpha - \theta$ to the turn method, where θ is the robot's current heading.

Chapter 12

The Traveling Salesman

A genetic algorithm for a computationally demanding problem

Imagine a situation where some friends or relatives have come to visit, and you want to take them on a tour to show them your campus or your home town. If you are going to be walking, you'd like to find the shortest tour so you don't wear yourselves out and you have some time to relax at each stop on the tour. For a tour with only a few destinations, the most efficient route may be obvious. But if the tour has more than four or five stops you might need to do some planning. A simple strategy, like moving from a location to the one closest to it, might be all you need (Figure 12.1).

As more and more destinations are added to the tour, however, the simple strategy of going to the nearest location you haven't been to yet begins to break down. Figure 12.2 shows a larger map, with 25 destinations, and a tour that starts at the location at the top of the map. After visiting the first few locations, the strategy of moving on to the closest point would select the location shown by the dashed line, but it turns out this choice would not lead to the shortest possible tour.

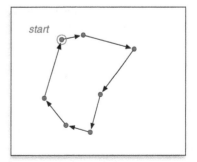

Figure 12.1: *A simple strategy for planning the shortest tour with only a few sites is to start at a random location, and then after each site move on to the nearest location not yet visited.*

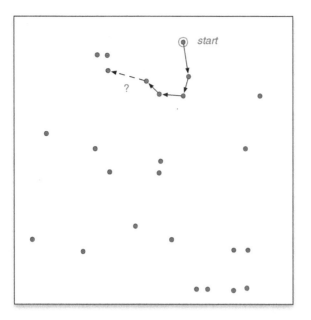

Figure 12.2: *With a larger set of destinations the strategy of moving to the closest point does not always lead to the shortest tour. Before looking at the solution (which is shown in Figure 12.18 at the end of the chapter) can you try to figure out on your own what the shortest tour is? Hint: The best tour does not include the line segment drawn with a dashed line in this picture.*

It's not likely you will ever be called upon to devise a tour that includes 25 destinations. But the same basic idea, of planning the optimal route that connects several different points, is an important part of several real-life situations, and mathematicians and computer scientists have devoted considerable effort to developing algorithms to solve this problem. Some familiar examples are a courier service that wants to plan the most efficient route for their delivery trucks at the start of each day, or a school district that needs to design the shortest route for a set of school buses that pick up and drop off students. A problem that has the same essential structure is figuring out how to connect circuits on the surface of a computer chip. In this case, the "destinations" are electronic components, and the goal is to figure out how to best place metal pathways on the chip to connect the components.

In computer science this problem is known as the **Traveling Salesman Problem**, or **TSP** for short. The input to an algorithm that solves the TSP is a map with a set of cities, where the distance between each pair of cities is defined in a table. The goal is to create the shortest tour that includes all the cities. A tour can start at any city, but it must visit every other city exactly once and then return to the starting point.

In applications of the TSP there are several ways to define the places to visit and the cost of moving from one place to another. For transportation problems, like finding the best route for a delivery truck, the "cities" are simply addresses where a package needs to be dropped off, and the cost can either be the driving distance between points or the expected amount of time it will take to drive between two addresses. For the computer chip layout problem, the "cities" are the components on the chip, and one of the factors in the the overall cost of the design is the total distance along the metal layers that make up the connections between the components.

A simple algorithm guaranteed to find the lowest-cost tour is to systematically check every tour. The objects we will use to represent maps in the projects for this chapter have a method named each_tour that creates every possible tour of the map. All we have to do to find the optimal tour is write a for loop that iterates over all possible tours and keeps track of the one with the lowest cost. This strategy is called an **exhaustive search**: it's exhaustive, in the sense of "exhausting all possibilities," and it's a search because the goal is to find a specific object from among a large number of alternatives.

But before we use the exhaustive strategy on a map with 25 cities it would be a good idea to do a few back-of-the-envelope calculations to estimate how long it might take to find the optimal tour. As we will see in Section 12.2, there are $(n-1)! / 2$ different ways to visit n cities and then return to the starting point. For a small map with 5 cities, there are only 12 different itineraries that visit each city exactly once, so the simple linear search would be a reasonable strategy. But if we add 5 more cities to the map, making a total of 10 cities, all of a sudden there are over 180,000 different tours. For a map like the one in Figure 12.2, with 25 cities, there are 3×10^{23} different tours. To put this in perspective, even if your computer could somehow compute the cost of one billion tours every second, it would take almost 1,000,000 years to iterate over all possible tours of 25 cities!

The Traveling Salesman Problem is another example of a problem that is at the boundary between what is computable and what is not computable. Like finding a winning move in chess (a problem described in Chapter 1), an exhaustive search for the best possible tour of a set of cities seems like a very straightforward computation. The computation is relatively simple when there are only a few cities, but is beyond the limits of what is possible for any machine when the map has more than a dozen or so cities.

This chapter introduces a new type of algorithm, called an **evolutionary algorithm**, to tackle problems like the traveling salesman where an exhaustive search is simply too time consuming. The algorithm begins by making an initial set of random tours and evaluating the cost of each one. A random tour is one that starts at an arbitrary city and then moves on to random destinations not yet visited, until arriving back at the starting point.

While we can't expect any one random tour to be a very good solution, some of them will be much shorter than the others. Evolution comes into the picture when we select the "fittest" of the initial tours and use them as the basis for forming new tours. The idea is to select the most efficient tours, throw out the rest, and then develop new tours that are similar to the "survivors." New tours are the same as they are derived from, except they have small "mutations" that change a tour, perhaps by visiting two of the cities in the opposite order. If we keep iterating this process, of tossing out the most inefficient tours and replacing

---

Factorials

The notation $n!$, pronounced "n factorial," stands for the product of the integers from 1 to n. The factorial function grows even faster than the exponential function introduced in Chapter 5. For every value of $n > 3$,

$$n! > 2^n$$

(see Exercise 12.1). As you can see from the table at right, $n!$ can be a very large number, even when n is as small as 30.

n	$n!$
3	$3 \times 2 \times 1 = 6$
4	$4 \times 3 \times 2 \times 1 = 24$
5	$5 \times 4 \times 3 \times 2 \times 1 = 120$
10	$10 \times 9 \times ... \times 1 = 3,628,800$
15	$15 \times 14 \times ... \times 1 \approx 1.3 \times 10^{12}$
30	$30 \times 29 \times ... \times 1 \approx 2.7 \times 10^{32}$

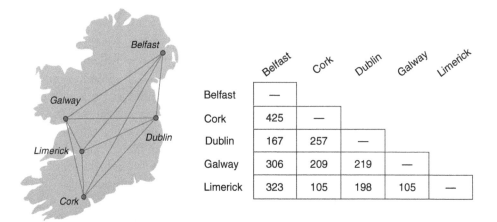

	Belfast	Cork	Dublin	Galway	Limerick
Belfast	—				
Cork	425	—			
Dublin	167	257	—		
Galway	306	209	219	—	
Limerick	323	105	198	105	—

Figure 12.3: *Matrix of driving distances in Ireland.*

them with slight modifications of the better tours, eventually a good solution—and in many cases the best possible solution—will emerge.

The project in this chapter will explore how an evolutionary approach can be used to solve the Traveling Salesman Problem. In the first section we will see how to implement a map as an object in Python and do some initial experiments that emphasize the point that an exhaustive search of all tours is not feasible. We will then see how to create random tours and how to make slight modifications to them so that eventually an optimal tour evolves from a primordial soup of random tours.

12.1 Maps and Tours

In the days when road maps were printed on paper, one could usually find a table of driving distances between major cities. The entry in row x, column y, would have the distance between cities x and y. Because the distance from x to y is the same as the distance from y to x the tables were often presented with a triangular layout, as shown in the example in Figure 12.3.

In mathematical terms, a rectangular table of numbers like the one shown in the figure is called a **matrix**. Driving distances are a special type of matrix, called a **symmetric** matrix, since the distances are the same for each direction.

TSPLab, the PythonLabs module we will use for this project, includes a type of object named Map that defines the distances between cities. When we create a map object, we can either pass it the name of a file that contains distances between real locations, or we can make up a random map when we want to test our algorithm on maps with more cities.

Here is how to create a map to represent the distances between the cities shown in Figure 12.3. This assignment statement creates a variable named m and makes it a reference to a Map object:

```
>>> m = Map(path_to_data('ireland.txt'))
```

As in previous projects, the argument passed to the constructor is a string containing the path to the data file. The path_to_data function, defined in PythonLabs, looks for the data file in the directory where PythonLabs was installed on your system.

We can use Python's index operator to look up the distance between a pair of cities. First we need to know how the cities are labeled on the map. Call a method named cities to get a list of city names:

```
>>> m.cities()
['belfast', 'cork', 'dublin', 'galway', 'limerick']
```

Then to find the distance between any two cities simply use the index operator. If m is a Map object, and x and y are cities on that map, the expression m[x,y] means "the distance between cities x and y." Here is how we can find the distance from Dublin to Galway:

```
>>> m['dublin', 'galway']
219.0
```

A method named make_tour will create an object that represents a tour between two or more cities. One way to create a tour object is to put the names of the cities, in order, in a list and pass the list as an argument to make_tour. For example, an object that represents a three-city tour from Dublin to Galway to Limerick and back is made by this expression:

```
>>> t = m.make_tour(['dublin', 'galway', 'limerick'])
```

Note that this example has only three cities, so it's not a potential solution to the problem of finding the shortest tour of all five cities. However, the ability to make shorter tours like this one will be useful for testing methods that operate on tour objects.

The cost of a tour is the sum of all the distances between destinations in the tour. The total cost of a tour is shown in parentheses at the end of the line when a tour is printed on the terminal. If we ask Python to show the value of t, the variable that refers to the tour we just made, this is what we'll see:

```
>>> t
<PythonLabs.TSPLab.Tour ['dublin', 'galway', 'limerick'] 522.000>
```

Because this map is based on driving distances, the number 522 represents a distance, in kilometers, but in other maps the cost could be based on travel time or ticket price or some other metric. An important note is that the tour is a "return trip" so the total cost of the tour includes the final leg from Limerick back to Dublin (which you can verify by adding the distances between these cities shown in the table in Figure 12.3).

If an algorithm needs information about a tour it can call one of several different methods defined for tour objects. Two of the attributes of a tour that we will be interested in are the path (the list of cities visited) and the cost. Using the example tour object from above:

```
>>> t.path()
('dublin', 'galway', 'limerick')
>>> t.cost()
522.0
```

Tutorial Project

Start an interactive session and load the module that will be used for projects in this chapter:

```
>>> from PythonLabs.TSPLab import *
```

T1. Create the map that has the driving distances (in kilometers) between five cities in Ireland:

```
>>> m = Map(path_to_data('ireland.txt'))
```

T2. A method named `display` will print the distance matrix on your terminal:

```
>>> m.display()
            belfast    cork   dublin   galway limerick
   belfast     0.00
      cork   425.00    0.00
    dublin   167.00  257.00    0.00
    galway   306.00  209.00  219.00    0.00
  limerick   323.00  105.00  198.00  105.00    0.00
```

Note: As a result of the way the city names are represented inside the object they may not be printed in the order you see here or in Figure 12.3, but each pair of cities should be in the matrix.

T3. Use the index operator (square brackets) to find the distance from Cork to Dublin:

```
>>> m['cork', 'dublin']
257.0
```

T4. You should get the same distance if you reverse the order of the cities:

```
>>> m['dublin', 'cork']
257.0
```

T5. Use the index operator to find distances between other pairs of cities. Do the results agree with the values in the matrix that was printed on your terminal?

T6. Make a three-city tour and save it in a variable named `t`:

```
>>> t = m.make_tour(['dublin', 'belfast', 'galway'])
```

T7. Get the total cost of the tour object by calling its `cost` method:

```
>>> t.cost()
692.0
```

T8. Verify this value is correct by summing the individual distances of each leg of the trip:

```
>>> m['dublin','belfast'] + m['belfast','galway'] + m['galway','dublin']
692.0
```

T9. Can you type an expression that makes a tour of the same three cities, but starting in Belfast instead of Dublin?

T10. Can you figure out the expression that makes a tour of the same three cities, but goes in the opposite direction, from Dublin to Galway to Belfast?

Try some more experiments on your own, including some tests with tours of four or five cities. Does it matter which city is the start of the tour, or is the cost the same as long as you visit all the cities in the same order? Do you get the same length tour if the cities are visited in the opposite order?

12.2 Exhaustive Search

The goal for the project in this section is to emphasize just how many tours are possible, even for a small map with as few as 10 cities.

In mathematical terms, the different tours are **permutations**. A permutation is one way of ordering the items in a list or a string. It's easier to see why tours are permutations if we represent a tour by a string of letters, using just the first initial of each city name. For example, the tour that starts in Dublin and goes to Cork, Galway, Limerick, and Belfast can be described by the string "DCGLB." A better tour, which goes to Limerick after Cork, is represented by the string "DCLGB." A tour that visits the cities in alphabetical order is "BCDGL." Each of these tours have all five letters, but the letters appear in a different order. They are simply different permutations of the string "BCDGL" (Figure 12.4).

One way to count the possible permutations is to start by considering the choices for the first city. In the case of the 5-city Ireland tour, there are 5 choices. After we choose the first city, there are 4 choices for the next one to visit. Since there are 5 starting points and 4 choices for the next step from each starting place there are $5 \times 4 = 20$ different ways to start a tour: the ones that start with B ("BC," "BD," "BG," "BL"), those that start with C ("CB," "CD," "CG," "CL"), and so on. Now each of the 2-city tours can continue in 3 different ways, since there are still 3 cities left to visit, so there are $5 \times 4 \times 3 = 60$ ways to make a tour that includes 3 of the 5 cities.

Continuing with this line of reasoning, it should be clear that the total number of strings made from 5 different letters is $5 \times 4 \times 3 \times 2 \times 1$, or 5!, which is 120. In the general case, for any string with n different letters, there are $n!$ different permutations.

The TSPLab module defines a function named each_permutation that we can use to experiment with permutations. Instead of creating all possible tours and returning them in a list, each_permutation will make one string at a time, as we need them. That means we can use a call to each_permutation in a for statement and have it create a new string at

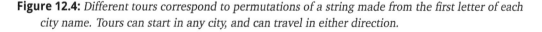

Figure 12.4: *Different tours correspond to permutations of a string made from the first letter of each city name. Tours can start in any city, and can travel in either direction.*

the start of each iteration. This example shows how to print all possible permutations of the
string "ABC":

```
>>> for s in each_permutation("ABC"):
...      print(s)
...
ABC
ACB
BAC
BCA
CAB
CBA
```

The function makes a single permutation and passes it back to the for statement, which
stores it in the variable s and evaluates the expression in the body of the loop. Because
there are $3! = 6$ possible permutations of the three letters, the print statement is executed
six times, eventually printing all different permutations on the terminal.

To relate permutations to tours, first consider that valid tours can start in any city. All that
matters is that a tour visits every city and returns to the starting place. Given a permutation,
we can "rotate" it so it starts at a different letter but otherwise has all the letters in the
same relative order. For example, "ABCDE" follows the same roads and has the same cost
as "BCDEA," it just starts with B instead of A. There are n ways to start a tour, so only
$n!/n = (n-1)!$ permutations correspond to different tours.

◆ Generators

The **each_permutation** function implemented in TSPLab is an example of special type of function that
produces a sequence of values. Instead of making a list of all objects in the sequence, a generator makes
them "on demand," one at a time.

A function that creates random numbers is a good example of when we might want to define a generator.
The exercises in Chapter 7 worked with a function named **prng_sequence** that created lists of random
values. This call produces a list of 1,000 numbers:

```
>>> prng_sequence(81,337,1000)
[0, 337, 634, 691, ... 749, 6, 823]
```

If we don't need all 1,000 numbers at once, we could write a new version of the function that works as a
generator, and then we can put a call to the generator in a for statement:

```
>>> for x in mcrng(81,337,1000):
...      # do something with x
```

When the code in the body of the loop has used all the values
it needs, it can execute a **break** statement to terminate the loop
early, and as a result the program makes only the numbers
used in the computation instead of all 1000 numbers.

*If you want to see how a generator
is implemented in Python download
random/mcrng.py from the book
website.*

```
1    def xsearch(m):
2        "Exhaustive search for the best tour"
3        best = m.make_tour()
4        for t in m.each_tour():
5            if t.cost() < best.cost():
6                best = t
7        return best
```

Figure 12.5: *A function that does an exhaustive search over all possible tours of a map.*

Download: tsp/xsearch.py

Note also that our hypothetical traveling salesman can travel in either direction on the tour. As far as TSPLab is concerned, the tour "ABCDE" is the same as "EDCBA." They both follow the same "roads," but the cities are visited in the opposite order. Because the costs of these tours are the same, we will consider them to be the same solution.

Putting this all together—the fact that $(n-1)!$ permutations correspond to unique paths and that tours can go in either direction—we end up with the formula given in the introduction: an algorithm that searches for the minimal cost tour of a set of n cities potentially has to consider $(n-1)!\,/\,2$ ways of ordering all n cities.

If we have a map object we can call a method named each_tour to generate tours. Like each_permutation, this method is a generator, meaning it will create the tour objects one at a time, as we need them. The main difference is that each_tour creates only those permutations that correspond to different tours. If the map has n cities, the generator makes only $(n-1)!\,/\,2$ orderings, corresponding to the different paths between the cities. Another difference is that each_tour makes tour objects where the paths are different permutations of the list of city names. Assuming the variable m refers to a map of Ireland, this example shows how to print the complete set of all possible tours:

```
>>> for t in m.each_tour():
...     print(t)
...
<PythonLabs.TSPLab.Tour ['belfast', ... 'limerick'] 1329.000>
<PythonLabs.TSPLab.Tour ['belfast', ... 'galway'] 1291.000>
...
```

The print statement is executed 12 times, because for a map with 5 cities the total number of tours is $(5-1)!\,/\,2 = 12$.

With the each_tour generator doing all the hard work it is very easy to do an exhaustive search for the minimal cost tour (Figure 12.5). Start by picking any tour to use as an initial value for the best tour. The statement on line 3 shows one way to make the initial tour: simply call the make_tour method without any arguments. It will return a tour with the cities in alphabetical order, which we can save in a local variable named best.

After we have an initial value for best we can use a for loop that calls each_tour to make every possible tour. Each time we get a new tour, check to see if it has a lower cost than best, and if so make it the new value of best. After the loop terminates simply return the reference to the best tour.

Here is an example of a call to xsearch, using the map of Ireland:

```
>>> t = xsearch(m)
>>> t.path()
('belfast', 'galway', 'limerick', 'cork', 'dublin')
>>> t.cost()
940.0
```

Of course this strategy won't work if we have a map with more than a few cities because there will be far too many tours to check them all. After experimenting with permutations and the each_permutation and each_tour generators we'll start looking into the evolutionary algorithm that will work on larger maps.

Tutorial Project

T11. Make a string for the permutation experiments:
```
>>> s = "1234"
```

T12. Before calling the iterator that generates all permutations of this string, can you predict how many permutations are possible?

T13. Use each_permutation to print every possible ordering of the letters in the string:
```
>>> for p in each_permutation(s):
...     print(p)
...
...
1234
1243
1324
...
```

Do you see how this list is organized? The first group of lines all start with 1, then there are lines starting with 2, and so on. The same sort of pattern occurs within a group, also. Did you notice that the six lines in the group that start with 1 are all the permutations of the digits 2, 3, and 4?

T14. The TSPLab module includes a method named factorial that will compute the factorial of a number. To compute 4!:
```
>>> factorial(4)
24
```

Did you get $4! = 24$ output lines from the previous expression?

T15. Another way to count permutations is to make a list of all permutations and than ask Python for the length of the list:
```
>>> a = list(each_permutation(s))
>>> a
['1234', '1243', '1324', ... '4321']
>>> len(a)
24
```

By looking at the pattern of the strings generated by each_permutation you should be convinced there are $n!$ different ways to arrange the letters in a string with n characters.

The next set of experiments will use `each_tour` to carry out an exhaustive search for the optimal tour. If you are not in the same interactive session from the previous section retype the statement that makes the object for the map of Ireland:

```
>>> m = Map(path_to_data('ireland.txt'))
```

T16. Call a method named `size` to get the number of cities in the map:

```
>>> m.size()
5
```

T17. A method named `ntours` will compute $(n-1)! / 2$, the number of tours in a map of a specified size. To compute the number of tours in the map of Ireland:

```
>>> ntours(m.size())
12
```

T18. Verify the claim that calling `make_tour` with no arguments returns a Tour object with the cities in alphabetical order:

```
>>> m.make_tour()
<PythonLabs.TSPLab.Tour ['belfast', ... 'limerick'] 1329.000>
```

T19. Type the `for` statement that uses `each_tour` to generate every tour:

```
>>> for t in m.each_tour():
...     print(t)
...
<PythonLabs.TSPLab.Tour ['belfast', ... 'limerick'] 1329.000>
...
```

It's not as easy to see a pattern in these tours as it was to see a pattern in the permutations of strings. The `each_permutation` and `each_tour` functions use different algorithms for reordering the items in a list, but you should see all twelve tour objects.

T20. Download the definition of `xsearch`, or use your IDE to create a new file named `xsearch.py` and type in the definition in Figure 12.5. If you create your own file you can copy and paste several of the statements from your interactive session.

T21. Load your definition of `xsearch` and call it with the map of Ireland:

```
>>> xsearch(m)
<PythonLabs.TSPLab.Tour ['belfast', ... 'dublin'] 940.000>
```

Is this the lowest-cost tour from the list printed previously in your terminal window?

If you want to try some more experiments to get a sense of how big a number $n!$ can be, even for strings with as few as 15 letters, try the following optional project, otherwise you can move on to the next section.

Before you do these exercises, however, a word of caution: figure out how to abort a computation in your interactive session. If you are using a Unix terminal window, for example, you can hit ^C to stop Python and return to the top-level prompt.

♦ Make a string with 15 characters:

```
>>> s = "ABCDEFGHIJKLMNO"
```

♦ Find out how many permutations will be printed if you use `each_permutation` as an iterator:

```
>>> factorial(len(s))
1307674368000
```

In scientific notation, that number is 1.3×10^{12}, or about one trillion.

♦ You can ask Python to print all the permutations of the new 15-letter string, but be ready to interrupt the computation by typing ^C:

```
>>> for p in each_permutation(s):
...     print(p)
...
ABCDEFGHIJKLMNO
ABCDEFGHIJKLMON
ABCDEFGHIJKLNMO
...
KeyboardInterrupt
```

As Python was printing these strings, did you see how the computation was progressing? The letters on the right side of the string were changing places, but the further to the left you looked, the slower the letters were changing.

12.3 Random Search

It should be clear by now that for all but the smallest sets of cities it is going to be impossible to find the shortest path connecting all the cities by evaluating every possible tour. The strategy used by the evolutionary algorithm outlined in the introduction is to start with a set of random tours and then apply small adjustments until the optimal tour is found. To prepare for experiments with the evolutionary algorithm, this section will explain in more detail what it means to make a random tour of the cities on a map.

In real-world terms, a salesman who wants to make a random tour of his cities could write the name of each city on a separate 3 × 5 card and then shuffle the cards. We can use this same basic idea in a search algorithm by making tours that are a random permutation of the list of city names. The method named permute, first described in Chapter 7, rearranges the items in a list and puts them in a random order. If we want to make a random tour, all we need to do is get a list of city names, pass them to permute, and then pass the result to make_tour:

```
>>> a = m.cities()
>>> permute(a)
>>> t = m.make_tour(a)
```

As a convenience, the make_tour method will carry out these steps for us. All we need to do is pass the the string 'random' in a call to make_tour and it will create a random tour:

```
>>> m.make_tour("random")
<PythonLabs.TSPLab.Tour ['belfast', ... 'limerick'] 1374.000>
```

The algorithm we will look at in this section tries to find the optimal tour simply by generating lots of random tours and selecting the one that has the lowest cost. We don't really expect the algorithm to find the best tour, or even a reasonably good tour. But this method does establish a baseline. When we start testing the evolutionary algorithm in the next section, we will want to know how well it performs, and one way to do this is to compare the tours it produces with the results of random searches.

In a random search we take a "random sample" of all the possible tours and keep track of the best one we find. The Python code is very similar to that for the exhaustive search function. The difference is that instead of using each_tour to examine all tours, we just

```
1    def rsearch(m, n):
2        "Random search for the best tour"
3        best = m.make_tour("random")
4        for i in range(n-1):
5            t = m.make_tour("random")
6            if t.cost() < best.cost():
7                best = t
8        return best
```

Figure 12.6: *A function that examines a random sample of possible tours of a map. The arguments are m, a reference to the map, and n, the number of samples to make.*

Download: tsp/rsearch.py

use a for loop with a range expression to create a specified number of random tours. The definition of a function named rsearch that implements this new algorithm is shown in Figure 12.6.

To test this algorithm and the evolutionary algorithm in the next section we need to make bigger maps. What we want are maps that have enough cities so there are too many tours to be able to find the best one by checking all possible orders, but small enough to be able to display a map and understand at a glance whether or not a tour is efficient.

The technique for making larger maps is the same one used in earlier chapters to make large lists of numbers for testing searching and sorting algorithms. If, instead of passing a file name to Map.new, we give it an integer n, we will get back a new map with n cities placed at random locations. For example, this assignment statement will make a map with 15 random cities and save it in the variable named m:

```
>>> m = Map(15)
```

In these maps the names of the cities will just be the numbers from 0 to $n - 1$:

```
>>> m.cities()
[0, 1, 2, 3, 4, 5, 6, 7, 8, 9, 10, 11, 12, 13, 14]
```

That means a random tour will simply be a permutation of these numbers:

```
>>> m.make_tour("random")
<PythonLabs.TSPLab.Tour [4, 6, 2, ... 5, 10] 3154.441>
```

A method named view_map displays a map on the PythonLabs canvas by drawing a small gray circle for each city (Figure 12.7). For these maps, the x coordinate is the distance from the left edge of the map, and the y coordinate is the distance from the top of the map, meaning the origin is in the upper-left corner. The distance between a pair of cities is the geometric distance defined by their map coordinates.

After displaying the map, we can call view_tour to see the connections between the cities defined by a particular tour. Each time we call view_tour, any tour that was displayed previously will be erased, and the path used in the new tour will be drawn on the map.

The experiments in this section will use a version of rsearch that is included with TSPLab. When we call this function, it will update the map on the screen each time it finds a new tour with a lower cost, and when it is done, it will return the tour object corresponding to the best tour it found.

Tutorial Project

T22. Start by making a map with 15 randomly placed cities:
```
>>> m = Map(15)
```

T23. Display the map on the PythonLabs canvas:
```
>>> view_map(m)
```

T24. A method named coords will give the (x, y) coordinates of a specified city. Type these expressions to see the locations of the first cities in your map:
```
>>> m.coords(0)
(390, 215)
>>> m.coords(1)
(173, 250)
```
As is always the case for experiments based on random numbers, the actual values you get will be different than the ones shown here. Locate cities 0 and 1 on your canvas, and verify the coordinates that were printed in your terminal window are accurate (remember the origin (0,0) is in the upper left).

T25. Ask Python to print the distance between cities 0 and 1:
```
>>> m[0,1]
219.80445855350615
```
Does the value you got seem accurate? Check the value printed by Python using the equation

$$d = \sqrt{(x_0 - x_1)^2 + (y_0 - y_1)^2}$$

T26. To see how a random tour is created, first make a list containing the names of the cities:
```
>>> a = m.cities()
```

T27. Call permute a few times:
```
>>> permute(a); print(a)
[0, 10, 9, 3, 11, 1, 13, 8, 6, 4, 5, 12, 7, 2, 14]
>>> permute(a); print(a)
[8, 14, 6, 1, 4, 0, 13, 12, 10, 3, 7, 11, 5, 9, 2]
```

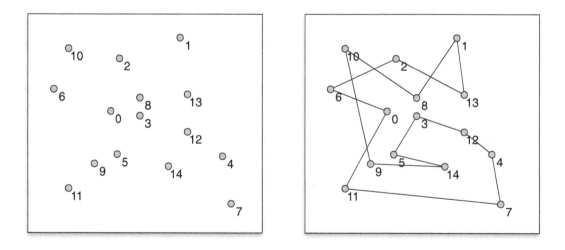

Figure 12.7: *A map of 15 cities placed at random locations by a calling Map(15), and a tour made by calling* m.make_tour("random")*.*

T28. Make a random tour of the 15 cities on your map m and print the path for the tour:
```
>>> t = m.make_tour("random")
>>> t.path()
(11, 13, 3, 10, 1, 0, 12, 7, 5, 14, 9, 8, 2, 4, 6)
```

T29. Tell Python to draw this tour on the map:
```
>>> view_tour(t)
```

The view_tour method should have connected the cities that were already on the canvas. Does the path shown agree with the one printed on your terminal window for the previous exercise?

Try making a few more random tours and displaying them on the canvas. Each call to view_tour should erase the previous tour. You should see a series of very different-looking tours, with no relationship between any tour and the one that follows it.

T30. Call the TSPLab version of rsearch, telling it to make 100 random tours. If you pass an optional argument named pause the function will wait for a short time so you can see the best tour each time it is updated:
```
>>> rsearch(m, 100, pause = 0.5)
<PythonLabs.TSPLab.Tour [6, 10, 7, ... 13, 5, 9] 2092.258>
```

T31. Repeat the experiment, but change the 100 to 1,000 so the search looks at 1,000 random tours:
```
>>> rsearch(m, 1000, pause = 0.5)
<PythonLabs.TSPLab.Tour [6, 12, 10, ... 13, 0, 8] 1795.936>
```

What you should see is that it's pretty easy for rsearch to find a better tour early on, but as the search continues, it becomes harder and harder to find a better tour, as indicated by the longer intervals between updates on the canvas.

Try running the rsearch algorithm a few more times, looking at as many as 10^5 tours (written as 100000 in Python) or even 10^6 tours (written as 1000000). Can you see how the best tour on the screen generally improves? Also, can you see that improved tours are harder to find the longer the search goes on?

You should also begin to notice something about the better tours as they are displayed. In general, the fewer the number of roads that cross over other roads, the lower the cost. In fact, the optimal tour will be a loop where no road crosses over any other road in the tour.

12.4 Point Mutations

The random search algorithm of the previous section is not likely to find the best tour on a map of 15 cities. There are $14!/2 \approx 3 \times 10^{10}$ different tours, so even if we let rsearch make 10^6 random tours, the probability of its picking the best tour is about $1/30,000$.

It may seem silly to even try to find the best tour just by making random guesses, but the reason we went through the exercise is that it sets the stage for the algorithm described in this chapter, which also works by generating lots of random tours. The difference between the random search algorithm and the evolutionary algorithm is that the latter tries to improve upon the tours it finds. Rather than simply record the fact that it has found a good tour and going back to guessing, this new algorithm will make minor changes to the good tours to try to make them even better.

Biologically Inspired Algorithms

The genetic algorithm described in this chapter is an example of what computer scientists call a *biologically inspired algorithm*.

The goal for these algorithms is not to simulate real biological systems, but simply to use data structures and operations that are similar to those found in nature to solve computational problems.

Biology	*Computation*
organism	tour object
DNA	sequence of city names
population	array of tour objects
natural selection	lower-cost tours survive
point mutation	rearrange order of cities
crossover	combine subtours

It is the idea of making a series of modifications to tours that gives rise to the term "evolutionary algorithm." The approach we are going to look at is a particular kind of evolutionary algorithm known as a **genetic algorithm**. The name comes from the fact that making a series of small changes to tours is reminiscent of the way changes in genes are passed from one generation to the next, as in a culture of yeast growing in a Petri dish.

At any one time, we will be working with a set of tours, which, to continue the analogy to genetics, corresponds to a small population. The main step of the algorithm weeds out the most expensive tours—the less fit organisms—and replaces them with new tours that are based on minor modifications of the least expensive tours. In other words, new tours are descended from those that survived the weeding out process in our virtual Petri dish.

In our experiments, a list of city names will serve as the "DNA" that defines a single tour. The technique for making a slight change to a tour is called a **point mutation**, based on the terminology used in molecular biology to describe the smallest possible change to the DNA in a real gene. In the TSP, a point mutation corresponds to selecting a city and then exchanging its place with the city that follows it.

As an example of a point mutation, suppose a tour named t visits five cities in this order:

```
>>> t.path()
('belfast', 'dublin', 'limerick', 'cork', 'galway')
```

The city in location 2 of this list is Limerick (remember the first item in an list is at location 0). A point mutation at location 2 would exchange Limerick with Cork, the city that follows it in this tour.

The method that makes this change is named mutate, and this is how we would call the method to make a mutation at location 2 in the path for tour t:

```
>>> t.mutate(2)
>>> t.path()
('belfast', 'dublin', 'cork', 'limerick', 'galway')
```

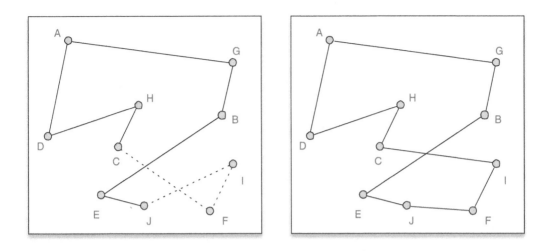

Figure 12.8: *The tour on the left has a small twist, where the road from C to F crosses the road from I to J. The tour can be improved by changing the segment CFIJ to a new segment CIFJ, as shown on the right. The change is the result of a single mutation that switches the position of the letter F with the letter that follows it, transforming the string "...CFIJ..." into a new string "...CIFJ..."*

Another example of a point mutation is shown in Figure 12.8. On this map, the city names are single letters, and the tour on the left includes a segment that visits cities C, F, I, and J, in that order. One can tell at a glance by looking at this map that the tour is not optimal, because the "road" that connects C to F intersects the road from I to J. A more efficient tour, shown on the map on the right, visits these cities in the order C, I, F, J, and because these paths do not intersect each other, the total length of this segment of the tour is shorter. The change in the list that represents this tour, from [C,F,I,J] to [C,I,F,J], is a single point mutation that swapped F with the city that followed it.

Figure 12.8 shows the general case of what we hope to achieve with point mutations. At any stage during the search for an optimal tour, there will be tours that have one or more segments like the one shown on the left, where the tour crosses over itself. If we imagine a tour as being a string or rubber band, the change that improves the tour is like untwisting or removing a "kink" from the tour. The optimal tour will be a single loop that has no kinks.

Of course there will be several false steps along the way. Not every mutation will improve a tour, and in fact it's probably the case that most will increase the cost, not decrease it. But tours with higher costs will eventually be weeded out, and it's the occasional improvement that will survive and be passed on to later iterations of the algorithm.

Tutorial Project

T32. Make a map object for the five cities in Ireland:
```
>>> m = Map(path_to_data('ireland.txt'))
```

T33. Make a new tour with the cities in this order:
```
>>> t = m.make_tour(["dublin","cork","limerick","belfast","galway"])
```

T34. Print the cost of this tour:
```
>>> t.cost()
1210.0
```

T35. Draw the map and tour on the canvas:

```
>>> view_map(m)
>>> view_tour(t)
```

You should see a map with five circles, and a tour that looks like a rubber band with a single "twist" in it.

T36. Note that Belfast is in location 3 in the list that represents the tour:

```
>>> t.path()
('dublin', 'cork', 'limerick', 'belfast', 'galway')
```

T37. If we exchange the order that this tour visits Belfast and Galway we will remove the twist. Call mutate to exchange the cities at locations 3 and 4:

```
>>> t.mutate(3)
```

T38. Call view_tour again to update the canvas to show the mutated tour:

```
>>> view_tour(t)
```

Notice how the mutation improved the tour. It is now a simple loop where no line segment intersects any other segment.

T39. Print the new path and cost:

```
>>> t.path()
('dublin', 'cork', 'limerick', 'galway', 'belfast')
>>> t.cost()
940.0
```

Do you see how the cities at locations 3 and 4 changed places? And that the cost was reduced from 1210 to 940?

Try some more experiments on your own, adding a point mutation at selected places in the list, until you are sure you understand what the mutate method does. Note that some mutations will make the cost higher; not every mutation is beneficial.

12.5 The Genetic Algorithm

The previous sections showed how to represent maps and tours as objects in Python, and we saw how to create random tours and how to slightly modify a tour. We're now ready to see how we can make a "population" of tour objects and how to implement a function that gradually changes tours until a good solution evolves.

The Python code for a function named esearch (the name stands for for *evolutionary search*) is shown in Figure 12.9. The arguments are the map object, which defines the cities that will be in the tour, the number of generations we want to simulate, and the number of tours we want to include in the population.

An obvious choice for how to represent the population is to create a list of Tour objects. The helper function named init_population, called from line 5, does just that. We need to pass it a reference to the map so it knows how to construct the random tours, and we also pass it the number of tours we want.

The main loop of esearch is an iteration that repeats the same operations for a specified number of generations. On each generation, some of the high-cost tours are removed from the population and replaced by new tours that are slight modifications of the survivors.

At the start of each iteration the tours are sorted. We will want the lowest-cost tour to be in location 0. The other tours will be arranged so that the more a tour costs the closer it will be to the end of the list. We can call sort (the built-in method defined in Python's list

```
1   from PythonLabs.TSPLab import *
2
3   def esearch(map, ngen, popsize):
4       "Evolutionary search for the best tour"
5       pop = init_population(map, popsize)
6       for i in range(ngen):
7           pop.sort(key = Tour.cost)
8           select_survivors(pop)
9           ns = compact_population(pop)
10          rebuild_population(pop, map, ns)
11      return pop[0]
```

Figure 12.9: *The top-level evolutionary search algorithm keeps the best tours in each generation and replaces the less fit tours by slightly different copies of the survivors.*

Download: tsp/esearch.py

class) to arrange the list for us if we supply an optional parameter that tells Python how to compare the items.

As an example of how to specify a sorting criterion, suppose we want to sort a list of strings, but instead of sorting alphabetically, which is the default, we want to sort by length, so the shortest strings are at the front of the list. Here is what the call to sort looks like:

```
a.sort(key = len)
```

The parameter tells Python to use the len function to compute the sort key, the value used to compare two items.

To sort our list of Tour objects according to cost we just have to pass the name of a function that looks up the cost of a tour. Recall that if t is a reference to a Tour object, a call to t.cost() returns the cost of the tour. The method named cost is defined inside the class named Tour, so the full name for this function is Tour.cost, and that is what we pass to sort (line 7 in Figure 12.9).

When removing tours from the population, an important question is how to decide which ones to delete from the list. One approach would be to remove a fixed percentage. For example, suppose the goal is to remove half the tours. If there are 10 tours in the population, we would keep the tours in locations 0 through 4 and delete those in locations 5 through 9.

A more subtle approach is one that mimics what happens in nature, where random selection plays a role. The goal is to allow some of the less fit tours to survive and to remove some of the stronger tours. The idea is that even if a tour has a relatively high cost, after a few generations one of its descendants might be radically different, and that descendant might be a better solution than one derived from one of the best tours in the founding population. This strategy helps maintain the "genetic diversity" of the population, and in most genetic algorithms it is an important strategy to help find the best solution.

The technique used in the PythonLabs implementation of the select_survivors helper function is to compute the probability of deleting a tour as a function of its location in the list. Let p be the size of the population. When iterating over the list, the tour in location i is deleted with probability i/p. For example, with $p = 10$ the first tour is always kept because $0/10 = 0$. The probability of removing the second tour is $1/10$, and so on (Figure 12.10). As a result, because the tours are sorted by cost, from lowest to highest, the best tour in any

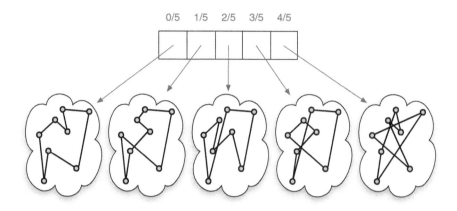

Figure 12.10: *The Tour objects in the population list are sorted according to cost, with the lowest-cost tours at the front of the list. The probability of removing a tour is based on its position in the list. For example, the probability of removing tour 3 is* $3/5 = 0.6$.

generation is always preserved, but the others are deleted with a probability that depends on their relative fitness.

When a tour is removed from the population the list does not shrink. Instead, the list is updated so there is a None object in place of the deleted tour. The helper function named compact_population rearranges the list by moving all the survivors to the front and all the None objects to the end (Figure 12.11). The function returns the number of tours that still remain. Note that if the list has q surviving Tour objects, they will be in locations 0 to $q - 1$, and the empty locations will be from q up to $p - 1$ (where p is the population size).

The final step, implemented by the rebuild_population helper function, is to replace all the None objects by new tours. The version we will experiment with in this section just chooses tours left from the previous generation at random. It "clones" these tours and calls mutate to make a random change somewhere in the path.

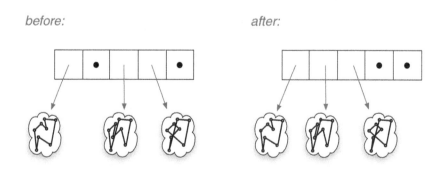

Figure 12.11: *A call to* compact_population *will move references to Tour objects to the front of the list and None objects (shown here as black dots) to the end of the list.*

Figure 12.12: *The helper functions that implement the genetic algorithm update the canvas to show the progress of the search for the best tour. The lowest-cost tour found so far is displayed on the map, and a histogram shows the cost of each tour in the population. In this snapshot, light gray bars correspond to tours that have been deleted; they will be replaced on the next call to* `rebuild_population`.

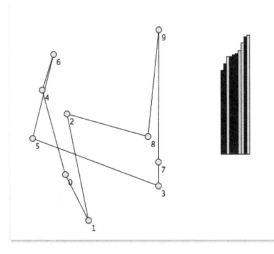

As we work through the tutorial project for this section, we will call each of the helper functions individually, to get a sense of how they work, and then call the PythonLabs implementation of esearch to run complete experiments that search for the optimal tour in maps of varying sizes.

The version of esearch implemented in TSPLab will display the progress of the algorithm on the canvas. If a map has been displayed previously, a call to init_population will draw a histogram, or bar chart, using one bar to represent each tour (Figure 12.12). The height of a bar is proportional to the cost of the corresponding tour.

The project will run experiments that compare tours found by esearch with tours found by calling the random search method rsearch for the same map. We'll make slightly larger maps than we did in the last section and have each algorithm look for the best tour in a map with 20 cities.

A natural question is, how many tours will be created by the esearch method? If it makes just as many tours as rsearch, and the results aren't that much better, then we might as well not bother, and just stick with a random search. What we will see is that not only does esearch produce much better solutions, it does so by looking at far fewer tours.

On each iteration, esearch replaces about half of the population. The tours at the front of the list have a low probability of being deleted, but this is balanced by the fact that tours at the end of the list have a higher probability of being removed. The number of tours deleted by the esearch function is thus approximately one half of the population on each iteration, and these tours are replaced with new tours by the rebuild_population method. At the end of the search, given a population of size p evolving over g generations, we can expect the number of tour objects made by the genetic algorithm to be roughly $(p/2) \times g$.

The genetic algorithm gives noticeably better results than the random search, but there is still room for improvement. With the strategy of making a series of small adjustments to tours, the small "kinks" in the paths will be smoothed out, but there will be many situations where evolution won't make any progress unless it is able to make more dramatic changes. We'll come back to this idea in the next section, where we introduce a second type of mutation that makes much larger changes to a tour.

Tutorial Project

T40. To test sorts with alternative search keys make a list of strings and print the list:

```
>>> a = RandomList(10, "cars")
>>> a
['infiniti', 'skoda', 'bmw', ... 'peugeot']
```

T41. If you sort the list with the default criterion the strings will be sorted alphabetically:

```
>>> a.sort()
>>> a
['acura', 'bmw', 'citroen', ... 'skoda']
```

T42. This call will sort the strings by length:

```
>>> a.sort(key = len)
>>> a
['bmw', 'acura', 'mazda', ... 'rolls-royce']
```

You should see short strings at the front of the list, longer ones at the end.

T43. Make a new map with 20 cities and display it on the canvas:

```
>>> m = Map(20)
>>> view_map(m)
```

T44. Call the init_population helper function to make a list of 10 random tours:

```
>>> pop = init_population(m, 10)
```

T45. The canvas should now have a histogram with 10 bars, one for each tour, along with the path for the first tour in the list.

T46. If you ask Python to print the list you will see the complete description of all 10 tour objects:

```
>>> pop
[<PythonLabs.TSPLab.Tour [6, 19, ... 1, 11] 3994.415>]
```

T47. That list is hard to read, so there is a function named summary in TSPLab that will print only the cost of each tour:

```
>>> summary(pop)
[5052, 4876, 4534, 4875, 4707, 5027, 4449, 4063, 5043, 3994]
```

Do these costs correspond to the heights of the bars in the histogram on the canvas?

T48. Sort the population by the cost of the tours:

```
>>> pop.sort(key = Tour.cost)
```

T49. Now when you print the population summary it should show the tours in order of increasing cost:

```
>>> summary(pop)
[3994, 4063, 4449, 4534, 4707, 4875, 4876, 5027, 5043, 5052]
```

The bars in the histogram should also be arranged in order of increasing height.

T50. Call the select_survivors helper function (it's part of TSPLab, and was defined when you imported TSPLab):

```
>>> select_survivors(pop)
```

T51. That call should have replaced roughly half the tours with None:

```
>>> summary(pop)
[3994, 4063, 4449, 4534, None, 4875, None, None, 5043, None]
```

Are most of the None objects at the end of the list? Have these tours been deleted from the histogram?

T52. Try making some more populations, using sizes of 20 or 30, and repeating the experiments that select random tours and print the results.

T53. Call the `compact_survivors` helper and save the result in a variable named ns (for "number of survivors"):

```
>>> ns = compact_population(pop)
```

T54. Print the summary again:

```
>>> summary(pop)
[3994, 4063, 4449, 4534, 4875, 5043, None, None, None, None]
```

Are all of the None objects in your list at the end of the list? On the canvas all the empty locations should be on the right side of the display.

T55. Print the value of ns:

```
>>> ns
6
```

This is the expected answer for the population shown in these examples. Notice how locations 0 to 5 in the list are references to Tour objects and locations 6 to 9 are None.

T56. Call the `rebuild_population` helper function:

```
>>> rebuild_population(pop, m, ns)
```

Note that the number of survivors in pop is passed as an argument to this helper function.

T57. Now when you print the population all the None objects should be replaced by new tours:

```
>>> summary(pop)
[3994, 4063, 4449, 4534, 4875, 5043, 5127, 4426, 3880, 4080]
```

Notice also how the display shows a full population again with new tours on the right.

T58. This for loop will print the paths of all the tours from the previous generation:

```
>>> for i in range(ns):
...     print(pop[i].path())
...
(5, 3, 7, 2, 17, ... 15, 1, 11)
...
```

T59. Print the path for the first new tour:

```
>>> pop[ns].path()
(6, 0, 12, 10, ... 17, 2, 16)
```

Can you find the "parent" of this new tour? If you look at the list of survivors printed from the previous exercise, you should find one that is identical to the new one except for one place in the path where two cities are in the opposite order. You should also be able to find the "parents" of all the remaining new tours.

We're now ready to run some experiments with esearch. At the start of each experiment, esearch calls init_population to make a set of random tours. It then repeatedly calls the helper functions to update the population. As we will see, we can also pass it options to tell it to vary the population size and control a number of other parameters.

T60. Type this expression to have esearch run for 100 generations to find a tour for map m:

```
>>> esearch(m, 100, 10)
<PythonLabs.TSPLab.Tour [0, 3, 9, ... 13, 19, 6] 2396.557>
```

Notice that esearch also displays some text on the canvas to tell you how the search is progressing.

T61. Call esearch a few more times, and record the costs of the tours you get.

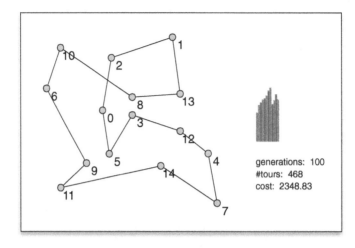

Figure 12.13: *A tour produced with point mutations only. These tours typically have one or two large loops that are hard to "untwist" by exchanging a city with the one next to it on the tour.*

One of the pieces of data displayed on the canvas is the number of tours created by esearch. If the population size is $p = 10$ and the algorithm runs for $g = 100$ generations, esearch should make roughly $(p/2) \times g = 500$ tours. Is that about how many you saw next to #tours on the canvas?

Next let's see how rsearch will do if it looks at the same number of tours.

T62. Use the random search algorithm to find the best tour out of 500 random samples:
```
>>> rsearch(m, 500)
<PythonLabs.TSPLab.Tour [10, 9, 3, ... 13, 0, 18] 3372.460>
```

T63. Repeat the previous expression a few times, and record the costs of the tours you get from rsearch.

What do you conclude from these experiments? Is esearch generally producing better tours than rsearch when both are generating the same number of tours?

The number of tours in a population has a big effect on the search. When there are more tours to choose from, the evolutionary algorithm has a better chance of finding a high-quality tour.

T64. Run the genetic algorithm for 100 generations on a population of size 50:
```
>>> esearch(m, 100, 50)
<PythonLabs.TSPLab.Tour [12, 17, 15, ... 5, 11, 7] 2145.054>
```

T65. Repeat the previous expression a few more times. Does a larger population size give better results?

How would you characterize the tours made by esearch? In most cases, they should appear less jumbled than the random tours found by rsearch, but they will still have a few large "twists" or "kinks" where one segment of the path crosses over another one (Figure 12.13)

♦ Try some more experiments on your own, perhaps with a map of 30 or 40 cities, or with a larger population size or running for more generations. In most cases, the genetic algorithm will make lots of improvements at first, and then make fewer and fewer improvements as the search continues, and eventually seem to get stuck on a tour with one or more large twists.

12.6 Crossovers

Most of your experiments with `esearch` probably ended with a tour that looks like the one in Figure 12.13, where there are several large regions that look like they might be part of the optimal tour along with other areas where there are twists or "kinks" in the path. The tour in Figure 12.13 would be improved if a mutation exchanged cities 2 and 8, making the tour one large loop with no roads that cross each other. But there are other cities in the path between 2 and 8. The `mutate` method cannot exchange 2 and 8 because they are not right next to each other in this tour.

The reason it is so difficult for `esearch` to work out this sort of wrinkle is that smaller changes that are a step in the right direction, such as exchanging 2 with 1 to make the loop one path shorter, often increase the overall cost, so any tour that has the change doesn't survive long enough to pass its "genes" to succeeding generations.

One approach to solving this problem is to use a second type of mutation, one that makes larger changes in the "DNA" of a tour. A common type of mutation, and one that is implemented in TSPLab, is based on another process that occurs naturally in real DNA in real cells.

In genetics, a **crossover** mutation occurs when two chromosomes (which are very long strands of DNA) break apart. When the pieces are brought back together, to form whole chromosomes again, there is often some mixing, and as a result new chromosomes can have a mixture of parts from the original chromosomes. This is the source of genetic recombination, and it is the reason we all have a combination of traits from both of our parents.

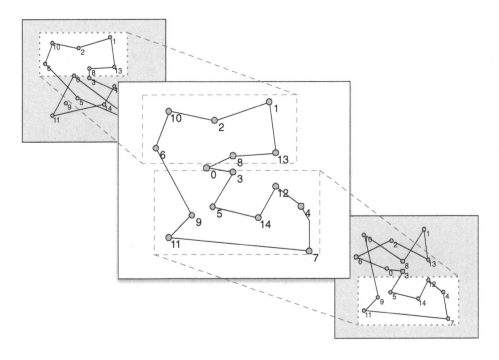

Figure 12.14: *In a crossover mutation, portions of two existing tours (shown in gray) are spliced together to form a new tour. The new tour often combines the best portions of its "parents."*

In a genetic algorithm for the TSP, a crossover is basically a "cut-and-paste" operation. Part of the list of city names from one tour is appended to part of the list from a second tour, resulting in a third tour that has large pieces from each of the original tours (Figure 12.14). We won't go into the details of how this operation works; for the tutorial project, it is sufficient to know that it is implemented by the same make_tour method that creates random tours and tours based on point mutations. If you are interested in how this operation is implemented, you can find a description in the TSPLab documentation section of the Lab Manual.

Having a second type of mutation to use when making new tours raises a new question, however: when the rebuild_population method creates new tour objects, how does it decide whether to make the new tour with a point mutation or with a crossover? The answer is that rebuild_population basically just "flips a coin." Each time it goes to make a new tour, the method uses a random number generator to decide which type of mutation to apply.

Figure 12.15 shows the definition of a simplified version of the function that rebuilds the population. The third argument passed to the function is the number of survivors from the previous generation. As we saw in the previous section, this value is a dividing line in the population list: locations 0 through ns − 1 have references to tours, and locations from ns to the end of the list are empty. The for loop iterates over all the empty locations, calling make_tour to create a new object to store in each location.

A new detail shown in this function is that we can pass a new type of argument to the function that creates Tour objects. In previous projects, we either passed a list of city names to get a Tour object with a specified path, or passed the string 'random' if we wanted a random tour. The calls on lines 6 and 10 in Figure 12.15 show make_tour also makes tours that are slight variations of existing tours. If we pass the string 'random' and a reference to another tour we will get back a copy of that tour with a point mutation at a random location in the path. If we pass 'cross' and reference to two other tours the new tour will be a "genetic recombination" of the other tours.

```
1   def rebuild_population(pop, map, ns):
2       "Add new tours to the end of pop so all locations contain Tour objects"
3       for i in range(ns, len(pop)):
4           if select_with_probability(0.75):
5               mom = pop[ randint(0, ns-1) ]
6               kid = map.make_tour( 'mutate', mom )
7           else:
8               mom = pop[ randint(0, ns-1) ]
9               dad = pop[ randint(0, ns-1) ]
10              kid = map.make_tour( 'cross', mom, dad )
11          pop[i] = kid
```

Download: tsp/rebuild.py

Figure 12.15: *Python implementation of the rebuild_population helper function.*

The `if` statement on line 4 determines whether to do a point mutation or a crossover. A new helper function named `select_with_probability` uses a random number generator to make the decision. This helper returns `True` with the specified probability. In this case, 75% of the new tours will be point mutations: the assignment on line 5 selects a random tour from the first part of the population list, and the statement on line 6 creates the new tour. The other 25% of the time the assignments on lines 8 and 9 select two survivors at random, and these tours are passed to `make_tour` to create a new tour using a crossover.

In the TSPLab implementation of `rebuild_population` the probability of each type of mutation is defined by a parameter called the **distribution**. By default, if we call `esearch` without any extra parameters, it uses a distribution that forces all the mutations to be point mutations, as we saw in the previous section. If we want to include crossovers, the easiest way is to use a distribution named `'mixed'`, as in this example:

```
>>> esearch(m, 200, 50, dist = 'mixed')
```

There are several different sorts of distributions defined in TSPLab, and it is possible to define new ones. For example, we can tell `rebuild_population` to make 90% point mutations and 10% crossovers (see the Lab Manual for details).

The project for this section will carry out some experiments on maps with 20 or more cities, using various combinations of the different types of mutations. What we will see is that both types of mutations are necessary. Without crossovers, as we have seen, the evolutionary search algorithm cannot make changes it needs to get "unstuck" from solutions that have large twists. On the other hand, using only crossovers will lead to wild changes almost like those seen in a random search, and the algorithm won't be able to "fine-tune" any solutions that are almost optimal. With a combination of both types of mutations, tours with very low costs, and perhaps even the optimal tour, will eventually emerge.

Tutorial Project

T66. If you still have the map with 20 cities from the previous section, you can use it for this project, otherwise make a new map and display it:
```
>>> m = Map(20)
>>> view_map(m)
```

T67. Call `esearch` to see what kind of tour it can find using point mutations only in 500 generations and a population of 50 tours:
```
>>> esearch(m, 500, 50)
<PythonLabs.TSPLab.Tour [7, 14, 13, ... 15, 17, 8] 2258.722>
```
It's possible this search may lead to a tour with no twists, but usually there will be one or two places where roads intersect each other.

T68. Repeat the previous expression, but this time tell `esearch` to use a combination of point mutations and crossovers:
```
>>> esearch(m, 500, 50, dist = 'mixed')
<PythonLabs.TSPLab.Tour [16, 9, 18, ... 17, 15, 2] 1607.581>
```
Was the algorithm able to find a better tour?

Try each of the previous searches a few more times to convince yourself that using crossovers helps `esearch` find better tours. You can also try each search (with and without the `:dist` parameter) on different maps of size 20. Some maps may be laid out in such a way that only point mutations are required, and some may have regions that are difficult even when both types of mutations are used.

T69. At this point you might be wondering, if crossovers are so effective, why even have point mutations? Type this expression to tell esearch to use only crossover mutations:

```
>>> esearch(m, 500, 50, dist = 'all_cross')
<PythonLabs.TSPLab.Tour [7, 17, 15, ... 14, 11, 8] 1679.313>
```

T70. Repeat the previous expression a few more times. Do the tours made using only crossovers have anything in common?

When esearch is not able to use any point mutations, it usually finds a tour that has the best overall "shape," but there will be several places where a point mutation that exchanges two cities right next to each other would make a better tour. But such small changes are not likely when the only way to make a new tour is to "cut and paste" from two different tours.

If you would like to try some more experiments on larger maps here are some suggestions:

♦ Make a map with 30 cities and draw it on the canvas:

```
>>> m = Map(30)
>>> view_map(m)
```

♦ Call esearch using the same parameters you used for the map with 20 cities:

```
>>> esearch(m, 500, 50, dist = 'mixed')
<PythonLabs.TSPLab.Tour [17, 26, 3, ... 10, 27, 1] 1947.363>
```

♦ One way to see if esearch can find a better tour is to let it run longer. Repeat the previous expression, but increase the number of generations to 1,000:

```
>>> esearch(m, 1000, 50, dist = 'mixed')
<PythonLabs.TSPLab.Tour [15, 9, 20, ... 13, 5, 12] 1820.638>
```

♦ Another possibility for getting a better tour is to use a larger population size. Type this expression to use the original number of generations (500) but twice the population size:

```
>>> esearch(m, 500, 100, dist = 'mixed')
<PythonLabs.TSPLab.Tour [3, 11, 22, ... 28, 25, 4] 1799.569>
```

♦ Do some more experiments on your own. Try increasing the number of generations up to 2,500, and the population size up to 300. In general, which strategy is more effective, letting the search run for more generations or increasing the population size?

♦ If you want to see how well this genetic algorithm works on larger maps, try making maps of up to 100 cities, and using population sizes of up to 1,000. The canvas only has room for 400 cities, so with more than 100 cities the map will be very crowded. The population size can be as big as you want, but obviously the larger the population size, the longer it will take to make each new generation.

12.7 ♦ TSP Helper Functions

The helper functions called by esearch provide an opportunity for several small programming projects. If you would like to work on your own versions of these functions the following sections provide more information.

Here is a suggestion for how to organize your project. Create a single file for all the helper functions, and call it something like tsp-helpers.py. The first line of the file should be a command that imports all the functions defined in TSPLab:

```
from PythonLabs.TSPLab import *
```

```
from PythonLabs.TSPLab import *

def select_survivors(p):            # your code for
    ...                             #   select_survivor

def compact_population(p):          # your code for
    ...                             #   compact_population
```

Figure 12.16: *When you work on your own versions of the helper functions put all the definitions in one file. When this file is loaded into an interactive session the definitions of the helpers will replace the versions imported from TSPLab.*

When you write your own version of a helper function add the definition to the end of this file. To test your function, start an interactive session and load the file. TSPLab will be imported, and then your definition will replace the one imported from TSPLab (Figure 12.16). Now you can call esearch, and when it calls a helper function, it will use your code instead of the function in TSPLab.

Before you call esearch, however, be sure to run some tests and try to debug your functions in an interactive session. For example, you can call your version of init_population to make sure it makes the requested number of tour objects, or you can make a population using the TSPLab version of init_population but then call your version of select_survivors to see if it is deleting tours the way you want it to.

♦ init_population

The projects in Chapter 7 introduced a Python construct called *list comprehension*. If we write an expression (typically involving a for statement) between a pair of brackets Python will evaluate the expression and create a list from the results. For example, if we want a list of multiples of 5 for the numbers between 0 and 9, we could use a for statement and a call to range to generate the numbers, and we then simply have to multiply the numbers by 5:

```
>>> [5*i for i in range(10)]
[0, 5, 10, 15, 20, 25, 30, 35, 40, 45]
```

You can do a similar thing if you want to create a list of tours. Use a for expression to specify how many times to call the make_tour method and collect all the new Tour objects in a list.

♦ select_survivors

According to the description in Section 12.5, the probability of removing the tour at location i in the population list is i/n, where n is the size of the population. Because the Tour objects are always sorted the decision of whether or not to delete can be based on the tour's location in the population list.

Assume the population is a list named p, and i is the name of the loop index variable. At each location i pass i/len(p) to select_with_probability. If the result is True store None in p[i].

◆ compact_population

The key to implementing the function that moves all the None objects to the end of the population list is realizing that this function is basically just creating a new permutation of the list. The same general strategy used by permute (Figure 7.11) applies here, as well. Write a loop that iterates over the list, and whenever a location contains a tour, find a None somewhere to its left and exchange the two. If we keep repeating this basic step eventually all the tours will be in the left part of the list and all the None objects on the right. The only question is where to find a None to exchange with a tour.

One simple way to solve this problem is to use a variable d, initially set to 0, to hold the distance between the items being exchanged. We'll use the same names as above, where p is the population list and i is the index variable. On each iteration, if there is no tour at p[i] (if p[i] is None) add 1 to d. Otherwise p[i] is a tour, so swap the items in p[i-d] and p[i].

Figure 12.17 has a diagram that explains why this algorithm works. The important observation is that at the start of each iteration, all the None objects to the left of the current location are all next to each other, and they are all immediately to the left of location i. Think of them as a "bubble" that is continually shifting to the right. When p[i] holds a tour, the step that swaps the tour with the None at p[i-d] simply shifts the bubble to the right by one location (Figure 12.17a). When p[i] is None, the bubble grows in size (Figure 12.17b). By the end of the main loop all the None objects will have been collected into a single bubble on the right side of the list.

One final detail to be aware of is that compact_population needs to return the number of tours in the population. But that's easy to figure out: when the for loop terminates, the variable d is the total number of None objects, so the number of tours is just len(pop)-d.

◆ rebuild_population

If the genetic algorithm is going to use only point mutations, the rebuild_population helper is simple to implement. The argument n passed to the function is the address of the first None in the population (p). Simply write a for loop that sets i to every value in the range from n to the length of p, and in the body of the loop set p[i] to a new tour.

To make the new tour, choose a random number between 0 and n-1 (Python's randint function, from the random library, will do this for you). Then call make_tour, passing the string 'mutate' and a reference to the selected tour, and you will get back a copy of that tour with a single point mutation added somewhere.

If you want to write a version of rebuild_population that randomly does point mutations or crossovers, you can use the code in Figure 12.15 as your starting point. You can try a different probability of doing a point mutation, or you can read the Lab Manual to learn about other types of mutations and add code that will randomly try these other mutations in addition to point mutations and crossovers.

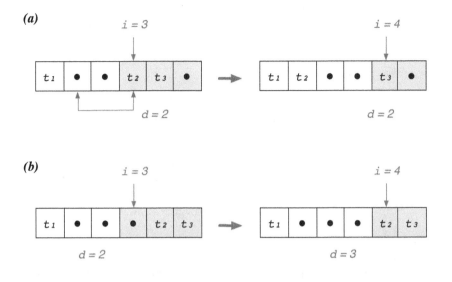

Figure 12.17: *This diagram shows an example of a single iteration of* compact_population. *At the start of the iteration the index variable* i *is 3. The white boxes show the part of the list that has been processed already. Because two* None *objects have been seen so far* d *is 2.* **(a)** *If the item at* p[i] *is a tour it is swapped with the item at* p[i-d] *and* d *remains 2.* **(b)** *If there is a* None *in* p[i] *just increment* d.

◆ select_with_probability

The functions that select survivors and rebuild the population are based on decisions made with a specified probability. For example, if we want an if statement based on an expression that is True 90% of the time we would write

```
if select_with_probability(0.9):
```

The first step is to call another function from Python's random library. The function, also named random, returns a floating point number between 0 and 1. Here is an example from in an interactive session:

```
>>> from random import random
>>> random()
0.4056570589472771
```

To implement a function that returns True with a certain probability p, all we need to do is call random and then compare the result with p. To see why, suppose $p = 0.9$. 90% of the numbers returned by a call to random will be less than 0.9, so our function should return True when the value returned by random is less than 0.9.

The obvious way to write the function is to use an if statement:

```
def select_with_probability(p):
    if random() < p:                # naive version
        return True
    else:
        return False
```

A better approach is based on the fact that we can save the result of a Boolean expression in a variable. We can write an assignment statement that has our Boolean expression on the right side:

```
x = random() < p
```

As a result of this assignment, x will be either `True` or `False`, and we just have to return the value of x. Here is a second version of the helper function:

```
def select_with_probability(p):
    x = random() < p            # better version
    return x
```

Better yet, we don't even have to save the Boolean value at all, just return it as soon as it is computed:

```
def select_with_probability(p):
    return random() < p         # best version
```

At this point you might be asking why we even have a helper function if all it's going to do is evaluate one expression. While a function call does add some overhead to a program, there are (at least) two good reasons to have a helper function. For one thing, it makes the program just a little easier to understand. If you show the definition of `rebuild_population` to a friend (or if you come back to this project in a month or so) it will be clear what the intention is.

A second reason is that if `select_with_probability` is called from several different locations in the program, it will be easier to modify its behavior, for example by updating to a different random number generator. You will have to change only one piece of code instead of having to search for all the places where the result of calling `random` is compared to another number.

As a final note, before you start writing the code, ask yourself how you will test your function. Just calling it once or twice and seeing you get back a Boolean value is a good start. But what do you need to do to make sure the function is working correctly?

12.8 Summary

This chapter introduced an important problem in computer science known as the Traveling Salesman Problem. Although it sounds like a simple puzzle or brain-teaser, it is actually a very challenging problem faced by professionals in a wide variety of different areas. Traveling to a set of cities on a tour that visits each city exactly once before returning to the starting point is the same basic problem as driving a delivery truck to drop off packages for customers, laying out components on a circuit board, and a variety of other problems in transportation and manufacturing.

What makes this problem so difficult is the huge number of alternatives to consider. There are $(n - 1)! / 2$ different itineraries for visiting n cities. This equation grows very quickly with increasing n, so that for as few as 20 cities there are far too many alternatives for an algorithm to examine each possible tour.

Concepts and Terminology Introduced in This Chapter

Traveling Salesman Problem	An optimization problem where the goal is to find the lowest-cost tour that visits each city on a map exactly once
random search	An algorithm that looks at random solutions; for the TSP, the algorithm evaluates random tours
genetic algorithm	A biologically inspired algorithm that begins with random tours and then tries to "evolve" better tours through a series of mutations
point mutation	A small change in DNA, or, in the TSP, in the order in which two cities are visited
crossover	A process that combines DNA from two sources, or, in the TSP, makes a new tour by combining large parts of existing tours

---

Instead of doing an exhaustive search that considers every tour, we tried two approaches that examine random tours. A method named rsearch simply makes random tours, hoping to find a reasonably good one. Building on that idea, a method named esearch implements an evolutionary search that tries to improve on any good tours it finds.

The idea behind a genetic algorithm is to mimic how a population of real organisms changes over time. The algorithm starts with a collection of random tours, and over a series of iterations throws out the more costly tours and replaces them with slight variations of the better tours. Under the right conditions, involving a mixture of the different kinds of "mutations" that modify tours and a large enough "population" to work with, a very good tour eventually emerges.

One of the curious things about the Traveling Salesman Problem is that it is possible to calculate the *cost* of the optimal tour without knowing the actual *path* for the best tour. There are algorithms that use distances between cities to compute a lower bound on the cost of a tour. One might think that the algorithm to compute the cost of the best tour could also be used to find the tour itself, but unfortunately that's not the case.

Knowing the lower bound for the cost of a tour allows different strategies for halting the algorithm. The simple technique used in the projects in this chapter was to specify a fixed number of generations ahead of time, and then just see what the method would produce in that number of generations. But if we know the cost of the optimal tour, we can tell the algorithm to run until it finds a tour with this cost.

In many cases, such as a courier company planning the routes for a delivery truck, it may not be necessary to have the very best tour. The company may be happy to know the driver will follow a tour that is almost as good as the best possible tour. In this case, they might be satisfied with a computer program that runs for half a minute to produce a tour that is within 5% of the lowest cost, instead of a program that runs for five hours to find the absolute best tour.

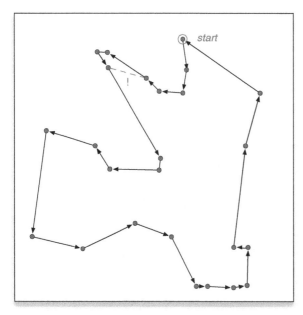

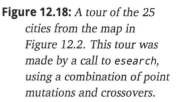

Figure 12.18: *A tour of the 25 cities from the map in Figure 12.2. This tour was made by a call to* esearch, *using a combination of point mutations and crossovers.*

If you took the challenge at the beginning of the chapter, to find the optimal tour of the map in Figure 12.2, you can compare your solution to the one found by esearch, which is shown here in Figure 12.18. After visiting four cities, it's tempting to move on to the closest city, as shown by the dashed line, but that would be a mistake. The best tour has to skip over this city and then visit it on the way toward the cities at the bottom of the map. How did your solution compare to the one found by Python?

Explore Further

A resource for learning more about the Traveling Salesman Problem, including the history of the problem and real-world applications, is http://www.tsp.gatech.edu, a website maintained by a research group at the Georgia Institute of Technology. This site also has a Java applet that will allow you to try to solve the TSP yourself. If you open this applet with your web browser, it will display a set of cities, like the one shown in Figure 12.2, and let you click on the cities to try to make the best tour.

Exercises

12.1. Prove that $n! > 2^n$ for every value of $n > 3$. Hint: How many terms are in $n!$? in 2^n?

12.2. How many possible tours are there for a map with 16 cities?

The next set of problems refer to the map in Figure 12.19. The distances between the cities on the map are defined by this matrix:

```
        A       B       C       D       E       F       G
A     0.00
B   151.54    0.00
C   202.99  187.45    0.00
D    82.02  199.41  284.85    0.00
E   142.56  185.95  325.58   89.55    0.00
F   167.36  302.42  242.33  202.43  290.11    0.00
G   253.62  404.88  383.51  239.02  323.36  144.51    0.00
```

12.3. What is the distance between these pairs of cities?

        A—E   F—C   B—E

12.4. What is the length of the three-city tour A—D—E?

12.5. What is the length of the seven-city tour shown by the solid lines in Figure 12.19?

12.6. Suppose a variable t contains the tour shown by the solid lines in the figure. Its path is:

```
>>> t.path()
('C', 'F', 'G', 'A', 'E', 'D', 'B')
```

What parameter would you pass in a call to t.mutate to "straighten out" the tour so it uses the connections shown by the dotted lines?

12.7. Draw the map that would result from calling t.mutate(1).

12.8. What is the cost of the path that follows the dotted lines in Figure 12.19? In other words, find the cost of the path

```
('C', 'F', 'G', 'A', 'D', 'E', 'B')
```

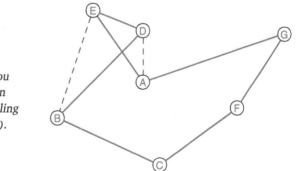

Figure 12.19: *This map with 7 cities is used in Exercises 12.3 to 12.8. If you want to check your answers you can make a map with these cities by calling* Map(path_to_data("test7.txt")).

12.9. Suppose you have a tour named t, and you call t.mutate(2) to exchange the cities in locations 2 and 3 of the path. What would happen if you called t.mutate(2) a second time?

12.10. In a tour of n cities the last city as at location n-1. Does it make sense to call t.mutate(n-1) to introduce a point mutation starting at the last city? Make a Tour object and use it to verify your answer.

12.11. What proportion of the possible tours of 16 cities would rsearch consider if it looked at 10^6 random tours?

12.12. Approximately how many tours of a map with 16 cities would be evaluated when esearch is called with the following sets of parameters?

  a) population size = 25, number of generations = 50
  b) population size = 50, number of generations = 100
  c) population size = 100, number of generations = 1,000

12.13. In the previous problem, will the number of tours evaluated by esearch change if a different size map is used? Or is the number of tours created independent of the map size?

12.14. ♦ A formula for estimating the number of tours created by a genetic algorithm was given on page 402. It turns out this formula overestimates the number slightly, because on each generation the best tour is never thrown out. Devise a formula that is a more accurate prediction of the number of tours that will be created, and compare the number of tours made in your experiments with this new formula.

Programming Projects

12.15. Write a function named factorial that takes a single argument, an integer named n, and returns the value of $n!$:

```
>>> factorial(6)
720
>>> factorial(7)
5040
```

12.16. Write your own version of the ntours function that computes the number of possible tours of a specified number of cities:

```
>>> ntours(7)
360
>>> ntours(15)
43589145600
```

12.17. Write a function named distance that will compute the distance between a pair of cities. The arguments will be a pair of tuple objects, where each tuple has an x and y map coordinate of a city:

```
>>> m = Map(10)
>>> m.coords(2)
(273, 355)
>>> m.coords(3)
(232, 372)
>>> distance(m.coords(2), m.coords(3))
44.384682042344295
```

12.18. Write a function named `average_distance` that will compute the mean distance between all pairs of cities in a Map object.

12.19. Write your own version of the `cost` method, in the form of a function named `tour_length`, that will compute the sum of the distances between all cities in a tour:

```
>>> t = m.make_tour('random')
>>> t.cost()
2490.3675289744338
>>> tour_length(t)
2490.3675289744338
```

The sidebar on page 6.3 explains how to access parts of a tuple.

12.20. In an interactive Python session, make a Map object with 7 cities. Use a watch (or a stopwatch app on a smartphone) to measure the amount of time it takes to print all possible tours of the map (see Exercise T19 on page 389). Use the result to predict how long it would take to print all possible tours of 15 cities.

12.21. Repeat the previous exercise, but instead of having Python print the tours, write a function named `xtime` that will measure the amount of time to generate all tours on a map of a specified size. A function named `time`, defined in a module also named `time`, returns the system "timestamp":

```
from time import time
```

If you call `time` once before the loop that generates tours, again at the end of the loop, and then take the difference, you will have the number of seconds it took to execute the loop:

```
t1 = time()
  # a loop that makes tours
t2 = time()
  # t2 - t1 is the time required to execute the loop
```

Here is an example, showing how long it took a laptop to make all tours of 10 cities:

```
>>> xtime(10)
7.20
```

12.22. Write your own version of the `mutate` method, in the form of a function that takes two arguments, a list of items and a location. The function should exchange the item at the specified location with the one before it:

```
>>> a = list('ABCDEFGHIJ')
>>> a
['A', 'B', 'D', 'C', 'E', 'F', 'G', 'H', 'I', 'J']
>>> mutate(a, 3)
>>> a
['A', 'B', 'D', 'C', 'E', 'F', 'G', 'H', 'I', 'J']
```

Make sure your function handles cases where the specified location is the beginning or end of the list:

```
>>> mutate(a, 0)
>>> a
['J', 'B', 'D', 'C', 'E', 'F', 'G', 'H', 'I', 'A']
```

12.23. Write a function named `greedy` that implements the "choose the closest city" strategy to construct a tour. The argument passed to the function will be a Map object. Make the tour by starting at the first city on the map. The next city will be the city closest to the first one. Keep iterating, at each step adding the city closest to the one chosen on the previous iteration, until all cities are added to the tour.

```
>>> m = Map(path_to_data('ireland.txt'))
>>> m.cities()
['belfast', 'cork', 'dublin', 'galway', 'limerick']
>>> greedy(m)
['belfast', 'dublin', 'limerick', 'cork', 'galway']
```

Hint: Start by making a list called `unvisited` that initially has all the cities. On each iteration remove one city from `unvisited` and append it to the tour. The `del` statement (described on page 75) will delete an item from a list.

12.24. We might improve the evolutionary algorithm by making sure a population does not contain duplicates of the same tour. One way to look for duplicates would be to make sure all cities start at the same location, making it easier to find tours that follow the same path. Write a function named `normalize` that will take a list of integers as a parameter and return the same list, but with the number 0 in the first location. For example, here is a list of 10 city numbers:

```
>>> t
[4, 1, 3, 7, 9, 6, 2, 0, 8, 5]
```

Passing the list to `normalize` should "rotate" the tour so it has the same numbers, and in the same order, but 0 will be in the first location:

```
>>> normalize(t)
[0, 8, 5, 4, 1, 3, 7, 9, 6, 2]
```

Hint: Review the section on using the slice operator in the implementation of merge sort (Chapter 5).

12.25. Perform an experiment to estimate the probability a point mutation improves a tour. Choose a reasonably large value for the number of tests, for example, choose $n = 1000$. Make n random tours, and for each one, call `mutate` to add a point mutation. Keep track of the number of times the mutation lowered the cost. After all the tests are complete, divide the number of improved tours by n. For example, if mutations lower the cost of 300 tours, the probability of improving is $300/1000 = 0.3$.

Index